AF528557

Secrets of

Similipal

A Wonderland of Elephants and Melanistic Tigers

THE AUTHOR

Dr Debabrata Swain was born in an idyllic village of Bajitpur in Balasore district of Odisha. He had hardly completed his primary education there when he and his entire family shifted to Khunta in Mayurbhanj district of Odisha, where his father was posted as Block Level Cooperative Extension Officer. Khunta was surrounded by sal (*Shorea robusta*) coppiced forest with a silhouette of Similipal which was distinctly visible on a clear sky day. As a young boy, Dr Swain along with his friends, often sneaked into the nearby forest to catch jewel beetles (*Sternocera chrysis*) or dragonflies (Odonates) and relish numerous berries like kendu (*Diospyros melanoxylon*), amla (*Phyllanthus emblica*), jamun (*Syzygium cumini*), and wild mango (*Mangifera indica*), to mention a few. In this forest they often encountered wild animals like wolf (*Canis lupus*), jackals (*Canis aureus*), several birds and snakes. Such playful incidents and interactions with nature sowed seeds of conservation that germinated and blossomed in later days of his life.

Incidentally, Dr Swain joined the Indian Forest Service in 1984 and undertook great journeys until his retirement in 2019. He retired as the Chief of Forest Service in his state. During his 35 years of career in forest services, he held several significant assignments in the state including State Siviculturist, Chief Conservator of Forests (Plan, Programme and Afforestation) and Special Secretary (Forest and Environment), besides Principal Chief Conservator of Forests and Head of Forest Force, Odisha. Nevertheless, his dream came true when he was assigned to be the Director, Similipal Biosphere and Tiger Reserve in April 2002 where he remained till November 2007. Being a native of the land, he was able to strike a right chord with the local people, and involved the community in conservation of this precious wealth of nature. He conceptualised the formation of Green Corps (Sabuja Vahinee) consisting of youths in the fringe villages of Similipal. By 2006, 37 Sabuja Vahinee groups, consisting of 323 volunteers (even women participated) were formed, and they were successful in convincing local tribal communities against Akhand Shikar (mass hunting of wild animals). They also protected forest from many threats such as tree felling and poaching. The dense forest cover of Similipal substantially increased during those years. During 2006, the Management Effectiveness Evaluation of the reserve rated it as one of the best-managed reserves in the country.

Moreover, he performed interesting assignments while being in central deputation with the government of India. He served as Inspector General, central region, Nagpur and Additional Director General-cum-Member Secretary of National Tiger Conservation Authority (NTCA), New Delhi. The Central Region of NTCA consists of Maharashtra, Madhya Pradesh, Chhatisgarh and Odisha, encompassing 17 tiger reserves. He extensively travelled and supervised the activities of all the tiger reserves and made several constructive suggestions for improvement of the reserves. Thus with the support of NTCA and hard work of field staff, Madhya Pradesh which occupied the third position (with 308 tigers) in the tiger census of 2014, rose to first (with 526 tigers) in the tiger population estimation of 2018. Similarly, Maharashtra that had 190 tigers (5th position) in 2014, escalated to 4th position with 312 tigers in 2018. Tiger population in Odisha and Chhatisgarh although remained stable. Regional IG, NTCA (Dr Debabrata Swain) and scientists of Wildlife Institute of India planned five underpasses and four minor bridges on 37 km stretch of NH 44 (earlier NH 7) passing through Pench tiger landscape in the Maharashtra part of the NH during 2014-15. Dr Swain monitored the location and construction of those underpasses from 2015 to 2018. Now these elevated stretches of the NH have proved to be a win-win situation for tiger conservation and development. During his incumbency of Member Secretary, NTCA, he initiated the fourth cycle of All India Tiger Estimation. He laid the foundation of introducing the latest technology and tools in tiger estimation in collaboration with Wildlife Institute of India.

Dr Swain's contribution to elephant conservation is also equally significant. His PhD thesis on elephants of Odisha, also published as a book titled 'Asian elephants: Past, present and future' is a reference book catering to elephant conservation and management. Following his passion for elephants, he founded the 'Save Elephant Foundation Trust' to reduce human-elephant conflicts by partnering with the Forest Department and the local community. He skillfully scripted a documentary film on elephants to create awareness among the communities, forest managers and others. Besides he has published over 60 scientific and popular articles on the forest, wildlife and climate change. He is also a recipient of some national and international fellowships, namely Charles Bullard Fellowship of Harvard University 2008-09, Fulbright Indo-American Environmental Leadership Programme Fellowship 2005, Smithsonian Institution (MAB) Programme Fellowship 2005, Indo-US Science and Technological Forum Fellowship 2005, ITTO Freezailah Fellowship 2009, and Fellow, Zoological Society of India 2001 and Orissa Environmental Society 2019. He also received Eminent Environmentalist of Odisha by Prasanna Kumar Dash Birth Centenary Memorial Committee in 2019.

Dr Swain is also known for his impeccable honesty and integrity. He leads a simple lifestyle following the principles of peace and serenity of Brahma Kumaris. He is deeply dedicated to biodiversity conservation, sustainable development and a corruption free society. He is currently a member of Odisha Lokayukta.

Secrets of

Similipal

A Wonderland of Elephants and Melanistic Tigers

Debabrata Swain

— Foreword by Sri Naveen Patnaik —

2021

Daya Publishing House®
A Division of
Astral International Pvt. Ltd.
New Delhi – 110 002

Save Elephant Foundation Trust, Odisha
Baripada – 757 003

ISBN: 978-93-90435-70-8 (HB)

Published by : **Daya Publishing House®**
A Division of
Astral International Pvt. Ltd.
– ISO 9001:2015 Certified Company –
4736/23, Ansari Road, Darya Ganj
New Delhi-110 002
Ph. 011-43549197, 23278134
E-mail: info@astralint.com
Website: www.astralint.com

Co-published by **Save Elephant Foundation Trust**
Debendrapur, Baripada-757003, Odisha
Ph. 9937623053/8917560418/7008807839
E-mail: saveelephantodisha@gmail.com
Website: http://saveelephant.net/

Dedicated to Supreme Almighty Authority, Supreme Soul for creating this earth with amazing landscapes like Similipal and I pray to bestow us sanity to protect it from onslaught of selfishness, narrowmindedness and lack of fellow-feelings. Only the Supreme can save us from impending disaster.

Acknowledgments

It is my proud privilege to express sincere gratitude to Shri Naveen Patnaik, Honourable Chief Minister of Odisha for sharing his love for Similipal and setting the tone for the book through an insightful foreword. I am grateful to Shri Suresh Chandra Mohanty, former Principal Chief Conservator of Forests and Head of Forest Force, Odisha for carefully reviewing the draft versions and providing a preface for this book.

I am deeply indebted to my revered teachers, Late Dr Basanta Kumar Behura, Professor (retired), Department of Zoology, Utkal University and Dr Peter Ashton, Professor (retired), Harvard University, USA for their guidance, and inspiration for putting together a book on Similipal.

I am thankful to Dr Kishore Kumar Mandal and Dr Truptirekha Kar for providing me updated list of plants discovered in Similipal and some rare photographs of those plants to be reproduced in this book. I am also grateful to Dr Manoj V Nair for his ungrudging help in preparing the list of avi-fauna of Similipal and providing photographs of elephants. I convey special thanks to Dr Lala Ashwini Kumar Singh, Senior Research Officer (Retired), Wildlife Department, Odisha for his advices while writing the book. I would also like to acknowledge Engineer Dr Sarat Misra, retired Chief Engineer (Works) and a reputed orchidologist; Shri Jitasatru Mohanty, retired Conservator of Forests; Dr Jagyandatta Pati, Deputy Director, Similipal Tiger Reserve; Shri Santosh Joshi, Divisional Forest Officer, Keonjhar division; Shri Ajit Satpathy, Divisional Forest Officer, Keonjhar (Wildlife) division; Shri Prasanna Kumar Behera, Divisional Forest Officer, Karanjia division; Shrimati Anshu Prangya Das, Deputy Conservator of Forests (Wildlife); Dr Sushil K Dutta, retired Professor, North Odisha University; Dr Pratyush P Mohapatra, Scientist, Zoological Survey of India; Dr Hopi Hoekstra, Professor, Harvard University; Dr Uma Ramakrishnan, Professor, National Centre for Biological Sciences; Dr Paul J Morris, Harvard University; Dr Hrudayanath Thatoi, Professor, North Odisha University; Dr Srustidhar Rout, Associate Professor, North Odisha University; Dr Tanveer Ahmed, Wildlife Institute of India; Shrimati Sunita Jee, Assistant Conservator of Forests; Kumari Swati Saini, GIS Specialist, Wildlife Institute of India; Kumari Medha Nayak, Research Scholar, National Institute of Science Education and Research; Shri Jayadev Ghadei, Graphic Designer; and Shri Rajkumar Behera, Artist, Nanadankanan Biological Park; for collaborating and providing necessary inputs in form of literatures, photographs, drawings and information for this book. I am also thankful to Dr Peter Ashton and photographer Dr Hans Hazebrock for sharing photographs from their visit to Similipal in the year 2008. Dr J P Das, eminent cardiologist had held a few health camps inside Similipal and he has shared some sketches drwan by him during his visit to the reserve. I am thankful to him.

My grateful thanks also go to the field staffs of Similipal Tiger Reserve without whose support and cooperation this work would not have been possible. I am also obliged to Save Elephant Foundation Trust, Odisha and Daya Publishing House, New Delhi for taking the initiative and responsibility to publish this book.

I express my gratitude to my mother Shrimati Sudhamani Swain and father (Late) Shri Chakradhar Swain for inspiring me to pursue my interest in wildlife conservation and study the rich biodiversity of Similipal.

I am grateful to Late Shrimati Chandramani Behura, wife of my teacher Prof Behura, for her constant encouragement to successfully carry out my work. My late youngest sister, Bidyut had avid interest in Similipal and she accompanied me on several of my trips to the reserve. She is close to my heart and is an inspiration even in her absence. My thanks are also due to my other sisters (Sarojini didi, Nistarini didi, Devi and Geeta), brother (Satyabrata), brother-in-laws (Shri Chittaranjan Sur and Dr Rajanikanta Jena), nieces (Puspita, Sujata and Sanjukta) and nephews (Anjan, Gautam and Rajib) for their help and aesthetic suggestions for this book. Last but not the least, I express gratitude to my colleagues, friends and well-wishers for their unwavering support and encouragement.

Dr Debabrata Swain

NAVEEN PATNAIK
CHIEF MINISTER, ODISHA

LOKASEVA BHAVAN
BHUBANESWAR

FOREWORD

Similipal in Mayurbhanj district is an amazing floral and faunal repository displaying a mosaic composition of plant and animal species representing northern, southern, western and north-eastern India. It is one of the most vibrant ecosystems on earth. Luxuriant forest traversed with numerous perennial streams and cascading waterfalls are the beauty and bounty of this biosphere. A treasure house of wide varieties flora and fauna, Similipal has 98 types of orchids, 552 varieties of medicinal plants besides a diverse class of ferns, lichens, mammals, reptiles, birds and amphibiotic animals. Prominent as the land of elephants and melanistic tigers, this vast and varied landscape has many fascinating and splendid features deserving wide publicity for benefit of humankind.

I am glad to know about the book " Secrets of Similipal: A Wonderland of Elephants and Melanistic Tigers" by Dr. Debabrata Swain, who has long experience in forest management. Apart from providing vivid scientific information, the author has sincerely tried to unfold many untold facts of this precious natural asset of our State. The book, I hope, will be very useful to promote research, conservation and environmental services on sustained basis. I am sure; this endeavour will certainly receive high appreciation from researchers, scholars as well as the general public.

(NAVEEN PATNAIK)

Phone : Office : 0674-2531100, 2531500, 2535100 (Fax)
Residence : 0674-2591099, 2590299, 2590833 (Fax)

e-mail : cmo@nic.in | twitter : twitter.com/COM_Odisha | Facebook : facebook.com/CMO.Odisha

Preface

For someone like me who has been to Similipal many times, not as a casual visitor but always on call of duty, there is also what you call a call of the wild, an enchantment with its every nook and corner. For me therefore it has been a fascinating journey to go through this manuscript.

Let me also say that I have worked closely with the author for a number of years as Chief Wild-Life Warden of Odisha when he was Field Director, Similipal Tiger Reserve. Both earlier, and in later parts of my career I have also shared in his keen sense of belonging to Similipal with its wildlands, near pristine wilderness, the majestic Sal trees, the fast-flowing perennial streams and waterfalls, and the scattered hamlets of tribal inhabitants.

Within the covers of this treatise the author has tried to pack a lot of facts on the state of Similipal forest. No one is more suited than the author for presenting this subject, since he is a rare Forester who has spent almost a decade in getting hands-on experience in managing Similipal in various capacities. The whole text is aptly organised into seven chapters: introductory, the flora, the fauna, the people, management issues, the corridor, and the epilogue. There is a sizeable number of references to various publications and websites; indeed the more broadbased the perception the closer it gets to the truth. A list of 1359 species of plants found in Similipal has been appended, as also the list of species among them which have medicinal applications. Further 141 species of lichens, 33 species of bryophytes, and 19 species of corticolous algae have been added. A checklist of 60 species of mammals, 385 species of birds, 23 species of amphibians, 70 species of reptiles, 66 species of fishes, and 133 invertebrates of these forests is also appended which will prove to be a standard source of reference.

In not too distant past Similipal was sought to be managed by an exclusive Corporation with the object of exploiting its timber and non-timber resources and providing gainful employment to the local forest dwellers in the process. Government reversed its policy bowing to the strong conservation lobbies, and a complete moratorium on felling of trees was announced in 1982. 2250 km^2 of this massif has since been finally notified as a sanctuary after settlement of rights, and 2750 km^2 of the forest has been notified as the tiger reserve (critical tiger habitat).The whole of the sanctuary and a belt of 10 kms width surrounding it has been designated a biosphere reserve with international recognition and status. The whole of Similipal forest is part of the notified Mayurbhanj elephant reserve. It is the focal area of conservation.

It is stated in the chapter on People that the decadal growth of population inside the tiger reserve is more than the rate of population growth in the state of Odisha. Obviously, the conditions are still quite congenial for hoodwinking the Law and eking out a livelihood inside these forests, even if these are lands meant to be (for more than four decades now) safe havens of tigers, elephants and the attendant biodiversity. In-bound migration may be more than out-bound migration as the social composition of the villages keeps changing. What is more alarming is that the extent of cultivated (agricultural) area has been increasing, and has its impact on the natural habitat: more water of the streams (meant for animals) is being diverted to grow crops.

Any conservation project of this nature will always have critics, many of them well-intentioned.The cognoscenti would swear that from all visible signs the forest ecosystem and the tiger habitat of Similipal is on continuous decline. The common refrain is to ascribe it to the rising tide of anthropogenic pressures. But, surely these are not such insurmountable issues that no solution can be found for them. One might tweak the policy or the strategies a bit to suit the local conditions. But in the ultimate analysis, there is no substitute for dedicated hard work at all levels with commitment to the overall goal. It is the managers and the people running the reserve who matter the most in delivery of success or failure. For all of them this book should be a valuable guidepost.

Suresh Chandra Mohanty,
IFS (Retired),
Ex – PCCF & HoFF, Odisha
Chairman, State EIA Authority , Odisha

Contents

Abbreviations

ADG	Additional Director General
AIC	Akaike Information Criterion
AIGF	Assistant Inspector General of Forests
CITES	The Convention on International Trade in Endangered Species of Wild Fauna and Flora
COP	Conference of Parties
DDR	Dung Decay Rate (also called Dung Disappearance Rate)
DDT	Dichloro-diphenyl-trichloroethane
DFO	Divisional Forest Officer
DLC	District Level Committee
EDC	Ecodevelopment committees
ER	Elephant reserves
EI	Ecological integrity
FAO	Food and Agriculture Organization of the United Nations
FD	Forest division
FRC	Forest Right Committee
GIS	Geographic Information System
GPS	Global Positioning System
IG	Inspector General
IGF	Inspector General of Forests
IRS-1C/1D	Indian Remote Sensing Satellites-1C/1D
IUCN	International Union for Conservation of Nature
IVI	Importance Value Index
LISS-III	Linear Imaging Self-Scanning Sensor 3
MAB	Man and Biosphere (Programme of UNESCO)
MDF	Moderately dense forest
MIKE	Monitoring of Illegal Killing of Elephants
MoEF&CC	Ministry of Environment, Forests and Climate Change
MSL	Mean Sea Level
NCBS	National Centre for Biological Sciences

NE	North-East
NF	Non-forest
NGO	Non-Governmental Organisation
NH	National Highway
NTCA	National Tiger Conservation Authority
NTFP	Non-timber forest product
OF	Open forest
PA	Protected area
PF	Protected forest
PMGSY	Pradhan Mantri Gram Sadak Yojana
RF	Reserve/Reserved forest
SBR	Similipal biosphere reserve
SECR	Spatially explicit capture-mark-recapture
Sp.	Species (single/plural)
Spp.	Species (plural)
STR	Similipal tiger reserve
TR	Tiger reserve
UBK	Upper Baraha Kamuda (Range of Similipal TR)
UNDP	United Nations Development Programme
UNESCO	United Nations Educational, Scientific and Cultural Organization
VDF	Very dense forest
VHF	Very high frequency (communication system)
WII	Wildlife Institute of India
WNBR	World Network of Biosphere Reserves
WWF	World Wide Fund for Nature

Chapter 1

The Landscape

"This is my delight,
thus to wait and watch
at the wayside
where shadow chases light
and the rain comes
in the wake of the summer."

— Rabindranath Tagore [1861-1941] in *Gitanjali* (1912) while describing summer rain. Summer rains are common in Similipal hills.

INTRODUCTION

Enthralled by the enchanting beauty of **Similipal** (also called **Similipahar**), the noted Odia poet Radhanath Ray has described this hill range as the 'Salmali Saila' meaning "moss covered hills" in his masterpiece 'Usha'. It is a densely forested hill range (**Plate 1.1**) in the heart of the Mayurbhanj district of Odisha (Orissa). A true representative of Mahanadian biogeographical region, Similipal is not only a compact mass of hills, valleys, streams, ravines and waterfalls, but a bewildering panorama of hundreds of varieties of trees, climbers, orchids, ferns, mosses, fungi, animals, birds, microorganisms and of course the indigenous people of these hills and forests. It is more than a worldly asset, as it linked to the heart-throb and emotions of the people of Odisha in general and Mayurbhanj in particular. It has the strong appeal of a sacred entity with a bond of assimilation into the age-old culture and livelihoods of the district and its adjacent areas.

Plate 1.1. Densely forested hill range of Similipal (near Joranda water falls), Odisha, India. Photo: B Mohanty.

THE NAME SIMILIPAL

The name Similipal (also called Similipahar) may well be rooted in charming red flowers of simul (*Bombax ceiba*) trees which are abundant throughout these hills. It is also the *mirabile visu* for its undulating hills with varied topography, cascading waterfalls, floral and faunal wealth of great diversity.

LOCATION

The entire area of Similipal hill range is a compact block of elevated plateau comprising of 2750 km^2 located in the central portion of the Mayurbhanj district of Odisha State (**Figure 1.1**). It lies between 20°17' and 22°34' north latitude and 85°40' and 87° 17' east longitude. The inter-state boundary of Odisha with Paschim Banga (West Bengal) is located very close in north-east direction and that with Jharkhand, in the north-west direction of this hill range. The Similipal hill range has a number of reserved forests (RF) and protected forests (PF), and is connected through some forest blocks to Hadgarh sanctuary of Keonjhar district and eventually to Kuldiha sanctuary of Balasore district in the south. The Kuldiha sanctuary is also connected to Noto RF of Similipal hill range through Sukhuapata hill, a vast chunk of revenue forest surrounded by Bhandar RF and Talagadia, Sarisua, Padmapokhari, Mankadapoda, Khadikhania and Jharanaghati PFs. Similarly, Hadgarh sanctuary is connected to Similipal through Satkosia RF. On the west, the forest connectivity extends to the forests of Keonjhar and beyond, and there is connectivity to even Jharkhand in the north. **Figure 1.2A** depicts terrain surrounding Similipal and **Figure 1.2B** shows the contours of Similipal sanctuary, sanctum sanctorum of Similipal landscape.

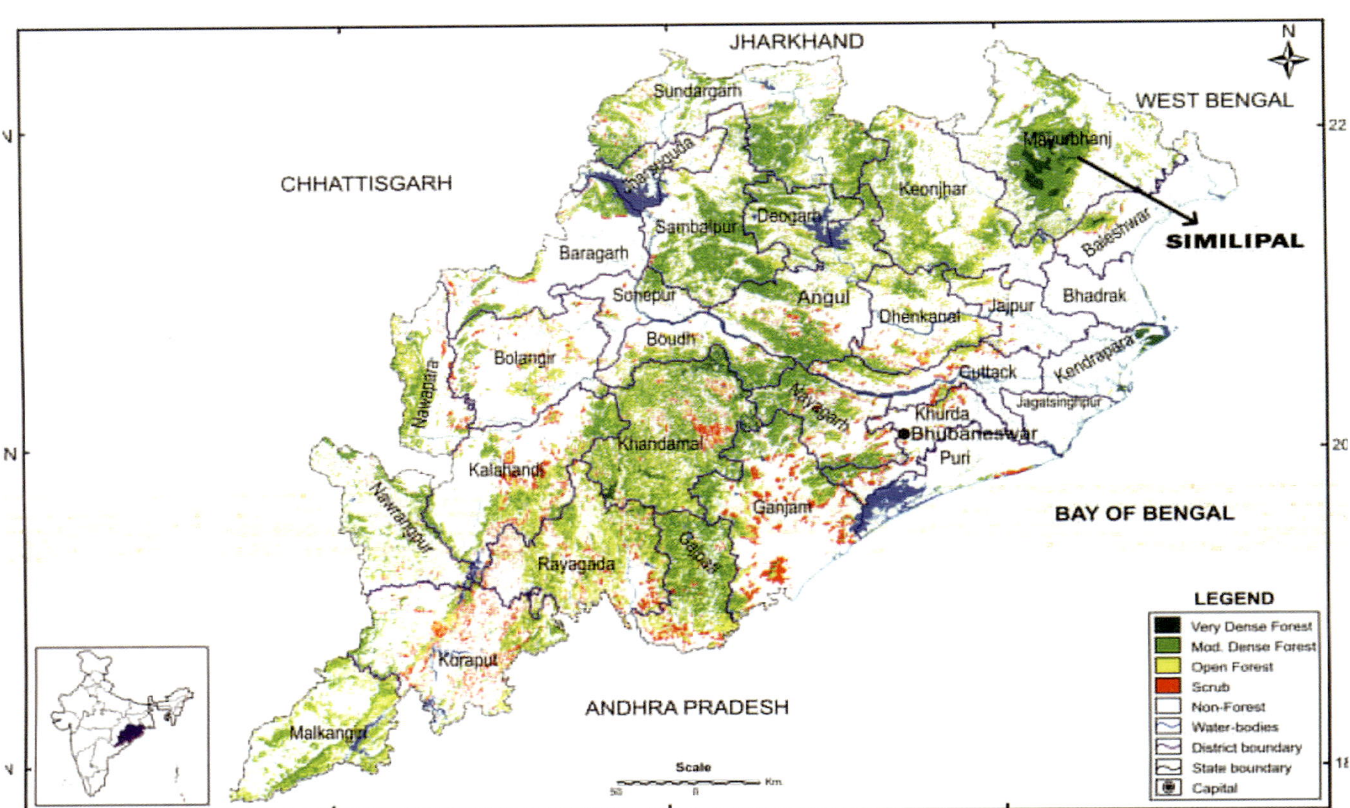

Figure 1.1. Location of Similipal forests, Odisha, India.

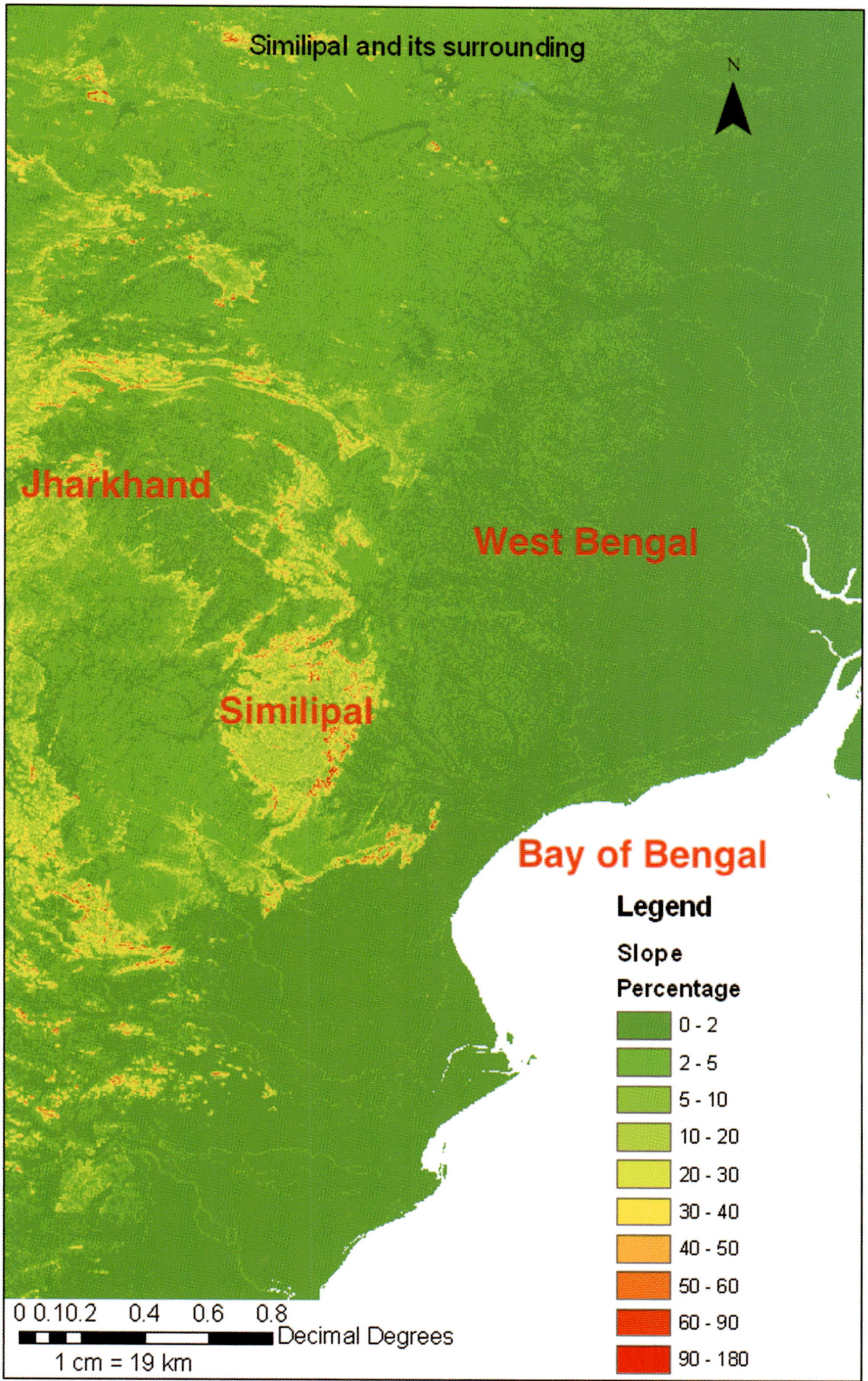

Figure 1.2A. Similipal hill range and its surrounding terrain, Odisha, India.

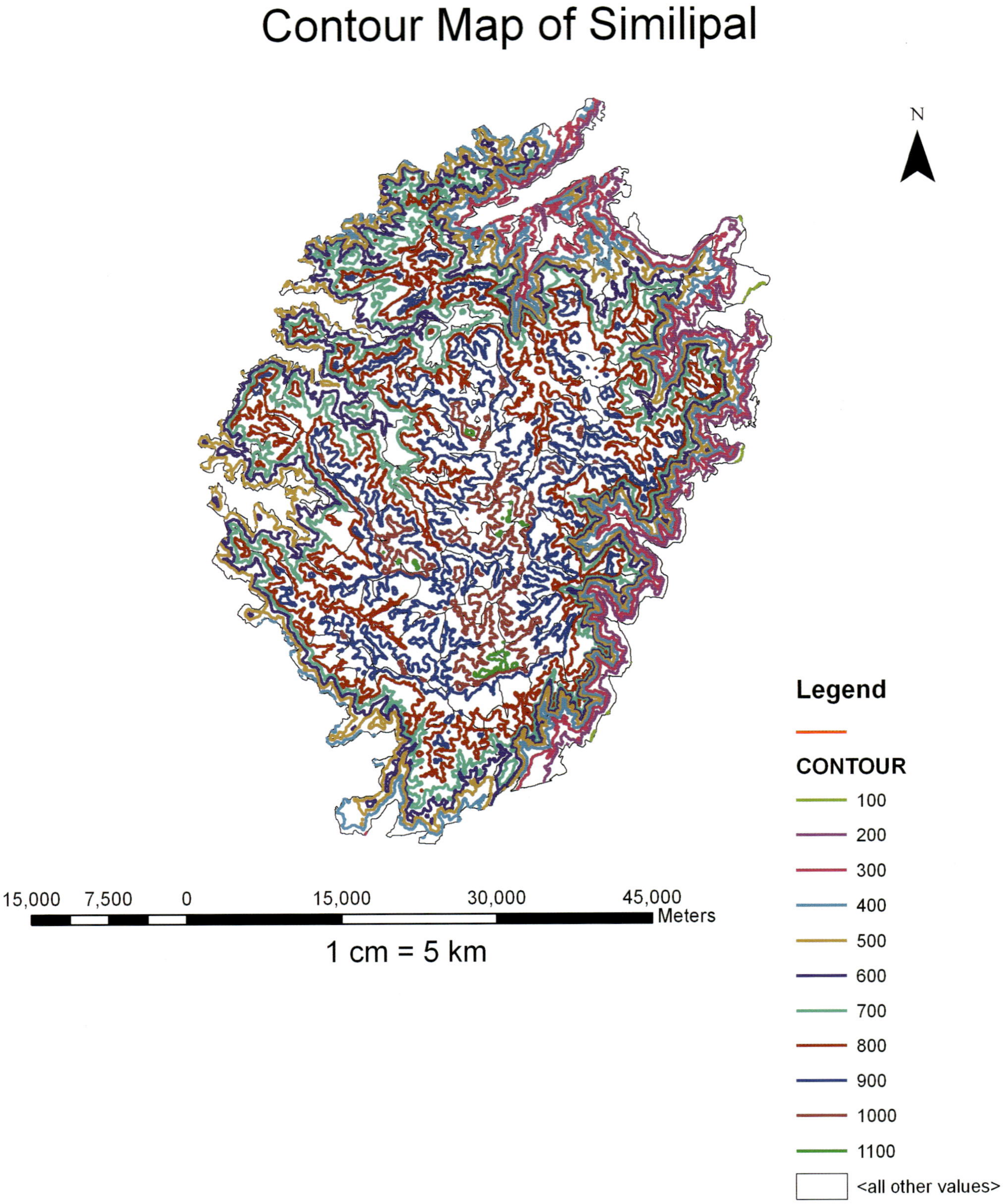

Figure 1.2B. Contour map of Similipal sanctuary, centrally located in the Similipal landscape, Odisha, India.

TERRAIN

The tract is mostly undulating and hilly **(Figure 1.2B)** interspersed with open grasslands and wooded areas. The inclined plateau has risen abruptly from the low coastal plains of the district. The steep side faces the Bay of Bengal and runs northwards to finally merge with Chhotanagpur (of average elevation of 500m). In the basin of the hills lie numerous valleys supporting meadows. The elevation of the highest point (Khairiburu) is 1168m above MSL **(Plate 1.2)** which stands along with Meghasani 1165m MSL as twin towers of natural grandeur. They are situated in south Similipal. The elevation of the central region of the plateau near Dhuduruchampa is 1010m and that of Chahala in the north is 770m. The elevations of a few other points in the locality are as follows:

Plate 1.2. Khairiburu (1168m MSL) and Meghasani (1165m MSL) of Similipal, Odisha, India. Photo: H Hazebrock.

Place	*Altitude in Metres (above MSL)*
Jenabil	865
Upper Barhakamuda	824
Bhanjabasa	706
Nawana	730
Dhuduruchampa	914
Bakua	926

GEOLOGY

The Similipal plateau is underlain by an assemblage of volcano-sedimentary rocks **(Plate 1.3)** consisting of alternating sequence of quartzites and spilitic lavas. Three quartzite formations alternate with three lava flows (Mahalik, 1997). The geological details of the area have been described by Iyengar and Banerjee (1964) and Iyengar and Alwar (1965). The geological formations are depicted in **Figure 1.3**. The age of the rocks is believed to be around 2,200 million years and hence belong to the early Proterozoic era. The main layout of the formation layers is of three concentric cups of metamorphic rocks inter-bedded with sub-metamorphic layers *i.e.*, volcanic with outer interspaces and igneous with inner interspaces. The metamorphic rocks are granitite gneiss, true gneiss and mica schist with pegmatite. The gneissic rocks are much intersected by dykes of basic and intermediate rocks. The sub-metamorphic rocks are shale, hematite, laterites, limestone, calcareous deposits, quartzite, phyllite and micaceous schist. Hematite, laterite and shale occur in extensive formations in central and south Similipal. Outcrops of sub-metamorphic and quartzite hematite occur all over Similipal hills.

Plate 1.3. Ganapati hill exposing basic lava flow of volcano-sedimentary rocks near Debasthali meadow in Similipal, Odisha, India. Photo: H Hazebrock.

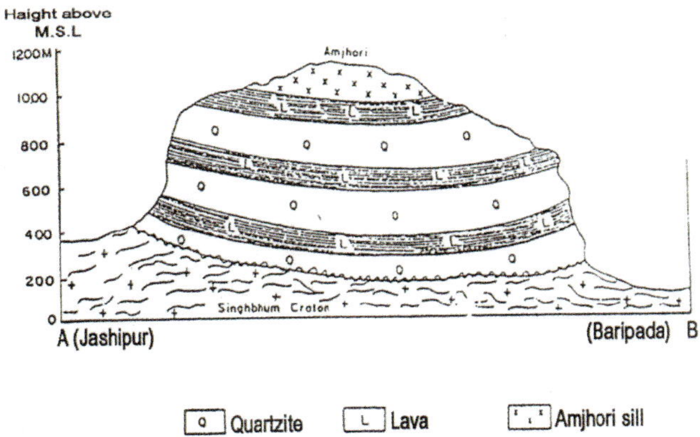

Figure 1.3. Geological and topographical cross section of Similipal (Mahalik, 1997)

SOIL

Extensive weathering of the volcano-sedimentary suite of rocks under the humid tropical climate has produced thick soil cover in Similipal. The soil is acidic in nature having a pH range of 4.8 to 6.8 in most areas. The main soil types are as follows:

(a) Derived from hematite, rich red loam soil is having intense biotic activity and dense tall woodlots.

(b) Laterites, produce deep soils, reddish in colour having the capability to support good tree growth. Where the depth of the soil is less, it supports poor tree growth, but good grass growth for animals.

(c) Shale, on weathering produces substantial deposits of clay and clayey loam soils good for biotic growth.

(d) Outcrops of sub-metamorphic sandstone and quartzite hematite on disintegration produce sandy soil. In deep layers, it supports good growth of both plants and animals. Where the depth of soil is thin it favours the growth of grasses.

Presently, soil erosion does not pose any serious threat except for Budhabalanga valley. The integrated complex of rock, soil and vegetation held in a stable formation provides little scope for soil degradation. However, sporadic denudation in small pockets remains, but it is not significant.

DRAINAGE

The unique geological formation of Similipal, thus, described as above, gives immense water retention capacity to the hills and makes it a land of perennial streams. Coupled with innumerable crests and valleys, clothed with lush forests, Similipal serves as the origin of a large number of streams flowing out in all directions to drain the area. It is, thus, the richest watershed in Eastern India and is the source of water to about 10 million people of North Odisha, South Jharkhand and South Bengal. Principal rivers originating from Similipal are Budhabalanga, East Deo, Palpala, Sanjo, Khadkei, Bhandan, Khairi, West Deo, Salandi, Thakthaki, and Telsim. The catchment area of each of the rivers forming part of the Similipal sanctuary is furnished in Table1.1. There is deep dissection of the Similipal plateau surface by each of these drainage systems, giving rise to narrow, deep valleys as well as waterfalls. From among a number of waterfalls in these hill ranges, two important ones attracting tourists from far and wide are Barehipani Fall (217m) **(Plate 1.4)** and Joranda Fall (181m) **(Plate 1.5)**.

CLIMATE

In this region, the year can be divided into three principal seasons: (1) Warm, wet season (July-October); (2) Cool dry season (November-February);

Plate 1.4. Barehipani waterfall in Similipal, Odisha, India. Photo: D Swain.

Plate 1.5. Joranda waterfall in Similipal, Odisha, India. Photo: B Mohanty.

Table 1.1. Catchment areas of rivers/streams constituting Similipal sanctuary

Sl. No.	*Name of River/Stream*	*Area of Catchment Constituting Similipal Sanctuary (in km²)*
1.	Budhabalanga	408.51
2.	Khairi	320.85
3.	West Deo	319.11
4.	Palpala	199.42
5.	Sanjo	177.61
6.	Salandi	172.07
7.	Bhandan	159.83
8.	East Deo	123.33
9.	Khadkei	115.65
10.	Thakthaki	95.77
11.	Tel and Sim	94.51
	Total	**2186.66**

(3) Hot, humid season (March-June). Temperature and precipitation for different seasons in Similipal forests based on data recorded in meteorological stations of Similipal tiger reserve are presented in Table 1.2.

Table 1.2. Temperature and Precipitation for different seasons in Similipal sanctuary (Swain and Morris, 2014)

	Warm, Wet Season	Cool, DrySeason	Hot, Humid Season
Maximum Temperature	17°C-44.5°C	6°C - 42°C	14°C - 49°C
Mean	28.6°C	25.3°C	32.7°C
SD	3.2°C	3.6°C	4.4 °C
Minimum Temperature	8°C - 32.5°C	1°C - 32°C	6°C - 37°C
Mean	22.2°C	19.1°C	21.4°C
SD	2.8°C	4.2°C	4.4°C
Precipitation	82.16cm-172.72cm	1.19cm-22.55cm	50.06cm-99.00cm
Mean	139.87 cm	7.74 cm	71.58 cm
SD	27.55cm	6.76cm	16.12cm
Months	July-October	November-February	March-June

The Similipal complex on land is located in such a way that this structure acts as a water tower of Odisha, Paschim Banga (West Bengal), Chhatisgarh and Jharkhand. The south-eastern part of Similipal gets more precipitation compared to the north and north-western parts. The atmosphere in the south-eastern portion which is the sea-ward side of the hill range normally remains humid due to the wind flow. The peaks of Khairiburu and Meghasani obstruct the wind flow. The micro-climate of the area assists in providing more precipitation on this side of Similipal. Thus, this landscape gets comparatively more precipitation due to the monsoon wind from sea coast side. The rainy season is sultry, humid and occurs between June-October. The South West Monsoon brings nearly 90 per cent of the total annual precipitation. North Similipal is comparatively drier relative to south Similipal. During winter, frost **(Plates 1.6A and B)** occurs in the Upper Barhakamuda valley and other adjoining valleys in south Similipal. Nawana valley in central Similipal also receives sporadic frost resulting in significant dip in temperature in these areas.

The air is quite humid with relative humidity always exceeding 70 per cent. No information on wind speed is available presently. During winter, a cold wind blows inside the Similipal sanctuary locally called as '*Kaliani Pabana*'. Spring breeze is quite common all through the sanctuary. Pujari (1997) examined rainfall data of places surrounding Similipal and drew some conclusions

Plate 1.6A. Icy frost deposited on the grassland at Upper Barhakamuda of Similipal, Odisha, India. Photo: D Swain.

Plate 1.6B. Frost pooling in winter scorched medium sized sal (*Shorea robusta*) trees in south Similipal, Odisha, India. Photo: D Swain.

about the cyclic patterns in the trend of annual rainfall. However, sufficient data in the matter are not available for arriving at any definite conclusion in regard to pattern of rainfall. The presence of numerous perennial streams, rivulets, and rivers indicates no significant drought in any year inside Similipal hill ranges.

CONSTITUTION OF SIMILIPAL TIGER RESERVE (STR)

Similipal was designated as one of the initial nine tiger reserves in the country during the year 1973 covering an area of 2750 km^2. The break up of the tiger reserve area in 1973 was as follows:

Total area: 2750 km^2.

Area of reserved forests (Similipal RF, notified in 1959): 2091 km^2.

Non-RF area: 659 km^2 (includes Protected Forests, Revenue Lands, village areas, cultivated lands, *etc.*).

Original (1st) core area: 304.57 km^2 (North Similipal).

In 1979 the State Government declared whole of Similipal RF as a sanctuary under Wlidlife (Protection) Act 1972. The core area of Similipal tiger reserve was further extended in 1985 to cover South Similipal (2nd core area) and the revised core area became 845.70 km². The core area of the sanctuary is a proposed national park as per notification issued in 1980 (1st core area) and in 1986 (2nd core area).

After amendment of Wlidlife (Protection) Act 1972 in 2006, Similipal tiger reserve was reconstituted in 2007 **(Figure 1.4)** as follows:

Revised total core area: 1194.75 km²

Buffer area: 1555.25km².

CONSTITUTION OF SIMILIPAL BIOSPHERE RESERVE (SBR)

Similipal was constituted as biosphere reserve (**Figure 1.5**) during the year 1994, the details of which are furnished as follows:

A. Core (= Core of STR)
   (Revised 2007) — 1194.75 km²

B. Buffer — 1555.25 km²

C. Transitional Area — 2819 km²

(Transitional area is within about 10 km radius from the outer boundary of the buffer)
TOTAL (SBR) ... 5569 km²

CONCLUSION

Similipal is one of the largest legally notified reserve forests in the country. It is also a tiger resrve and a biosphere reserve. The landscape is unique and replete with rolling hills, exquisite waterfalls, perennial streams, and diverse flora and fauna. Following chapters of this book unfold the mysteries of amazing Similipal, a wonderland of fairly pristine ecosystem, that merits preservation for posterity.

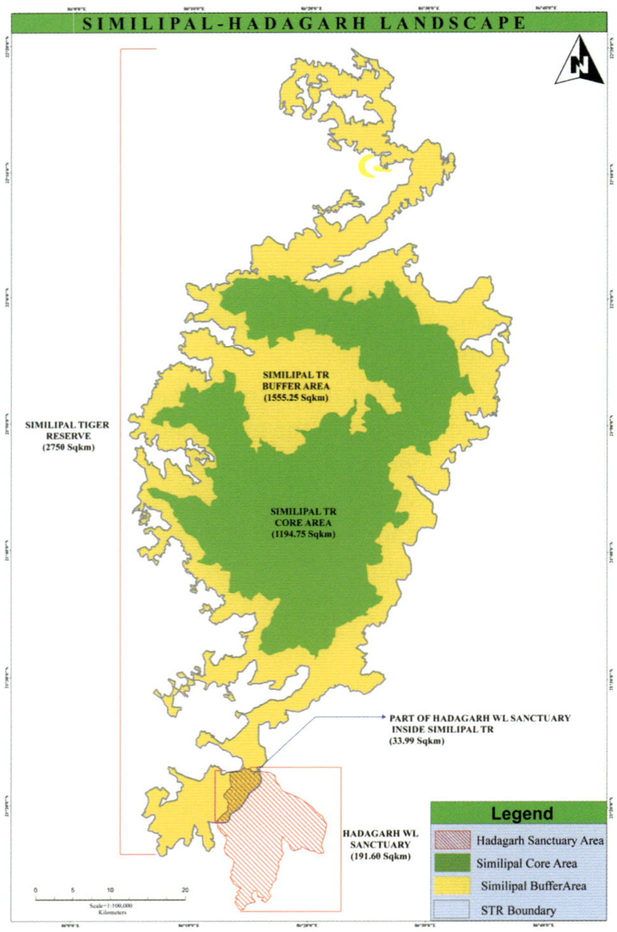

Figure 1.4: Constitution of Similipal tiger reserve, Odisha, India

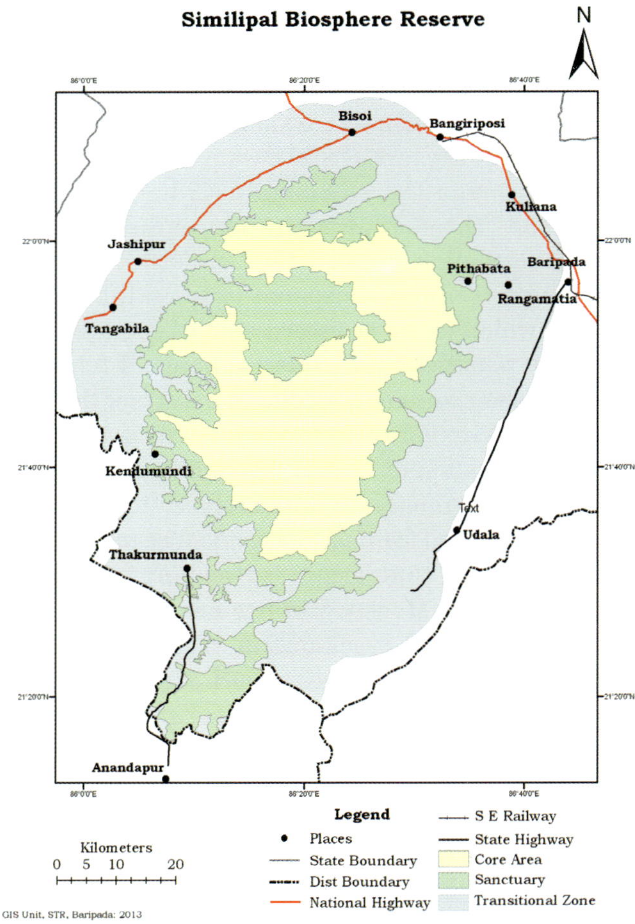

Figure 1.5: Constitution of Similipal biosphere reserve, Odisha, India.

Chapter 2

The Forest

"The forest is a peculiar organism of unlimited kindness and benevolence that makes no demand for its sustenance and extends generously the products of its life activity; it affords protection to all beings, offering shade even to the axe-man who destroys it."

– Gautama Buddha (c 563 BCE to 483 BCE) in: John Alexander, James D. Lazell *Ribbon of Sand: The Amazing Convergence of the Ocean and the Outer Banks*, UNC Press Books, 2000, p. 91.

INTRODUCTION

A few years back, I made a presentation on Similipal before Professor Peter Ashton at Harvard University. Professor Ashton is an authority on South Asian forests and he had travelled widely for his studies in the forests of South Asian countries. Though he had occasions to visit forests of Hazaribag and Singbhum in India, unfortunately, he could not visit Similipal earlier. When he learnt that Similipal, a sub-tropical forest with a repository of large flora having frostbitten grass lands/savannahs was a mixture of south Indian, north Indian and Andaman species, having the largest patch of sal (*Shorea robusta*) forest in the sal-teak (*Tectona grandis*) transition zone and having similarities with elements of Western Ghats and North-east India in floral and faunal composition, he was amazed and requested me to facilitate his visit to this unique forest of India. He visited Similipal thereafter and states in his book "On the forests of tropical Asia: Lest the memory fade" that Similipahar (Similipal) massif is the foremost island of deciduous lowland and lower montane forest remaining between Meghalaya and the Western Ghats (Ashton, 2014). On frostbitten valleys he writes that it is unexpected to see the lowland tropical sal annually reshooting (from biting frost) in grassland dominated by temperate grasses and herbs.

FOREST TYPES

The forests of Similipal (Similipal tiger reserve of 2750 km^2) comprise of an intermingling and a juxtaposition of different forest types, and according to Champion and Seth (1968) classification of forest types, there are the northern tropical moist deciduous forest, the northern tropical semi-evergreen forest, the dry deciduous hill forest, the high level sal forest, and lastly, grasslands and savannahs. Ashton (2014) classifies moist deciduous forest as 'tall sal forest' and dry deciduous forest as 'short sal forest'.

The village Jamuani (488m above MSL) is situated just outside the northern boundary of Similipal and when one travels from Jamuani to Chahala which is located at 770m above MSL, deep inside Similipal, she/he comes across northern tropical moist deciduous **(Plate 2.1A)** forest (type- 3C/C2e) or tall sal **(Plate 2.1B)** forest (according to Ashton, 2014) which is found all over Similipal **(Plate 2.1C)** except the moist valleys and on the southern and the eastern aspects of the hills. It extends over an area of about 2100 km^2.

Sal (*Shorea robusta*) **(Plate 2.2)** forms 50 per cent to 90 per cent of the standing crop. The quality of sal (based on height and girth of a sal tree) is 'IV' (inferior) on the steep drier aspects and 'II' on gentler slopes with deep soil on cooler aspects. Quality-I (superior) sal occurs in small pockets under good soil and climatic conditions.

Plate 2.1A. Northern tropical moist deciduous forest (type- 3C/C2e) in Similipal, Odisha, India. There are from left jamun (*Syzygium cumini*), mango (*Mangifera indica*), asan (*Terminalia tomentosa*), bauhinia (*Bauhinia semla* syn. *B. retusa*) and kusum (*Schleichera oleosa*) with other associates seen in the picture. Photo: H Hazebrock

Plate 2.1B. A large Sal (*Shorea robusta*) tree stands on the way from Kaliani to Gudgudia in Similipal, Odisha, India. Photo: H Hazebrock.

Plate 2.1C. Upper mixed moist deciduous forest on summit ridge at Meghasani (1160m altitude) in Similipal, Odisha, India. Photo: D Swain

Plate 2.2. Sal (*Shorea robusta*) has many economical and ecological uses. Cattle gather under sal trees to consume dropped sal flowers during flowering of sal. Above a sal flower. Photo: H Hazebrock.

The common associates of sal in this type of forests are: *Terminalia* spp. **(Plate 2.1A)**, *Anogeissus latifolia*, *Pterocarpus marsupium* **(Plate 2.3A)**, *Adina cordifolia* **(Plate 2.3B)**, *Schleichera oleosa* **(Plates 2.1A, 2.3B and C)**, *Toona ciliata* (rare), *Magnolia champaca*, *Mangifera indica* **(Plate 2.1A)**, *Bombax ceiba* **(Plate 2.3A)**, *Careya arborea*, *Mallotus philippensis*, *Dillenia pentagyna*, *Gmelina arborea*, *Garuga pinnata*, *Lannea coromandelica*, *Syzygium cumini* **(Plate 2.1A)**, *Desmodium oojeinense*, *Xylia xylocarpa*, *Kydia calycina*, *Lagerstroemia parviflora*, *Bridelia retusa*, *Mitragyna parvifolia*, *Phyllanthus emblica*, *Ziziphus* spp., *Cassia fistula*, *Buchanania lanzan*, *Sterculia villosa*, *Miliusa velutina*, *Helicteres isora*, *Indigofera* spp., *Croton* spp., *Melastoma malabathricum* **(Plate 2.4)**, *Colebrookea oppositifolia*, *Flemingia* spp. **(Plate 2.5A and B)**, *Strobilanthes* spp. **(Plate 2.6)**, *Wendlandia tinctoria*, *Imperata cylindrica*, *Themeda caudata*, *Cymbopogon martini*, *Eulaliopsis binata*, *Thysanolaena latifolia*, *Curcuma aromatica*, *Cheilocostus speciosus* **(Plate 2.7)**, *Hedychium coronarium*

Plate 2.3A. Piasal or Bija (*Pterocarpus marsupium*) in the meadow (artificially created for viewing animals) of Chahala and simul (*Bombax ceiba*) to right in Similipal, Odisha, India. Photo: H Hazebrock.

Plate 2.3B. Kusum (*Schleichera oleosa*) plant in a bloom of red leaves and kuruma or haldu (*Adina cordifolia*), 35m tall to left in Similipal, Odisha, India. Photo: H Hazebrock.

Plate 2.3C. View over forest from Meghsani hill top looking NE, with red *Schleichera oleosa* crowns. Photo: H Hazebrock.

Plate 2.4. *Melastoma malabathricum* in Similipal, Odisha, India. Photo: D Swain.

(Plate 2.8), *Bauhinia vahlii, Millettia extensa, Smilax zeylanica, Combretum album, Dioscorea* spp., *Asparagus racemosus, etc.*

Plate 2.5A. *Flemingia chappar* in Similipal, Odisha, India. Photo: D Swain.

Plate 2.5B. *Flemingia macrophylla* in Similipal, Odisha, India. Photo: H Hazebrock.

Plate 2.6. *Strobilanthes lupulinus* (flowers every 7 years) occurring at 1000 m MSL in the Meghasani hill of Similipal, Odisha, India. Photo: H Hazebrock

Plate 2.7. *Cheilocostus speciosus* in Similipal, Odisha, India. Photo: D Swain

Plate 2.8. *Hedychium coronarium* in Similipal, Odisha, India. Photo: D Swain

Plate 2.8A. *Cyathea gigantea* (tree fern) in Similipal, Odisha, India. Photo: D Swain.

Ferns and orchids are found in moist places. Ferns found are *Adiantum* spp., *Dryopteris* spp., *Cyathea gigantea* **(Plate 2.8 A)**, *Cyclosorus* spp., *etc.* Common orchids in this type of forest belong to the genera *Acampe* **(Plate 2.9A and B)**, *Bulbophylum* **(Plate 2.10)**, *Cymbidium* **(Plate 2.11)**, *Dendrobium* **(Plate 2.12)**, *Liparis* **(Plate 2.13)** and *Vanda* **(Plate 2.14).**

Plate 2.9A. Orchid *Acampe ochracea* in Similipal, Odisha, India. Photo: K K Mandal.

Plate 2.9B. Orchid *Acampe rigida* in Similipal, Odisha, India. Photo: S C Misra.

Plate 2.10. Orchid *Bulbophyllum crassipes* in Similipal, Odisha, India. Photo: S C Misra.

Plate 2.11. Orchid *Cymbidium aloifolium* in Similipal, Odisha, India. Photo: T Kar.

Plate 2.12. Orchid *Dendrobium regium* in Similipal, Odisha, India. Photo: S C Misra.

Plate 2.13. Orchid *Liparis viridiflora* in Similipal, Odisha, India. Photo: S C Misra.

Plate 2.14. Orchid *Vanda testacea* in Similipal, Odisha, India. Photo: T Kar.

At Chahala, in the upper reaches of the damp valleys of perennial streams and rivers, northern tropical semi- evergreen forests (type- 2B/C3) is seen.

Plate 2.15A. Northern tropical semi-evergreen forests (type- 2B/C3) of Similipal, Odisha, India. Photo: H Hazebrock.

Plate 2.15B. Evergreen understory in Northern tropical semi-evergreen forests of Similipal, Odisha, India. Photo: H Hazebrock.

This type of forest (**Plate 2.15A and B**) extends over an area of about 160 km^2 of southern and eastern part (Upper Barhakamuda, Jenabil and Pithabata Ranges) of Similipal and is characterized by a large number of dominant species having climbers (**Plate 2.16A**) and abundant epiphytes (**Plate 2.16B**). A few of the species of the top canopy and most of the under storey are evergreen and occur intimately mixed with each other. The species found under this type depending upon the soil and micro climatic conditions are as follows.

Plate 2.16A. Climber (*Combretum album* syn. *C. roxburghii*) tangles in semi-evergreen forests of Similipal, Odisha, India. Photo: D Swain.

Plate 2.16B. Epiphytes in semi-evergreen forests of Similipal, Odisha, India. Photo: K K Mandal.

On Stream Beds

Rheophytic *Salix tetrasperma* **(Plate 2.17 A and B),** *Ardisia solanacea* **(Plate 2.18),** *Trewia nudiflora, Macaranga peltata, Aphanamixis polystachya, Symplocos cochinchinensis, Glochidion* spp., *Bischofia javanica, Syzygium cumini, Pongamia pinnata, Diospyros malabarica, Saraca asoca* and at places *Terminalia arjuna.*

Plate 2.17A. Rheophytic *Salix tetrasperma* on stream bed of Similipal, Odisha, India. Photo: H Hazebrock.

Plate 2.17B Inflorecence of *Salix tetrasperma*. Photo: K K Mandal.

Plate 2.18. *Ardisia solanacea* in Similipal, Odisha, India. Photo: T Kar.

A little Higher up in Damp Areas

Bombax ceiba, Alstonia scholaris, Ficus spp., *Polyalthia cerasoides, Anthocephalus chinensis, Dillenia pentagyna, Litsea* spp., and *Citrus* spp.

Still Higher up

Magnolia champaca, Artocarpus lacucha **(Plate 2.19A)**, *Toona ciliata, Mangifera indica, Xantolis tomentosa, Mesua ferrea, Stereospermum chelonoides, Xylia xylocarpa* and *Bridelia retusa. Lasiococca comberi* **(Plate 2.19B)** endemic to Eastern Ghats (Sandhyarani *et al.*, 2007) occasionally occurs in semi-evergreen forests at Balidarh, Tarinibila, Hatisal and Bhanjabasa of Similipal.

Plate 2.19A. *Artocarpus lacucha* in semi-evergreen forests of Similipal, Odisha, India. Photo: T Kar.

Plate 2.19B. *Lasiococca comberi* endemic to Eastern Ghats occasionally occurs in semi-evergreen forests of Similipal, Odisha, India. Photo: D Swain.

On the steep and exposed slopes particularly in the northern aspect interspersed with moist deciduous forest there are dry deciduous hill **(Plate 2.20A)** forests (type- 3C/C3) or according to Ashton (2014)

Plate 2.20A. Dry deciduous hill forests (type- 3C/C3) of Similipal, Odisha, India. Photo: H Hazebrock.

Plate 2.20B. Dhatuki (*Woodfordia fruticosa*) in the understory of dry deciduous hill forests of Similipal, Odisha, India. Photo: K K Mandal.

short sal forest. It is extended over an area of 80 km^2. Both upper and lower canopies in these forests are composed of deciduous species. This type of forest has sal as its major species up to 30 per cent of the crop. Other associates are *Anogeissus latifolia, Sterculia urens, Boswellia serrata, Dalbergia latifolia, Cleistanthus collinus, Gardenia gummifera, G. latifolia, Erythrina suberosa, Cochlospermum religiosum, Helicteres isora, Woodfordia fruticosa* **(Plate 2.20B)**, *Nyctanthes arbor-tristis, Coffea benghalensis, Mimosa himalayana* **(Plate 2.20C),** *etc.*, with an abundance of many herbs, shrubs and grasses as ground cover.

Plate 2.20C. *Mimosa himalayana* in the understory of dry deciduous hill forests of Similipal, Odisha, India. Photo: D Swain.

At an elevation above 850m on the plateaus near Dhudurchampa, Bakua and between Jenabil and Upper Barhakamuda, high level sal **(Plate 2.21A)** forests [type- 3C/C 2e(i)] occur in patches of area of about 250 km^2. Pure stands of poor quality sal are found with *Dillenia pentagyna, Syzygium cerasoides, Pterocarpus marsupium* and grasses like *Imperata cylindrica* and *Themeda caudata* co-existing as associates. Large patches of *Phoenix acaulis* **(Plate 2.21 B)** occur as the usual open forest cover.

Plate 2.21A. High level sal forests [type- 3C/C 2e(i)] of Similipal, Odisha, India. Photo: D Swain.

Plate 2.21B. *Phoenix acaulis* (Khajuri) occurring in High level sal forests. Photo: D Swain.

Grassland and savannah (type- 3C/DS-I) extends over about 40 km^2 and it is seen between Jenabil and Upper Barhakamuda. Altitudinally ubiquitous in nature, grasslands are found on hill tops over 900 metres high as well as in the frosty valleys **(Plate 2.22)** and along the bank of the streams **(Plate 2.23).** In the former it is perhaps a climax type whereas in the latter, it seems to be of seral origin, a stable "pre-climax" under the combined influence of edaphic and climatic factors, mode of origin and intensity of biotic effects.

Plate 2.22. Grassland and savannah (type- 3C/DS-I) of Similipal, Odisha, India. Sal (*Shorea robusta*) and other frost tender woody plants are annually bitten back to field level whippy growth. Photo: H Hazebrock.

In the frosty open valleys, once cleared by fire during late January/early February, the sal and other frost tender woody plants are annually bitten back to field level whippy growth **(Plate 2.22)**. The species include

Plate 2.23. Grassland along the bank of the streams of Similipal, Odisha, India. Bottomland frost ponding (Ashton, 2014) scorched peripheral sal (*Shorea robusta*) unaffecting willow (*Salix tetrasperma*) allong the stream and also sal on the distant slopes. Photo: D Swain.

Syzygium cerasoides, Symplocos racemosa, Dillenia pentagyna, etc. The common members of Poaceae are *Imperata cylindrica, Saccharum spontaneum* and *Themeda* spp. *Aeginetia indica* **(Plate 2.24),** a holoparasitic herb is seen occasionally in the grassland at higher altitudes.

Plate 2.24. *Aeginetia indica*, a holoparasitic herb seen occasionally in the grassland at higher altitudes of Similipal, Odisha, India. Photo: D Swain.

Exacum tetragonum **(Plate 2.25)**, endemic to peninsular India is frequently found in these grasslands. Microtherm plants like *Hypericum* species **(Plate 2.26)** are also observed.

On hill tops: *Cymbopogon caesius, Eulaliopsis binata, Pogonatherum paniceum, Sehima nervosum, etc.*

Plate 2.25. *Exacum tetragonum*, endemic to peninsular India is frequently found in the grasslands of Similipal, Odisha, India. Photo: D Swain.

Plate 2.26. Microtherm shrub *Hypericum gaitii* in Eucalyptus plantation near Tinadiha (between Patbil and Upper Barhakamuda) of Similipal, Odisha, India. Photo: D Swain

In moist areas: *Apluda mutica, Arundo donax, Eragrostis atrovirens, Phragmites karka, Sporobolus indicus, Sacciolepis indica, etc.*

In other grasslands: *Bothriochloa bladhii, Cymbopogon flexuosus, Cynodon dactylon,*

Heteropogon contortus, Imperata cylindrica, and *Themeda* spp., etc.

FLORAL DIVERSITY

Saxena and Brahmam (1989) report the occurrence of 1076 species of plants representing 168 families in Similipal. Misra (1997) reports 93 species of orchids in Similipal. The works of Haines (1921-1925), Mooney (1950), Panigrahi *et al.* (1964), Saxena and Brahmam (1989, 1994-96), and Misra (1997) account for 1097 plant species of Similipal. Misra *et al.* (2011) added occurrence of 118 species in this reserve. Work of Biswal *et al.* (2013), Kar *et al.* (2014a,b,c,d), Kar *et al.* (2017), Kar *et al.* (2018), Mandal *et al.* (2012), Mandal *et al.* (2014), Misra and Kar (2017), Misra and Sahoo (2014), Nayak *et al.* (2014) and Nayak (not dated) brought the total number of plant species occurring in Similipal to 1359 **(Appendix I).** The ranges of flora existing and evolving are the elements of evergreen forests of medium to high tablelands as well as alpine and savannah elements. The flora at hilltops in Similipal presents very interesting phytogeographical features as a number of Himalayan species, both temperate and subtropical and some south Indian hill species are found here. There are four species of gymnosperms of which only two namely, *Gnetum ula* and *Cycas circinalis*, are native to Similipal. Saxena and Brahmam (1989) state that the percentage ratio of dicotyledonous and monocotyledonous species in the flora of Similipal is 71.5:28.5 as against 68.9:31.1 in the flora of Odisha and 81.3:18.7 in the world flora. This shows higher percentage of monocotyledons in the flora of Similipal compared to that in the world flora, as is also the case with the Odisha flora. This is due to the large number of species of grasses in Similipal.

The Similipal forest is topographically and climatically located to form dense vegetation. The high rainfall in these hill forests creates a humid atmosphere favourable for growth of orchids. Botanical survey conducted in Similipal by Sarat Misra (Misra, 2014) showed the occurrence of a large number of orchids not known earlier from here, Odisha or India; 94 (41 terrestrial and 53 epiphytic) species were recorded. This included *Eria meghasaniensis* **(Plate 2.27)**, a new species endemic to Similipal, and *Goodyera thailandica* a new record for India.

Two terrestrial species, *Disperis neilgherrensis* **(Plate 2.28)** and *Eulophia mackinnonii* **(Plate 2.29)**, and one mycotrophic species, *Epipogium roseum* have been recorded recently (Misra and Kar, 2017; Kar *et al.*,2017). Further *Zeuxine gracilis*, a terrestrial orchid reported from nearby Rebana forests of Keonjhar district (Misra, 2014), is noticed to occur in Similipal (Truptirekha Kar *pers. comm.*) to make a total of 98 (44 terrestrial, one mycotrophic and 53 epiphytic) species known from this reserve (Appendix-I).

Plate 2.27. *Eria meghasaniensis*, a new species of orchid endemic to Similipal, Odisha, India. Photo: J D Pati.

Plate 2.28. *Disperis neilgherrensis*, a terrestrial orchid species in Similipal, Odisha, India. Photo: T Kar.

Plate 2.29. *Eulophia mackinnonii*, a terrestrial orchid species in Similipal, Odisha, India. Photo: T Kar.

A drastic decline in the orchid population of Similipal has been observed (Misra, 2004). Shrinkage of vegetation cover is the single factor that can be attributed to this cause. Orchids survive and regenerate best in undisturbed forests; it cannot survive in isolation unlike many other plants. Orchids are therefore considered as climatic indicators of a healthy eco-system. Simlipal is a tiger reserve. It is therefore expected that its vegetation cover now remains intact. Several forest compartments, especially in South Similipal, with rich orchid population should be identified as protectorates. Favourable measures like plantation of native host trees along the bank of perennial streams in open areas may be taken up. This will help in spreading of population of the epiphytic species. Orchid species which have now become rare, or the species found only in the Similipal forest, should be reintroduced to their habitats. Example may be cited of one epiphytic species, *Dendrobium regium* (**Plate 2.12)**, bearing large and magenta coloured flowers, that has suffered over-exploitation from its habitats. One terrestrial *Tainia penangiana* **(Plate 2.30A)**, a Himalayan element found only in one locality – near a stream at Jenabil, in a very small area too is facing depletion from the wild. Several orchids have therapeutic value in the Indian System of Medicine (Behera *et al.*, 2013). These are collected discreetely from Similipal forest. To avoid this, farming of orchid plants with medicinal value should be encouraged in areas outside Similipal. Being the most advanced plant family, orchids may be put to use in the near future in genetic engineering. For all these usefulness, besides their immense floricultural value, orchids should be conserved *in situ* in the wild and *ex situ* in botanic gardens for the posterity **(Plate 2.30B)**.

Plate 2.30A. *Tainia penangiana,* a terrestrial orchid species and a Himalayan element found only in one locality – near a stream at Jenabil, in a very small area is facing depletion from the wild in Similipal, Odisha, India. Photo: K K Mandal.

Plate 2.30B. Orchidarium at Gurguria, assemblage of 93 species of orchids occurring in Similipal, Odisha, India. Photo: D Swain.

Singh and Kumar (2012) records 141 species of lichens **(Photo 2.31A and B)** from Similipal biosphere reserve **(Appendix II**), out of which 129 species are

Plate 2.31A. A lichen and an orchid on a Sal tree in Similipal, Odisha, India. Photo: K K Mandal.

Plate 2.31B. A lichen and an orchid on a rock on the way to Meghasani summit of Similipal, Odisha, India. Photo: D Swain.

new records for the state of Odisha, while three species *viz. Conotrema lumbricoides*, *Megalotremis biocellata*, and *Ocellularia lankaensis* are new records for India. Similipal with its varied topography is an excellent habitat for rich growth of lichens. It was unexplored lichenologically until Singh and Kumar (2012) recorded 141 species of lichens during May 2010. It is expected that the number may increase following a detailed explorations to the interior of Similipal. Further Alam *et al.* (2013) reports occurrence of 33 taxa of bryophytes in this biosphere reserve and its neighboring areas **(Appendix III)**. Bhakta *et al.* (2014) documented 19 species of diverse algal forms collected from tree bark surfaces at 10 locations inside Similipal, 18 being cyanobacteria and only one belonging to green algae. The major dominant cyanobacterial taxa were *Nostoc* (4) followed by *Gloeocapsa* (2), *Chroococcidiopsis* (2), *Aphanocapsa* (2), *Phormidium* (2), *Scytonema* (2), *Myxosarcina* (1), *Stigonema* (1), *Tolypothrix* (1), *Fischerella* (1) and a single species of green algae named *Desmococcus olevaceus* recorded during their study **(Appendix IV).**

Medicinal plant wealth of Similipal is as varied as its plant diversity. Saxena and Brahmam (1989) list out 552 plants of medicinal importance inside Similipal. Of these the commercially important ones having potentiality for commercial exploitation are 70 in number. Medicinal plant wealth of Similipal biosphere reserve has been reported by Pandey *et al.* (2000, 2002) and Rout (2004). Rout's study (Rout 2004) on the medicinal plants in Similipal reveals 195 species belonging to 67 families. Some of the medicinal plants species, the parts used and their medicinal values have been listed in **Appendix VA.**

Similipal is a good source of variety of mushrooms **(Plate 2.32A,B,C and D)** used as food and medicine by the local community. Sachan *et al.* (2013) recorded traditional uses of some wild edible mushrooms by ethnic tribes of Similipal. Sachan and Thatoi (2017) found 18 species of mushrooms from Similipal which belong to three orders, seven families and 10 genera, of which 14 are edible and four are inedible. Among the collected mushrooms, genus *Russula* followed by *Termitomyces*, *Volvariella*, *Coprinus* and *Pleurotus* were found to be abundant in their occurrence. Thatoi (2014) in his project report submitted to Department of Science and Technology, Government of Odisha listed 61 species of mushroom from Similipal biosphere reserve and Panda (2019) further added 20 species of mushrooms available in the region. Therefore so far 81 species of mushrooms are recorded in Similipal and adjoining localities **(Appendix VB)**, many of which have been identified up to genus level.

The vegetation study in a newly created preservation plot located in an area of 23 ha inside the core zone of Similipal biosphere reserve (Khairi Compartment no. 25 and 27 of Similipal RF) shows that there are 23 species in the over wood, 33 species in the middle storey and 19 species in the ground floor, besides 20 species of climbers and 23 species of orchids representing 43 families (Swain and Nanda 1997). The dominant species in the over wood in order of high to low importance value index (IVI) are of *Shorea robusta*, *Magnolia champaca*, and *Mallotus philippensis*. In the middle storey it is *M. philippensis* being commonly present followed by *Colebrookea oppositifolia*, *S. robusta*, and *Casearia tomentosa*. Among the herbs it is *Cyperus rotundus* followed by *Adiantum incisum* and *Sida acuta* being common to

Plate 2.32A. *Volvariella volvacea*, most common edible mushroom in Similipal, Odisha, India. It is a nutritious and delicious food item used by local people and believed to be lowering high blood pressure. Photo: H N Thatoi.

Plate 2.32B. Mushroom *Calvatia* sp. in Similipal, Odisha, India, used by local tribal people as vegetable and it has several medicinal properties. Photo: H N Thatoi.

Plate 2.32C. *Coprinus disseminatus, a* non-edible mushroom available in Similipal, Odisha, India. Photo: H N Thatoi.

Plate 2.32D. *Ganoderma chalceum,* a non-edible mushroom available in Similipal, Odisha, India. Photo: H N Thatoi.

locate. Climbers are mainly *Millettia extensa*, *Smilax* species, *Dioscorea* species, *Bauhinia vahlii*, *Acacia pennata etc.* Orchids are comprised of *Acampe* species, *Dendrobium* species, *Luisia* species, *Aerides* species, *Cymbidium* species, *etc.* The study further states that *S. robusta*, *Anthocephalus chinensis*, *Protium serratum*, *Bridelia retusa* and *Litsea monopetala* have adequate regeneration, whereas *Syzygium cumini*, *Actinodaphne angustifolia*, *Ficus semicordata*, *Ochna squarrosa*, *Anogeissus latifolia* and *Phyllanthus emblica* are having intermediate or higher girth classes with absence of seedling. The population structure of *M. champaca*, *M. philippensis*, *Grewia tiliifolia*, *Trema orientalis*, *Kydia calycina*, *Toona ciliata*, *Garuga pinnata*, *Xylia xylocarpa*, *Dillenia pentagyna*, *Pterocarpus marsupium*, *Cassia fistula* and *Desmodium oojeinense* indicates interrupted regeneration resulting from adverse climatic and/or bioadaphic factors inhibiting the regeneration completely for certain period of time.

Mishra *et al.* (2008) carried out vegetation analysis at 10 locations across Similipal forests to study changes in structure and composition in plant community distributed in the core and buffer areas of the reserve. They report a higher number of herbs and shrubs and a lower number of trees in the buffer area indicating greater anthropogenic disturbance. The total tree basal area varied from 48.7 to 78.61 m^2/ha in the buffer (disturbed) area and 81.4 to 104.9 m^2/ha in the core (undisturbed) area. Though the density of saplings and seedlings was nearly equal both at the disturbed and undisturbed sites, the rate of conversion of saplings to trees was greater at undisturbed sites. The lower rate of conversion at disturbed sites is due to the removal of seedlings of most of the tree species. The buffer area was comprised of high herb diversity (2.14 – 3.50) and low tree diversity (2.14 – 2.98) resulting from openings in the over wood which facilitated opportunity for the recruitment of herbs and shrubs.

FOREST COVER

Forest cover plays an important role in soil and water conservation in the catchments of rivers and streams (FAO, 2005). Forests also sequester and store more carbon than any other terrestrial ecosystem (FAO, 2006). Similipal tiger reserve consisting of 2750 km^2 of area has a core area of 1194.75 km^2 and buffer area of 1555.25km^2. Forest cover density maps of Similipal tiger reserve were studied based on India State of Forest Reports (ISFR) of Forest Survey of India, Dehradun for the year 2005, 2009, 2013 and 2017, and placed in Table 2.1 and 2.2. The satellite data utilised were of IRS-1D, sensor LISS-III, spatial resolution 23.5m*23.5m and spectral resolution 4 bands. The forest cover and land use have been classified as follows:

(i) Very dense forest (VDF): Canopy density >70 per cent
(ii) Moderately dense forest (MDF): Canopy density between 40 – 70 per cent
(iii) Open forest (OF): Canopy density between 10 – 40 per cent
(iv) Scrub: Canopy density <10 per cent
(v) Non-forest (NF)

There are little changes in different categories (very dense, medium dense and open) forest cover of Similipal tiger reserve from ISFR 2005 to ISFR 2017 as evident from Table 2.1 and Table 2.2. According to ISFR 2017, dense forest (very dense forest **(Plate 2.33)** and medium dense **(Plate 2.34)** forest together) constitutes

Table 2.1: Forest cover figures in buffer area of Similipal tiger reserve based on India State of Forest Reports (ISFR)

Sl. No.	ISFR Year	Data Year	Forest Cover (Buffer) in km²						
			A1	B1	C1	D1	E1	F1	G1
			VDF	MDF	OF	Total Forest	Scrub	NF	Total Area (D1+E1+F1)
1	2005	Nov-04	523.02	671.55	196.25	**1390.82**	1.89	162.54	**1555.25**
2	2009	Dec-06	523.02	671.55	196.25	**1390.82**	1.89	162.54	**1555.25**
3	2013	Nov, Dec-2011	521.31	673.31	196.20	**1390.82**	1.88	162.55	**1555.25**
4	2017	Nov-15	521.07	671.8	197.95	**1390.82**	1.97	162.46	**1555.25**

Table 2.2: Forest cover figures in core area of Similipal tiger reserve based on India State of Forest Reports (ISFR)

Sl. No.	ISFR Year	Data Year	Forest Cover (Core) in sq km						
			A2	B2	C2	D2	E2	F2	G2
			VDF	MDF	OF	Total Forest	Scrub	NF	Total Area (D2+E2+F2)
1	2005	Nov-04	687.77	474.88	20.67	**1183.32**	0	11.43	**1194.75**
2	2009	Dec-06	687.77	474.88	20.67	**1183.32**	0	11.43	**1194.75**
3	2013	Nov,Dec-2011	685.67	475.35	22.30	**1183.32**	0	11.43	**1194.75**
4	2017	Nov-15	686.82	474.86	21.64	**1183.32**	0	11.43	**1194.75**

85.62 per cent area of Similipal tiger reserve, whereas open forest **(Plate 2.35)** constitutes 7.98 per cent of the reserve. Scrub forest is negligible and present only in the buffer. In my earlier study (Swain, 2014) it was found out that grassland (meadow), crucial for wild herbivores **(Plate 2.23 and 2.36)** was only 0.04 per cent and it was available in the core area. No grassland was present in the buffer area of Similipal. Barren rocky/stony waste **(Plate 2.37)** constituted 2 per cent of Similipal forests. Water bodies, mostly in the form of streams **(Plate 2.38)** flowing inside Similipal formed 0.23 per cent. Tree-clad area was 0.12 per cent. Agricultural land and built-up land was 0.9 per cent.

Plate 2.33. Very dense forest (canopy density >70 per cent) as seen from top of Meghasani peak in Similipal, Odisha, India. Photo: D Swain.

Plate 2.34. A smooth canopy of medium dense forest (canopy density between 40 – 70 per cent) in Similipal, Odisha, India. Photo: H Hazebrock.

Plate 2.35. Open category forest (canopy density between 10 - 40 per cent) constitutes only 7.98 per cent in Similipal, Odisha, India. Photo: H Hazebrock.

Plate 2.36. Grassland (meadow), crucial for wild herbivores constitutes only 0.04 per cent in Similipal, Odisha, India. Photo: D Swain.

Plate 2.37. Barren rocky/stony waste constitutes 2 per cent of forests in Similipal, Odisha, India. Photo: D Swain.

Plate 2.38. River Palpala inside Similipal, Odisha, India. Parennial water bodies, mostly in the form of streams constitute 0.23 per cent of Similipal forest. Photo: D Swain.

The forests found in Odisha broadly fall under five out of the 16 major forest types of India (Champion and Seth, 1968), *i.e.*, tropical semi evergreen forests, tropical moist deciduous forests, tropical dry deciduous forests, subtropical broad leaved hilly forests and littoral swamp forests. Tropical dry deciduous forests form a major biome in the Indian peninsula by covering about 46 per cent of the total forest cover of the country (Singh and Singh, 1988). In Odisha, the dry deciduous forest being one of the conspicuous type of vegetation, contributes as much as 12 per cent to the country's deciduous forest cover (Mohanty, 2001). But in Similipal tropical moist deciduous forest constitutes 70 per cent of the forest cover. The chief feature of this moist deciduous forest is a leafless period in the dry season (March-May). The unique geological formation of Similipal, gives immense water retention capacity to the hills, as a result of which an appreciable number of deciduous trees get clothed in new flush of leaves and often flower before the onset of monsoon. The floor in most places is covered by grasses and annual fire is a common phenomenon resulting in the spread of grass, which becomes heavy and continuous especially where the canopy is open resulting into grassland and savannah.

CARBON SEQUESTRATION

Globally, each cubic meter (cum) of forest growing stock equals on an average, one ton of above ground biomass, 1.3 tons of total biomass and 0.7 tons of carbon (FAO, 2006). It is estimated that the world's forests store 283 gigatonnes (Gt) of carbon in their biomass alone and 638 Gt of carbon in the ecosystem as a whole (to a soil depth of 30 cm). The main carbon pools in tropical forest ecosystems are the living biomass of trees and understorey vegetation and the dead mass of litter, woody debris and soil organic matter. The carbon

Table 2.3: Biome-average tropical forest biomass carbon stock estimates (t C/ha) of Asia. All values are for above- and belowground forest biomass carbon. Table adopted from Gibbs *et al.*, 2007

Forest Type	*Houghton (1999)/ DeFries et al. (2002)*	*Brown (1997)/Achard et al. (2004)*	*Gibbs and Brown (2007a, 2007b)*	*IPCC (2006)*
All forests	-	151	-	-
Tropical equatorial forest	250	-	164	180/225[a]
Tropical seasonal forest	150	-	142	105/169
Tropical dry forest	-	-	120	78/96

a Values here are for continental and insular Southeast Asia, respectively.

Table 2.4: Forest biomass carbon stocks estimates for India, Odisha and Similipal. Units expressed as million tonnes of carbon (M t C)

Region	*Forest Cover Ctudy*	*Houghton (1999)/DeFries et al. (2002)*	*Brown (1997)/ Achard et al. (2004)*	*Gibbs and Brown (2007a, 2007b)*	*IPCC (2006)*	*Based on all Estimates*
India (Gibbs *et al.*, 2007)	GLC, 2000	8997	7333	8560	5085	5085–8997
Odisha (Swain, 2014)	FSI, 2013	733	738	694	513	513-738
Similipal (Swain, 2014)	FSI, 2013	31	31	29	16	16-31

stored in the aboveground living biomass of trees is typically the largest pool and the most directly impacted by deforestation and degradation. Thus, estimating aboveground forest biomass carbon is the most critical step in quantifying carbon stocks and fluxes from tropical forests (Gibbs *et al.*, 2007).

Nearly all estimates of emissions from tropical deforestation are based on a handful of biome-average datasets where a single representative value of forest carbon per unit area (*e.g.* tonnes of C per hectare) is applied to broad forest categories or biomes (*e.g.* Achard *et al.*, 2002, 2004; DeFries *et al.*, 2002; Fearnside, 2000; Houghton, 1999; Houghton *et al.*, 2001; Ramankutty *et al.*, 2007). Various authors (Achard *et al.*, 2004; Brown, 1997; DeFries *et al*,2002; Gibbs and Brown, 2007a; 2007b; Houghton, 1999; IPCC, 2006) provide biome-average tropical forest biomass carbon stock estimates (t C/ha) for different regions of the world. Carbon stock of tropical forest of Asia has been furnished in Table 2.3. Forest biomass carbon stocks estimates for India, Odisha and Similipal derived from Table 2.3 are set in Table 2.4.

CONCLUSION

Similipal is a unique repository of large flora consisting of 1359 vascular which include 98 species of orchids, 64 species of cultivated plants and 60 species of ferns and fern allies. There is also the occurrence of 552 species of medicinal plant. Though no systematic study has been made on mosses, lichens and fungi, but there are occurences of atleast 141 species of lichens, 33 species of bryophytes and 19 species of corticolous algae reported.

The unique geological formation gives immense water retention capacity to the Similipal hills, for which tropical moist deciduous and semi-evergreen forests together constitute most of the Similipal forest. The dominant species is sal in most of the area. Dense forest (canopy density >40 per cent) constitutes 85.62 per cent area of Similipal, whereas open forest (canopy density 10-40 per cent) constitutes 7.98 per cent of the reserve. Scrub forest (canopy density <10 per cent) is negligible and present only in the buffer. Grassland, crucial for wild ruminants is only 0.92 km^2 (0.04 per cent) and it is available in the core area. No grassland is present in the buffer area of Similipal. Barren rocky/stony waste constitutes 44.42 km^2 (2 per cent) of Similipal forests. Water bodies, mostly in the form of streams flowing inside Similipal form 5.06 km^2 (0.23 per cent). Tree-clad area is 2.64 km^2 (0.12 per cent). Agricultural land and built-up land is 19.63 km^2 (0.9 per cent). Forest biomass carbon stock of Similipal has been estimated to be 16-31 million tonnes of carbon and the dense forest cover of Similipal having understory and ground flora has high potential to mitigate carbon dioxide emission. Singh (2019) claims that conservation of old forest is 40 times more effective than planted forest in term of sequestering carbon. It is therefore obligatory on our part to protect Similipal at any cost.

Chapter 3

The Fauna

"The greatness of a nation and its moral progress can be judged by the way its animals are treated"

– Mahtma Gandhi [1869-1948]

INTRODUCTION

On 01 April 2006 at 10.30 PM, while I was patrolling inside Similipal reserve noticed that a herd of sambar in the grassland of Compartment No. 24 (Tinadiha beat of Upper Barhakamuda range) was frightened, and looking towards a particular point at the edge of the grassland. A female adult sambar's leg was terribly trembling. Focusing the light in the direction of the sambar's stance, I observed a pair of red eyes inside grasses just about a foot (0.3 meter) above the ground, confusing to be that of a nightjar (*Caprimulgus asiaticus*). Though the eyes of nightjar are amber in colour when seen through headlight of a vehicle, the position of red eyes under the grass cover confused that of a nightjar. But focusing the light on these eyes, I discovered that it was a tiger taking position to attack the sambar. Because of torchlight it got disturbed, stood up and slowly moved away from the site.

I saw a king cobra which was regularly basking in the grasses coming out of its den located in the stone cervices of a wooden bridge over a stream near to the beat house of Patbil. Being an ectothermic animal, the king cobra used to laze under sun around noon even in summer season, as Patbil is situated in the midst of dense forest at an altitude of 900m and the atmosphere is still cool during day time. Mugger crocodiles present in the river system of Similipal are seen to hunt barking deer coming to the streams to drink water. There are many marshy lands locally called '*daldali*' that even trap mega herbivores like elephants and gaurs. One can easily imagine that the diverse topography and soil conditions coupled with climatic variation and rich diversity of autotrophs is sure to support a fascinating faunal composition in Similipal.

The presence of grassland and savannah are unique to Similipal. Sal and other frost tender woody plants are annually bitten back to field level whippy growth **(Plate 3.1A)**. Such typical naturally occurring meadows attract varieties of animals which depend on grazing and browsing **(Plate 3.1B)**.

Plate 3.1A. Sal (*Shorea robusta*) and other frost tender woody plants are annually bitten back to field level whippy growth and that creates unique meadows in Similipal, Odisha, India. Many wild animals love to browse tender shoots of sal. Photo: D Swain.

Plate 3.1B. Droppings of sambar (*Rusa unicolor*) in a meadow in Similipal, Odisha, India. Photo: D Swain

According to biogeographic classification by Rodgers and Panwar (1988) the three hierarchical levels of planning units under which Similipal has been classified are as follows:

Biogeographic zone	: Deccan Peninsula
Biogeographic Province	: Chhotnagpur
Biogeographic Region	: Mahanadian

There are 60 species of mammals, 70 species of reptiles, 23 species of amphibians and 164 species of butterflies in Similipal (Dutta *et al.*, 2009; Mohapatra *et al.*, 2017; Mohapatra *et al.*, 2019; Pratyush P Mohapatra *pers. comm.*). During the rains of 2005, a gangetic dolphin (*Platanista gangetica*) was for the first time sighted in the river system of Budhabalanga **(Plate 3.1C)**. **Appendix VI** provides a list of mammals in Similipal. Nayak and Naik (2014) report 361 bird species in Similipal biosphere reserve, whereas Manoj V. Nair recorded 385 species of avifauna **(Appendix VII)** in the reserve (Manoj V Nair *pers. comm.*). Baliarsingh *et al.* (2013) report 66 species of fishes in Similipal biosphere reserve. There are also many species of insects and micro fauna present in this unique habitat.

Plate 3.1C. During the rains of 2005, a gangetic dolphin (*Platanista gangetica*) was for the first time sighted in the Budhabalanga River in the Similipal biosphere reserve, Odisha, India. The dolphion was stranded in the shallow water, for which it was shifted to a deep water of the same river. Photo: D Swain.

MAMMALS

The primary heterotrophs among the mammals are represented by sambar (*Rusa unicolor*) **(Plate 3.2)**, elephant (*Elephas maximus*), gaur or Indian bison (*Bos gaurus*) (**Plate 3.3**), cheetal (*Axis axis*) **(Plate 3.4)**, chausingha or four horned antelope (*Tetracerus quadricornis*), barking deer (*Muntiacus vaginalis*) (**Plate 3.5**), common langur (*Semnopithecus entellus entellus*), Rhesus macaque (*Macaca mulatta*), porcupine (*Hystrix indica*) **(Plate 3.6)**, Malabar giant squirrel (*Ratufa indica*), Indian pangolin(*Manis crassicaudata*), Rufus-tailed hare(*Lepus nigricollis ruficaudatus*) etc. Among the secondary heterotroughs are tiger (*Panthera tigris*), leopard (*Panthera pardus*) **(Plate 3.7)**, wild dog

Plate 3.2. A sambar (*Rusa unicolor*) in Similipal forest, Odisha, India. Photo: D Swain.

Plate 3.3. Gaur (*Bos gaurus*) in Similipal, Odisha, India. Photo: J D Pati.

Plate 3.4. Cheetal (*Axis axis*) at Debasthali inside Similipal forest, Odisha, India. Photo: J D Pati.

Plate 3.5. Barking deer (*Muntiacus vaginalis*) in an artificial salt-lick at Upper Barhakamuda inside Similipal. Odisha, India. Photo: D Swain.

Plate 3.6 Porcupine (*Hystrix indica*) of Similipal forest, Odisha, India. Photo: Similipal TR archive.

Plate 3.7. Leopard (*Panthera pardus*) captured in camera trap inside Similipal forest, Odisha, India. Photo: Similipal TR archive.

Plate 3.8. Leopard cat (*Prionailurus bengalensis*) captured in camera trap inside Similipal forest, Odisha, India. Photo: Similipal TR archive.

(*Cuon alpinus*), leopard cat (*Prionailurus bengalensis*) **(Plate 3.8),** jungle cat (*Felis chaus*) **(Plate 3.8A)** *etc.* A palm civet, often sighted in Similipal having white stripe on its back **(Plate 3.9)**, was identified earlier as *Paradoxurus jorandensis* (Ali *et al.*, 1988) and later it was said to be an albinistic form of *P. hermaphroditus* (Jairajpuri, 1991), which is now considered as a distinct subspecies as *P. hermaphrodites nictitans* and also available at other places in Odisha, eg. Kandhamal, Barbara RF and Boudh districts (Mohapatra *et al.*, 2019).

Plate 3.8A. Jungle cat (*Felis chaus*) captured in camera trap inside Similipal forest, Odisha, India. Photo: Similipal TR archive.

Plate 3.9. *Paradoxurus hermaphrodites nictitans* camera trapped in Similipal, Odisha, India. Photo: Similipal TR archive.

AVIFAUNA

Some common birds are peafowl (*Pavo cristatus*), painted spur fowl(*Galloperdix lunulata*), black partridge (*Francolinus francolinus*), Shahin falcon (*Falco peregrinus peregrinator*), fly catcher (*Muscicapa muttui*), hawk(*Spizaetus cirrhatus*), Malabar pied hornbill (*Anthracoceros coronatus*), grey hornbill (*Ocyceros birostris*), red jungle fowl (*Gallus gallus*) **(Plate 3.10)**, hill myna (*Gracula religiosa*), golden oriole (*Oriolus oriolus*), common snipe (*Gallinago gallinago*), owls **(Plate 3.11)**, crested serpent eagle (*Spilornis cheela*), doves, parakeets, pigeons and sun birds **(Plate 3.12)**. Besides, water birds like white ibis (*Threskiornis melanocephalus*), red-wattled lapwig (*Vanellus indicus*), teal (*Nettapus coromandelianus*), plover (*Pluvalis fulva*),Indian river tern (*Sterna aurantia*),sand piper (*Actitis hypoleucos*), herons, egrets and king fishers are met with in artificial and natural water bodies in and around Similipal.

Plate 3.10. Red jungle fowl (*Gallus gallus*) captured in camera trap inside Similipal forest, Odisha, India. Photo: Similipal TR archive.

Plate 3.11. Brown hawk owl (*Ninox scutulata*) of Similipal, Odisha, India. Photo: D Swain.

Plate 3.12. A male purple sunbird (*Cinnyris asiaticus*), a pollinator, in full breeding plumage sucking honey from the flower of a noi palaso plant (*Butea superba*) in Similipal, Odisha, India. Photo: H Hazebrock

HERPETOFAUNA

Dutta *et al.* (2009) reported that the herpetofauna of Similipal was represented by 21 species of frogs and toads, and 60 species of reptiles, comprising one species of crocodile, six species of turtles and tortoise, 20 species of lizards and 33 species of snakes. Later, Mohapatra *et al.* (2017) and Pratyush P Mohapatra *pers. comm.* further added two species of frog and 10 species of reptiles, enriching the herpetofaunal diversity to 93 species in Similipal. **(Appendix VIII)**.

The amphibian fauna of Similipal have five families. The family Bufonidae (toads) is represented by three species in Similipal, of which *Duttaphrynus stomaticus* has a very patchy distribution in the Similipal landscape and has been reported from the Karanjia forest division. The family Dicroglossidae (true frogs) is represented by nine species under four genera in Similipal. An interesting species of Fejervaryan frog, mentioned by Mohapatra *et al.* (2017) as *Fejervarya* sp. was later described from Eastern India is *Minervarya kalinga* (Raj *et al.*, 2018) which breeds during winter seasons in the hill streams of Similipal. The family Microhylidae (narrow mouthed frogs) is represented by five species belonging to two genera now. Earlier report of painted frog (*Kaloula taprobanica*) and variagated ramanella (*Ramanella variagata*) are now placed under genus *Uperodon* Duméril and Bibron, 1841. Furthermore, the earlier report of the Ornate narrow-mouthed frog (*Microhyla ornata*) from Similipal by Dutta *et al.* (2009) and Mohapatra *et al.* (2017) is now considered as *Microhyla mymensinghensis* Hasan *et al.*, 2014 (Pratyush P Mohapatra *pers. comm.*). There is also another species of *Microhyla* sp. seen in Similipal, which needs further taxonomic study for species level identification (Sushil K Dutta *pers. comm.*). The family Ranidae includes two species, the widespread fungoid frog (*Hydrophylax bahuvistara*) and Theobald's ranid frog (*Hylarana tytleri*, Pratyush P Mohapatra *pers. comm.*). The family Rhacophoridae (tree frogs) is represented by four species under three genera. In a recent taxonomic amendation of the bush frogs of Peninsular India, Mirza *et al.* (2019) synonymized *Philautus similipalensis* Dutta, 2003 under *Raorchestes sanctisilvaticus* (Das and Chanda, 1997) (earlier referred as *Philautus sanctisilvaticus* Das and Chanda 1997 in their paper Das and Chanda, 1997) based on molecular study. Hence, the Similipal bush frog **(Plate 3.13)** along with the sacred grove bush frog (*R. sanctisilvaticus*) and the Eastern Ghats bushfrog (*R. terebrans*) are now considered to be one widespread species distributed in the Peninsular India.

Plate 3.13. A bush frog *Raorchestes sanctisilvaticus* synonym *Phillautus similipalensis* in Similipal, Odisha, India. Photo: P P Mohapatra.

Among reptiles, the only crocodilian species found in this reserve is Mugger (*Crocodylus palustris*) **(Plate 3.14)**.

Plate 3.14. A marsh crocodile (*Crocodylus palustris*) on the bank of a stream inside Similipal, Odisha, India. Photo: J D Pati.

Out of six species of turtles and tortoise found in Similipal, there are four species of hard-shell fresh water turtles (Family: Geoemydidae) (**Plate 3.15**), one species of land tortoise (Family: Testudinidae) and one species of soft-shelled fresh water turtle (Family: Trionychidae). Lizards of Similipal include garden lizards (Family: Agamidae), chamaeleon (Family: Chamaeleonidae), geckos (Family: Gekkonidae) and fat-tailed gecko (Family: Eublepharidae), snake lizard (Family: Lacertidae), skinks (Family: Scincidae) and monitor lizards (Varanidae).

Plate 3.15. Tricarinate hill turtle (*Melanochelys tricarinata*), one of six species of turtles of Similipal, Odisha, India. Photo: D Swain.

The snake fauna of Similipal comprises members of family Boidae, Pythonidae, Colubridae, Homalopsidae, Lamprophiidae, Sibynophiidae, Elapidae, Viperidae and Typhlopidae. Boas (Family: Boidae) and pythons (Family: Pythonidae) are the primitive snakes. There are two species of boas and one species of python found in Similipal. Twenty-three species of Colubrids (Family: Colubridae) under three subfamilies and 15 genera have been recorded. Interestingly Mohapatra *et al.* (2017) and Deepak *et al.* (2019) report that the genus *Ahaetulla* is represented by three species in Similipal, namely common vine snake (*A. nasuta*) (**Plate 3.16A**), variable coloured vine snake (*A. anomala*) (**Plate 3.16B and C**) and brown vine snake or laudankia vine snake (*A. laudankia*) (**Plate 3.16D**). Among the potentially

Plate 3.16A. Common vine snake (*Ahaetulla nasuta*) in Similipal, Odisha, India. Photo: D Swain.

Plate 3.16B. Variable coloured male vine snake (*Ahaetulla anomala*) in Similipal, Odisha, India. Photo: P P Mohapatra.

Plate 3.16C. Variable coloured female vine snake (*Ahaetulla anomala*) in Similipal, Odisha, India. Photo: P P Mohapatra.

Plate 3.16D. Brown vine snake or laudankia vine snake (*Ahaetulla laudankia*) in Similipal, Odisha, India. Photo: P P Mohapatra.

dangerous snake species recorded from the reserve are the five species of Elapids (Family: Elapidae) under three genera which include king cobra, cobras and kraits. Other than these, *Rhabdophis subminiatus* reported from the reserve during 2012 by Sahoo *et al.* (2012) is also considered to be a venomous snake. Typhlops or Blind snakes (Family: Typhlopidae) are all fossorial, living either underground or in the leaf litter. There are three species of Typhlops under three genera. Similarly three species under three genera of vipers (Family: Viperidae) are also found here.

FISHES

There are 66 species of fishes (**Appendix IX**) belonging to 42 genera, 19 families and six orders (Baliarsingh *et al.*, 2013). The mahasheer (*Tor tor* and *T. Putitora*) fish though much depleted in number now is notable among the rare fishes of Similipal. *T. Putitora* is the only endangered fish according to IUCN status present in Similipal.

INVERTEBRATES

The invertebrates are much more diverse in Similipal and equally significant in their ecological role compared to vertebrates. But they have been left unstudied to a large extent. All the major phyla including annelids, arthropods and molluscs occur in Similipal but taxonomic studies are yet to be taken up. Honeybees, wasps (**Plate 3.17A**), termites, butterflies and moths are found in abundant numbers. 'Modal' ecorace of silkworm (*Antheraea mylitta*) **(Plate 3.17B)** is unique to Similipal (Rao and Satpathy, 2003).

Plate 3.17A. A wasp nest collected from Similipal, Odisha, India. Photo: D Swain.

Plate 3.17B. Silk cocoons of 'Modal' ecorace of silkworm (*Antheraea mylitta*), unique to Similipal, Odisha, India. Photo: P K Kar

Presence of certain insectivorous birds like the Indian Roller, flycatchers, bee eaters and drongos in large numbers indicate the richness of insect population. Earthworms and ground leeches form the prominent annelids. Other invertebrates are also available across Similipal with varying congregations.

Ramakrishna *et al.* (2006) report that field studies, observations and collections of permissible and unscheduled fauna from as many as 39 sampling sites, in buffer, core and transition areas of Similipal biosphere reserve yielded 217 species of invertebrate fauna (**Appendix X**) under following major groups : arthropods-194 species, malacofallna-20 species and annelids- 2 species besdes few interesting undetermined specimens (myriapods/spiders). Payra *et al.* (2016) recorded a total of 136 species of butterflies (**Plate 3.18 to 3.20**) belonging to 87 genera representing five families from two selected study sites in the fringe area of Similipal biosphere reserve during January 2014 to November 2015 (**Appendix XI**). They noticed 15 species protected under the Indian Wildlife (Protection) Act (1972). Only 28 species are common with respect to 41 species of butterflies cited by Ramakrishna *et al.* (2006). Similarly Ramakrishna *et al.* (2006) mention only 11 species of odonates (dragonflies and damselflies) in Similipal, where as Das *et al.* (2012) report 58 species of odonates (**Appendix XII**) in buffer area only of Similipal biosphere reserve, of which seven are common in both the studies. Therefore there is a lot of scope for study of invertebrate fauna in Similipal.

Plate 3.18. Banded peacock (*Papilio crino*) in Similipal, Odisha, India. Photo: P P Mohapatra.

Plate 3.19. Commander (*Moduza procris*) in Similipal, Odisha, India. Photo: P P Mohapatra.

Plate 3.20. Common Nawab (*Polyura athamas*) in Similipal, Odisha, India. Photo: P P Mohapatra.

THE TIGER

The tiger is highly endangered throughout its Range. The remaining five subspecies, namely, *Panthera tigris tigris* in India, Bhutan, Nepal, and Bangladesh; *Panthera tigris corbetti*, in eastern Myanmar, south-eastern China, Laos, Thailand, Cambodia, and Vietnam; *Panthera tigris jacksoni* in Malayasia; *Panthera tigris sumatrae* in Sumatra; and *Panthera tigris altaica* in north-eastern China, North Korea and Far Eastern Russia are all highly endangered. *Panthera tigris amoyensis* in South China has been extinct in wild. Tiger populations have been extirpated from Afghanistan, Turkmenistan, Iran, Turkey, Java, and Bali with the extinction of three subspecies, the Caspian tiger *Panthera tigris virgata*, the Javan tiger *Panthera tigris sondaica,* and the Bali tiger *Panthera tigris balica* (Maza'k, 1996; Nowell and Jackson, 1996; Kitchener, 1999; Seidensticker *et al.*, 1999; Sunquist *et al.*, 1999; Luo *et al.*, 2004; Dinerstein *et al.*, 2007). However according to Kitchener *et al.* (2017), there are only two subspecies of tigers: mainland Asia, *Panthera tigris tigris* having larger size, paler pelage with fewer stripes; and Sunda islands, *Panthera tigris sondaica* having smaller size, darker pelage with more frequent stripes. The primary causes of disappearance of tigers from their habitats are poaching, habitat loss and degradation, prey depletion, and conflict with humans (Nowell and Jackson, 1996; Ramakrishnan *et al.*, 1999; Seidensticker *et al.*, 1999; Johnsingh and Negi, 2003;

Graham *et al.*, 2005; Wang and Macdonald, 2006; Ranganathan *et al.*, 2008).

Similipal is one of the nine tiger reserves in India initially constituted during 1973. There were 17 tigers counted in 1972 based on a technique invented by the then Field Director (Padma Shri Late S. R. Choudhury), popularly known as 'Pugmark census'. Pugmark census (Swain *et al.*, 2005) was followed until 2004 **(Plate 3.21)**, and the tiger numbers based on Pugmark census year wise are set in Table 3.1.

Plate 3.21. Pugmark census in progress during January 2004 in Similipal, Odisha, India. Photo: D Swain.

Table 3.1. The year wise counting of tiger in Similipal based on Pugmark census

Year	*Male*	*Female*	*Cub*	*Total*	
1972				17*	*These figures are incomplete because the entire area of STR wasn't covered
1975				30*	
1976	46				
1977	23	29	4	56	
1979	22	39	4	65	
1984	31	43	7	81	
1986	32	51	6	89	
1989	21	51	21	93	
1990	22	52	20	94	
1991	24	50	22	96	
1992	24	49	22	95	
1993	24	49	22	95	
1995	25	47	25	97	
1997	26	48	24	98	
1999	28	44	26	98	
2000	29	39	29	99*	*Includes 2 of unknown sex
2002	29	43	27	99	
2004	28	41	32	101	

BOX-3.1 TIGER POPULATION ESTIMATION (SECR)

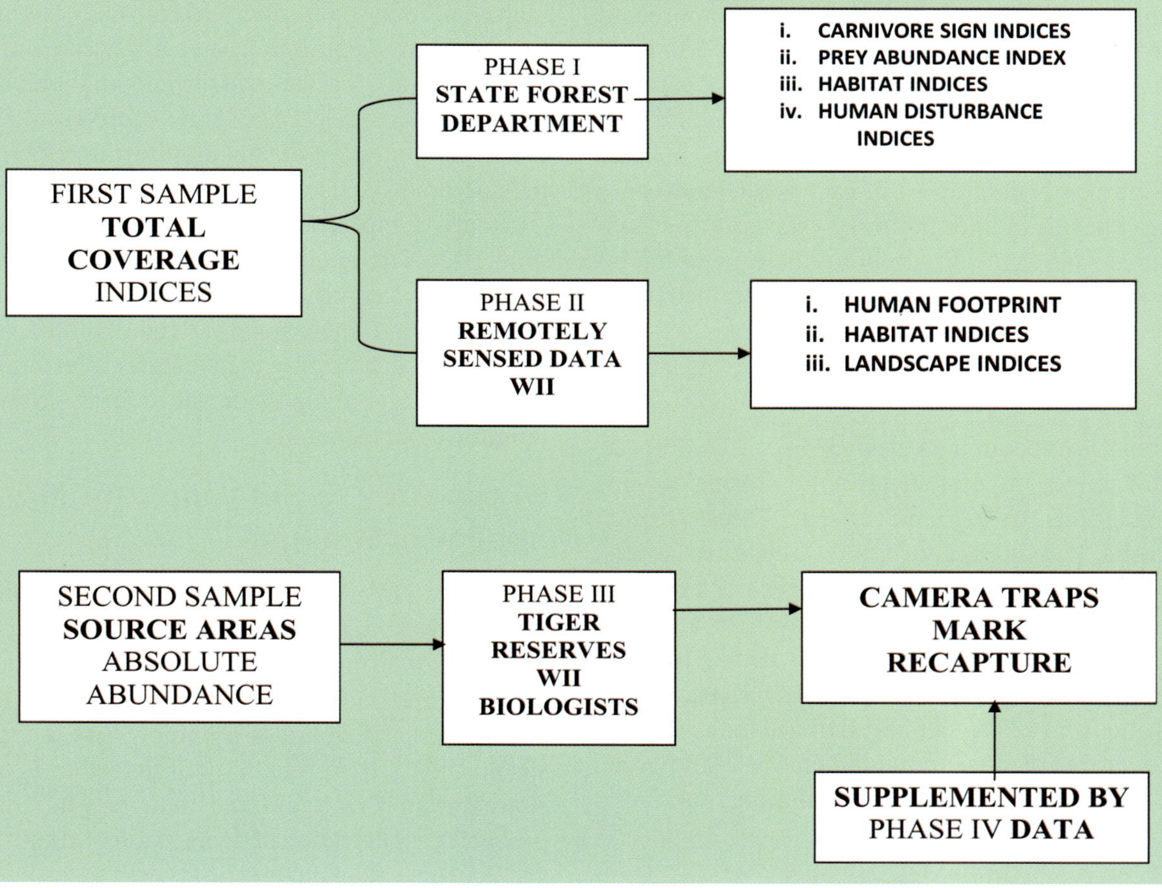

FIRST SAMPLE **TOTAL COVERAGE** INDICES

PHASE I **STATE FOREST DEPARTMENT**

i. CARNIVORE SIGN INDICES
ii. PREY ABUNDANCE INDEX
iii. HABITAT INDICES
iv. HUMAN DISTURBANCE INDICES

PHASE II **REMOTELY SENSED DATA WII**

i. HUMAN FOOTPRINT
ii. HABITAT INDICES
iii. LANDSCAPE INDICES

SECOND SAMPLE **SOURCE AREAS** ABSOLUTE ABUNDANCE

PHASE III **TIGER RESERVES WII BIOLOGISTS**

CAMERA TRAPS MARK RECAPTURE

SUPPLEMENTED BY PHASE IV **DATA**

From 2006, spatial data on individual tiger photo-captures are used in combination with spatial data on prey habitat, and anthropgenic factors as covariates in a joint likelihood spatially explicit capture-mark-recapture (SECR) framework to arrive at tiger population estimates for each tiger landscape. The method is illustrated in **Box 3.1** and briefly as stated below.

The method, described by Jhala *et al.* (2019), entails estimating spatial covariates of relative abundance of tigers, co-predators, and ungulates, human impact indices, and habitat characteristics across all potential tiger habitat in India, at a fine spatial resolution of a forest beat which is on an average of about 15 km^2 in area (Phase I and II). Subsequently an adequate area within each landscape was sampled using camera traps at a high spatial density of one double camera location in 4 km^2 during 2006, 2010 and 2014 and in 2 km^2 during 2018 (Phase III). The concept is similar to that of double sampling wherein indices or raw counts of abundance obtained from the entire sample space are calibrated against absolute density obtained from limited samples. The difference between double sampling and spatially explicit capture-mark-recapture (SECR) approach is that double sampling uses ratio or regression to calibrate indices, while tiger population estimation uses spatial information on capture-mark-recapture (that accounts for detection correction) in a joint likelihood with spatial covariates of tiger sign intensity, prey abundance, human disturbance and habitat characteristics. This approach estimates tigers directly within camera trapped areas and extrapolates it to areas with tigers, but not camera trapped based on joint distribution of covariates. Intensive monitoring of tiger source populations in tiger reserves and protected areas in each tiger landscape complex (Phase-IV) is done every year, and there is maintenance of a centralized photo-database of tigers at National Tiger Conservation Authority obtained from camera traps deployed across all tiger reserves.

'Pug Mark' vrs 'SECR'

One of the impediments to effective conservation is the lack of simple and cost effective monitoring of wild tiger populations. In 1970, a renowned Forest Officer from Odisha, Padma Shri Saroj Raj Choudhury (1926-1982) invented 'tiger tracer' to bring tiger track marks, popularly known as 'pugmarks' from nature to 'analysis room' (Choudhury 1970). Pug impression pads (PIPs) are laid on hard soils to receive good tracks to be used for analysis and from each pugmark so traced male, female and cub were segregated and individual tiger and even leopard numbers were arrived at using pugmark size and stride length of an animal (Singh 1999). Dr L A K Singh, senior research officer of Similipal reserve and his team worked in tendem to figure out tiger numbers in Similipal. They standardized census season, census unit and route, laying of PIPs, and proceeded linking cubs to mothers, and females to males, and spatial distribution of tigers in Similipal. But the reliability of the pugmark census technique has often been questioned by Karanth (1987, 1993 and 1999), Karanth and Nichols (1998), and Karanth *et al.* (2003) owing to its subjectivity and lack of validation on populations of known free-ranging tigers. Critics of the technique believed that an individual tiger's pugmark changes in shape and size over different substrate (soil texture, moisture and depth). Another source of variability, they believe, is the variation between different tracers' abilities to trace the features of the pugmark on the tracing sheet.

Realizing the shortfalls of the pugmark monitoring system in keeping pace with modern conservation biology needs for a monitoring scheme, the Project Tiger Directorate commenced a project in collaboration with the Wildlife Institute of India and the Forest Department of Madhya Pradesh in 2003 to evolve a monitoring programme for "Tigers, Co-predators, Prey and their Habitat" in the Satpura-Maikal Landscape (Jhala *et al.,* 2008). This pilot project evolved field friendly data collection protocols in consultation with field managers and scientists. The monitoring programme uses remote sensing, geographic information system, and global positioning system technology in combination with high resolution spatial data and field data, based on sign surveys, camera trapping, and distance sampling, to effectively monitor tiger and prey populations. After the Sariska crisis, the Tiger Task Force (2005) recommended the implementation of this monitoring scheme for all tiger occupied landscapes. The Project Tiger Directorate (currently the National Tiger Conservation Authority) synergized this mammoth task by liaisoning with the State Forest Departments to generate the required field data in appropriate formats and the Wildlife Institute of India to impart training in field data collection, and for estimating tiger and prey densities for the Nation wide monitoring programme.

Occupancy Estimation and Modeling Applied in Similipal

Tigers are cryptic and elusive animals. The dense tropical forest vegetation of their habitats coupled with highly undulating landscapes hampers the sighting, trapping and radio tracking of tigers. Non-detection of a species at a site does not imply that the species is absent unless the probability of detection is one. The characteristics of tigers and their habitats make the probability of detection of tigers in the wild substantially less than one. MacKenzie *et al.* (2002 and 2006) propose a model and likelihood-based method for estimating site occupancy rates when detection probabilities are less than 1. This method provides a flexible framework enabling covariate information to be included and allowing for missing observations. Their method provides good estimates of occupancy rates, generally unbiased for moderate detection probabilities (*e.g.* >0.3).

Occupancy estimation and modeling approach of MacKenzie *et al.* (2002 and 2006) was applied to observations of tigers and environmental covariates in Similipal tiger reserve. Following questions were asked: First, what are the conservation implications of the application of occupancy estimation and modeling of MacKenzie *et al.* (2006) to a tiger reserve? Second, is occupancy estimation and modeling a rapid, cost effective, and statistically robust methodology that can be performed by the management staff of a reserve to take quick management decisions?

Methods

Beat, the smallest administrative unit of the reserve, was chosen as a sampling unit. There are 139 beats covering an area of 2210.75 km² of Similipal sanctuary. The average size of a beat is 15.90 km² with the smallest and largest beats at 6.80 km² and 49.61 km² respectively. Of the 139 beats in the sanctuary, 75 are in the core area and 64 in the buffer area. Beats were sampled as part of the All India tiger and co-predator monitoring programme, and the protocol described below was developed for that programme (Jhala *et al.*, 2008).

Summary of Sampling Protocol

In each beat, three individuals local to the area walked three different transects of about 5 km each, one on each of three consecutive days in January (2006 Jan 18-20). The cool dry season is the most favourable for the formation, preservation, and detection of tiger sign, as tigers actively travel open paths during this season, leaving detectable scrape marks, scat, and kills, and leaving tracks in the moist sediments of natural track traps. The sampling team counted the number of signs of carnivores (tracks, scrape marks, kills, and scat) that they saw on each transect. Since tigers and leopards have a tendency to use dirt roads, trails, foot paths, river beds and nullahas, these landscape features within the beat were targeted. Three persons, one beat forest guard and two local residents were engaged for conducting survey for tiger and other carnivore signs in each beat, with another set of different local residents comprising the sampling team in each beat. The local residents and the forest guard knew the terrain and habitat features of their respective beat areas. The forest guard was the beat guard for that particular beat, familiar with the area and the signs of animals occupying that area, and having training on scat and sign identification. The GPS coordinates of the beginning point of each search path was recorded. Each search path can be thought of as a winding, non-linear, transect through a portion of the beat. Each sampling team was told that the total minimum distance to be covered in the three days in each beat while searching for tiger and other carnivore sign was 15 km (with each transect being 4 to 6 km). Tiger and leopard signs were classified into: (1) Pugmark trails **(Plate 3.22A)**, (2) Scats **(Plate 3.22B and C)**, (3) Scrapes **(Plate 3.23)**, (4) Scent marks (spray, rolling), (5) Rake/nail marks on trunks **(Plate 3.24)**, (6) Actual sighting, (7) Roaring (vocalization), (8) Kills (Predation on wild prey). Record was also kept for signs of any other carnivore that were encountered. The other carnivore species recorded were wild dog, bear, hyena, wolf and jackal. A brief description of the topography and forest type was recorded for each sign.

Plate 3.22A. Pugmark trails. Photo: D Swain.

Plate 3.22B. Tiger scat. Photo: D Swain.

Plate 3.22C. Tiger scat containing hair and bone. Photo: D Swain.

Plate 3.23. Scrape mark of a tiger. Photo: D Swain.

Plate 3.24. Nail mark of a tiger. Photo: D Swain.

A separate transect of 2 km each in a beat was walked on three subsequent mornings (2006 January 21-23), and on these the survey teams recorded the numbers of mammals (primarily ungulates) they saw. After considering the shape, size, vegetation, and terrain type of the beat, a transect line of 2 km was marked for sampling. The transect line was so marked as to traverse similar habitat (broad vegetation types) as far as possible, and the broad forest type and terrain type that the transect traversed were recorded. Care was taken that a line transect was not located near a busy road nor run parallel to a river or other features of the landscape that could create bias in sighting of ungulates. The GPS coordinates of the beginning and end points were recorded for each transect. Each transect was walked by three persons during the early morning hours (6:30 AM to 8:30 AM). Preferably one of the persons walking was a good field person who was able to spot wildlife. Each line transect was walked on three different mornings for estimating ungulate encounter rates.

Sign Detection Training

Intensive training was conducted for the master trainers who also acted as supervisors in the monitoring process. The master trainers imparted training including practical demonstration to beat forest guards in the Range level. One day prior to the beginning of monitoring programme, a mock exercise was conducted in each beat. Therefore, the Beat forest guards who were the leaders of each sampling unit, had received training of about 16 hours to identify animals, animal signs and recording of data on data sheets. They were provided with booklets giving detailed procedure and identification cards for animals and animal signs. As such forest guards working in the Similipal forests day in and day out, are highly skilled in identifying carnivore scats and signs. As director of the reserve, I personally supervised the training and implementation of the monitoring protocol.

Classification of Beats

Table 3.1 provides list of potential site covariates examined in this study and abbreviations mentioned for each site covariate are used in Table 3.2. For land use classification, GIS data generated from 2006 satellite imagery of IRS-1C/1D, sensor LISS-III were used. In this classification, dense forest is forest canopy cover with more than 40 per cent, open forest represents canopy cover between 10 per cent and 40 per cent, and degraded forest contained the canopy cover less than 10 per cent. Distance from habitation was expressed as the distance from the edge of the beat to the edge of the nearest inhabited place in km.

Table 3.1: List of potential site covariates examined in this study. Abbreviations are used in Table 3.2.

Site Covariate	*Abbreviation*	*Description*
Area	Area	Area of beat in hectares
Proportion degraded habitat	Degr	Area in hectares of the beat with a land use category of agricultural land, built-up land, degraded forest and open forest clubbed together divided by the total area of the beat.
Distance from Habitation	hdist	Distance from the edge of the nearest inhabited place to the edge of the beat (km)
Total prey	Prey	Sum of observed individuals of tiger prey animals observed in the beat from the ungulate observation protocol (sambar, chital, wild pig, and barking deer).
Proportion Dense Forest	Dfor	Area in hectares of the beat with a land use category of dense forest divided by the total area of the beat.
Proportion Grassland	Grass	Area in hectares of the beat with a land use category of grassland divided by the total area of the beat.
Prey: sambar	Sambar	Individuals of sambar observed in the beat from the ungulate observation protocol.
Prey: chital	Chital	Individuals of chital observed in the beat from the ungulate observation protocol.
Prey: wild pig	Pig	Individuals of wild pig observed in the beat from the ungulate observation protocol.
Prey: barking deer	Bdeer	Individuals of barking deer observed in the beat from the ungulate observation protocol.

Table 3.2. Summary of models that fit to Similipal tiger reserve data. Δ AIC is the difference between the AIC of a model and the AIC of the most parsimonious model, ** marks models with substantial support having a Δ AIC of less than 2, and * models having some support with a Δ AIC of less than 10. Untransformed estimates of some covariates are listed in the last three columns. Untransformed estimates of β, coefficients for covariates in models with a ΔAIC from the most parsimonious model of less than 10. Only values of β that are larger than the standard error are reported here. "——" indicates that a covariate is not present in a model, "<SE" indicates that the β value for that covariate was less than the standard error.

Model	*#Par*	*AIC*	*ΔAIC*	*AIC weight*	*Likeli-hood*	*-2logl*	*degr* β	*hdist* β	*Dfor* β	*Grass* β	*Sambar* β	*Prey* β
Ψ (degr, hdist, prey)	5	433.53	**0.00****	0.3672	1.0000	423.53	-0.81	0.81	——	——	——	1.17
Ψ (dfor, grass, prey)	5	435.29	**1.76****	0.1523	0.4148	425.29	——	——	0.85	86.1	——	1.31
Ψ (degr, prey)	4	435.31	**1.78****	0.1508	0.4107	427.31	-0.98	——	——	——	——	1.52
Ψ (degr, dfor, hdist, prey)	6	435.43	**1.90****	0.1420	0.3867	423.43	-1.08	0.80	< SE	——	——	1.19
Ψ (degr, hdist, sambar)	5	437.48	3.95 *	0.0783	0.2133	427.48	-0.90	< SE	——	——	5.02	——
Ψ (degr, dfor, gras, prey)	6	436.62	3.09*	0.5100	0.1388	424.62	< SE	——	< SE	84.67	——	1.37
Ψ (all)	8	437.60	4.07 *	0.4800	0.1307	421.60	< SE	0.72	< SE	80.15	——	1.29
Ψ (hdist, prey)	4	440.96	7.43 *	0.0089	0.0244	432.96	——	1.13	——	——	——	1.24
Ψ (samabar)	3	445.58	12.05	0.0009	0.0024	439.58	——	——	——	——	—-	——
Ψ (prey)	3	447.30	13.74	0.0004	0.0010	441.30	——	——	——	——	——	1.80
Ψ (deg, dfor, grass)	5	450.39	16.88	0.0001	0.0002	440.39	< SE	——	——	——	——	——
Ψ (hdist)	3	451.31	17.76	0.0001	0.0001	445.31	——	2.06	——	——	——	——
Ψ (degr)	3	455.53	22.00	0.0000	0.0000		-1.33	——	——	——	——	——
Ψ (pig)	3	457.7	24.17	0.0000	0.0000	451.70						
Ψ (bdeer)	3	472.18	38.65	0.0000	0.0000	466.18						
Ψ (chetal)	3	482.18	48.56	0.0000	0.0000	476.18						
Ψ (area)	3	482.89	49.33	0.0000	0.0000	476.89						
Constant p	2	483.45	49.89	0.0000	0.0000	479.45						
Survey-Specific p	4	484.05	50.49	0.0000	0.0000	476.05						

Site covariates were normalized [this doesn't affect the ranking of models, and only slightly affects the numeric results, for example, AIC for a model of Ψ (proportion of degraded habitat, distance from habitation, total prey) was 433.53 for normalized covariates, and 433.56 for raw covariates].

Summary of the Data

Figure 3.1 shows the relationship between each of the six potential covariates for tiger occupancy and the number of surveys in each beat in which tiger sign was observed. Of the potential covariates, the proportion of the beat that consists of degraded habitat, the proportion of the beat that consists of dense forest, the distance of the beat from human habitation, and the total number of prey (Sambar, Cheetal, Wild Pig, and Barking Deer, all known to be tiger prey in Similipal, and fitting the typical pattern of terrestrial prey half or more the large body weight of large felid carnivores, Gittleman 1985) observed in the ungulate surveys all show patterns suggestive of a relationship with tiger presence.

Occupancy Analysis

A data set consisting of binary observed presence of tiger sign on each of three successive samplings of a beat, and a set of site covariates was analyzed with the software package PRESENCE Version 2.2 (Hines 2006) (numeric results reported here using build <090330.1129> running under wine on a Debian Linux system were consistent with data exploration using bootstraps of 100 on a Windows XP system). All analyses used a bootstrap parameter of 2000.

The AIC (Akaike Information Criterion) is a measure of how well a model (a combination of parameters) explains the data, with a penalty added for more parameters. Its purpose is to find the most parsimonious combinations of parameters that explain the data. The model with the smallest AIC value is viewed as the best and simplest explanation of the data. The difference between AIC values (ΔAIC) between this minimum AIC and other models allows identification of models that also provide good and parsimonious explanations of the

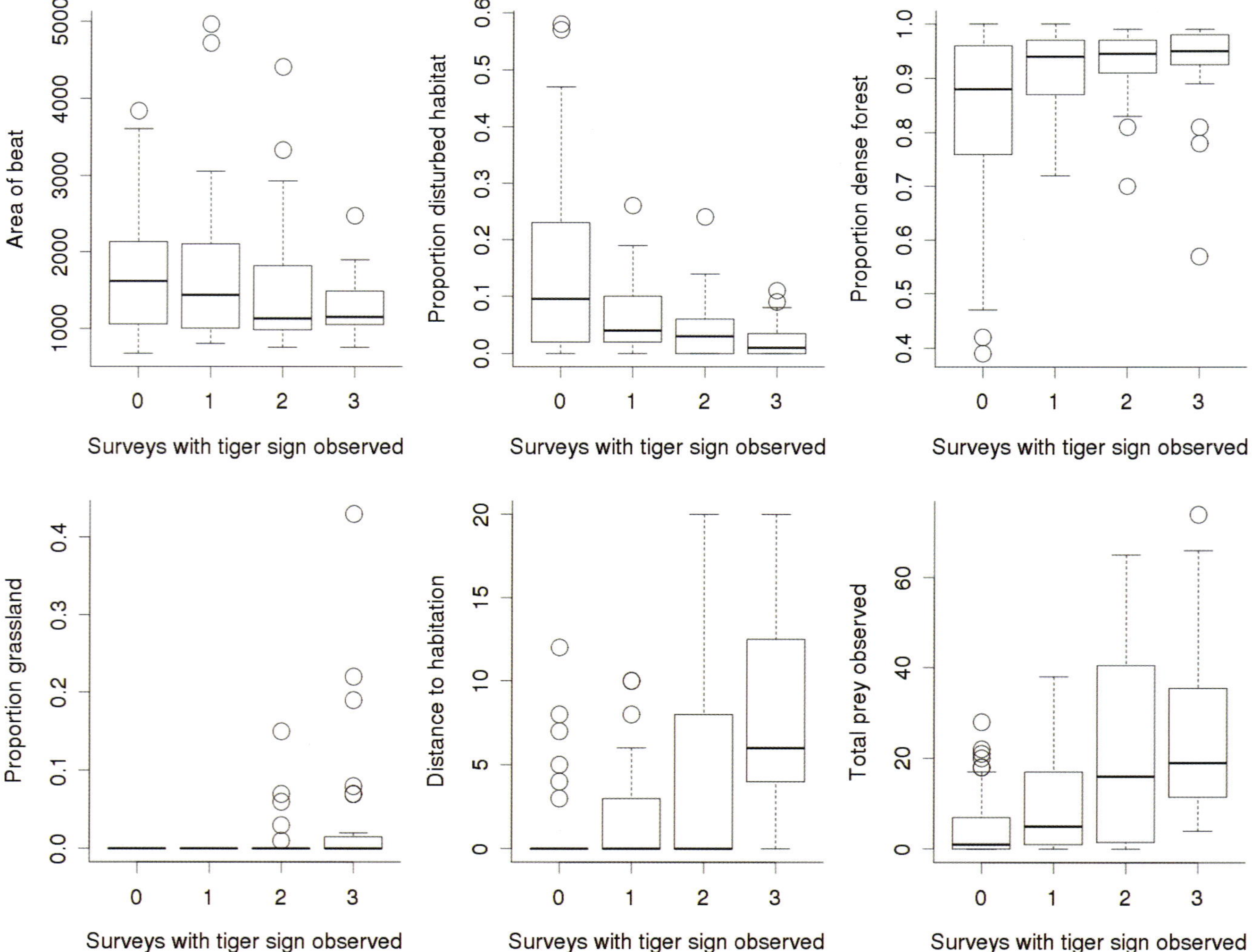

Figure 3.1. The number of surveys (out of three) in each beat in which tiger sign was observed compared with environmental characteristics of that beat in Similipal tiger reserve, Odisha, India.

data. Following Burnham and Anderson's (2002, p446) rule of thumb, the models were tagged with a ΔAIC of less than 2 as having substantial support, and those with ΔAIC values from about 2 to 7 as having some support, and those with ΔAIC values larger than 10 as poorly explaining the data.

Three null models (area, constant p, survey specific p) all have ΔAIC values larger than 10. Area doesn't correlate with tiger occupancy, a concern given the presence of larger beats in the buffer area. Likewise, Models of constant probability of detection alone (constant p) and consistent differences between surveys of a site (survey specific p) have very low explanatory value.

The single best model (the model with the fewest parameters that best explains the data), is Ψ(degr,hdist,prey); occupancy explained by a combination of the proportion of the beat that is composed of degraded habitat, the distance of the beat from human habitation, and the total number of prey animals observed in the beat during the ungulate surveys of that beat (**Figure 3.2**).

A map of the estimated values of Ψ (degr, hdist, prey) for each beat is shown in **Figure 3.3**; the highest tiger occupancy is in the South central core of the reserve.

The models that best explain the data are (1) the combination of proportion of degraded habitat, the distance to habitation, and total prey, (2) the proportion of habitat composed of forest and grassland, along with the total prey, (3) the proportion of degraded habitat and the total prey, and (4) the proportion of degraded habitat, the proportion of forest, the distance to habitation, and the total prey (Table 3.2). No one model stands out, following Burnham and Anderson's (2002 p. 184) rule of thumb of AIC weight of >0.9 for identification of a single most favourable model, with Ψ(degr,hdist,prey)

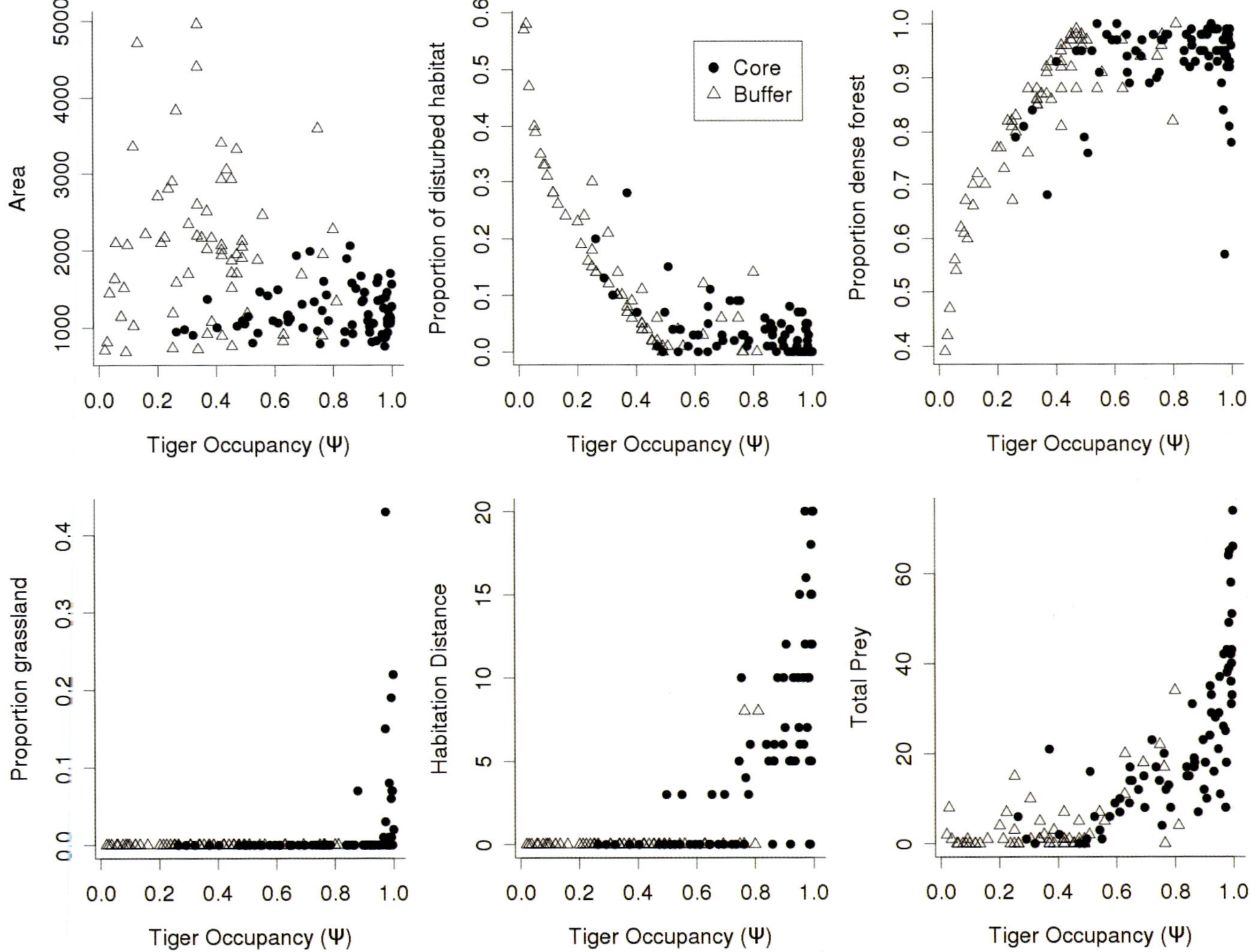

Figure 3.2. Correlations between site covariates and the tiger occupancy for a site as predicted by the psi (dist, hdist, prey) model in Similipal tiger reserve, Odisha, India. Solid circles are beats in the core area; open triangles are beats in the buffer area.

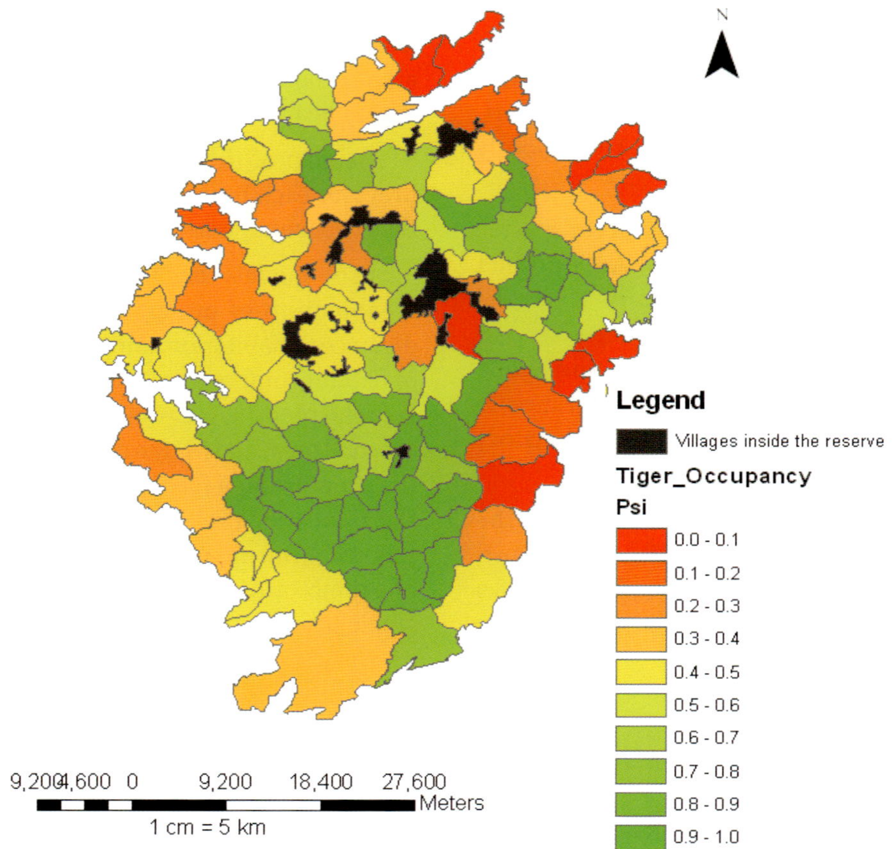

Figure 3.3. Tiger occupancy in Similipal tiger reserve, Odisha, India.

having an AIC weight of 0.37, the most important site covariates, that indicate the most important predictors of tiger occurrence, are the proportion of the beat that has disturbed habitat, the distance of the beat from human habitations, and the total prey observed, with grassland, when present also being an important predictor.

Sampling and Assumptions

MacKenzie *et al.* (2006) list the assumptions behind a single-species single-season occupancy model. Two of these assumptions are independent of potential covariates, and their validity is critical to the applicability of occupancy estimation. These critical assumptions are: (1) Closure, that is the occupancy of a site, does not change over the course of the sampling. A set of sampling surveys (a "season") provides a snapshot of the population unaffected by birth, death, immigration or emigration, and (2) independence of detection history of a species at a site, that is one survey of a site does not alter the occupancy probability or detection probability for a species in a subsequent survey at a site (along with independence of detection from site to site, a survey at one site doesn't alter the detection probability at another site). In more general terms, the application of a sensor to detect the target species should not affect the detection probability for subsequent applications of that sensor. For tigers, three repeated surveys of the same transect would be likely to violate the assumption of independent detection history.

The presence of human scent and disturbance from a survey would be likely to deter tigers from approaching that transect, reducing detection probabilities on subsequent resurveys of that same transect. Thus it is important that the three tiger surveys of each beat in this study were on distinct transects within the beat rather than repeated visits to the same transect over three days. These two critical assumptions need to be correct in order for the method of occupancy modeling to produce valid results.

Two other key assumptions of single-species single-season models involve covariates, and can be evaluated by the comparison of different models. These are: (3) that either occupancy probability is constant or the differences in occupancy probabilities are effectively modeled with covariates and (4) that either detection probability is constant or that detection probability is effectively modeled with covariates. Another assumption of occupancy analysis is that there is no substantive unmodeled heterogeneity in detection probabilities. All of these assumptions can be assessed from the analysis.

The sampling units of this study (the beats) are effectively arbitrary spatial units, although beat boundaries often follow linear natural features such as ridge lines or drainages. Area of beats vary (from 6.80 km^2 to 49.6 km^2, mean 15.9 km^2), with about half the beats in the buffer area being larger than the largest beat in the core area. Core area beats have a median area of 11.3 km^2 with first and third quartiles at 10.0 and 13.8 km^2 in contrast to the buffer beats, which have a median area of 19.8 km^2 and first and third quartiles of 13.0 and 24.8 km^2. Given the large ΔAIC value for area, the area of beats does not appear to explain the tiger occupancy patterns.

Management Implications

A naive conclusion from the observation that distance to habitation correlates with tiger occupancy would be that removing villages from the reserve would increase tiger occupancy. Closer examination of the data, however, leads to much more nuanced conclusions, particularly from the observation that both beat 51, Jenabil, compartment KH-26, and beat 40, Balikhal, compartment P-10 are adjacent to habitation, but both had tiger sign observed on two of three surveys and both have predicted Ψ values of about 0.98. These outliers indicate that human habitation in the reserve can be consistent with high tiger occupancy. Indeed 5 beats with a distance to habitation of zero have predicted Ψ values greater than 0.80, and predicted Ψ values for beats adjacent to villages span the range from 0 to 0.98. Distance to habitation alone is a poor predictor of tiger occupancy.

Examination of the relationships amongst tiger occupancy, proportion of each beat with disturbed habitat and the number of total prey leads to the conclusions that the proportion of disturbed habitat is a good predictor for the tiger occupancy of a beat only for tiger occupancies below about 0.50 (or when the proportion of disturbed habitat is above about 15 per cent), and that the abundance of prey animals is a good predictor of tiger occupancy for tiger occupancies above 0.50, and a poor predictor below that. This observation has clear management implications, first, for areas with low psi, reduce amount of disturbed habitat to below 10 per cent of total area. Second, for areas with moderate psi, increase prey animal abundance, which suggests an increased focus on enforcement against poaching on prey animals. Abundance of prey is a well known determinant of tiger population size (*e.g.* Miquelle *et al.*, 1999; Karanth and Stith, 1999).

The results of the occupancy analysis lead to divide the beats into four categories, each with its own management strategy: (1) Beats of greatest concern, with psi < 0.10, need extensive remediation of multiple problems. (2) Beats for targeted reduction of habitat disturbance (*e.g.* reduction of tree felling, reduction of grazing, reduction of human activity, increase of canopy cover). For these beats, it is proposed a target of reduction of the proportion of disturbed habitat for each beat to less than 10 per cent of the area of the beat. The proportion of beat area coded as disturbed habitat for these beats ranges from just over 10 per cent to about 30 per cent. (3) Beats for targeted reduction of human effect on prey population: These beats show lower availability of prey animals and a relatively small proportion (less than about 15 per cent) of degraded habitat. Target for these beats is increase of the populations of tiger prey species, through reduction of hunting of these prey species, such as through increased enforcement to prevent poaching on prey species. Miquelle *et al.* (2009 p. 98), recommend a management technique for increasing prey populations: "give local people an incentive to support higher populations of key prey species". Participatory wildlife management may be introduced, by motivating residents of buffer villages to protect prey species, and in return allowing collection of non-timber forest produce (NTFP) from the buffer. (4) The remainder of the beats in the reserve has high tiger occupancy and the least human disturbance. The strategy for these is, maintenance of current management procedures, including patrols and deployment of local youth as green brigades, a concept introduced in Similipal with great success particularly during the season of community hunting by indigenous people. **Figure 3.4** illustrates these management implications.

Continued Monitoring

Suggestion: In the next round of sampling, each of the largest beats should be subdivided into two samples, and the three transect surveys for sign in each half be performed. The procedure discussed herein is simple, cost effective, engages both staff and local populace in monitoring conditions of the reserve. A beat forest guard and for that matter a member from the local community who participates in search of signs of carnivores and direct sighting of ungulates can easily know during the survey the impacts of management strategy undertaken in the beat. The procedure is very simple as monitoring staff have to collect some basic data in the course of their patrols. As it is cost effective, it can be repeated every year and this process of monitoring is participatory (Evans and Guariguata, 2008) as it involves ground level staff and local community.

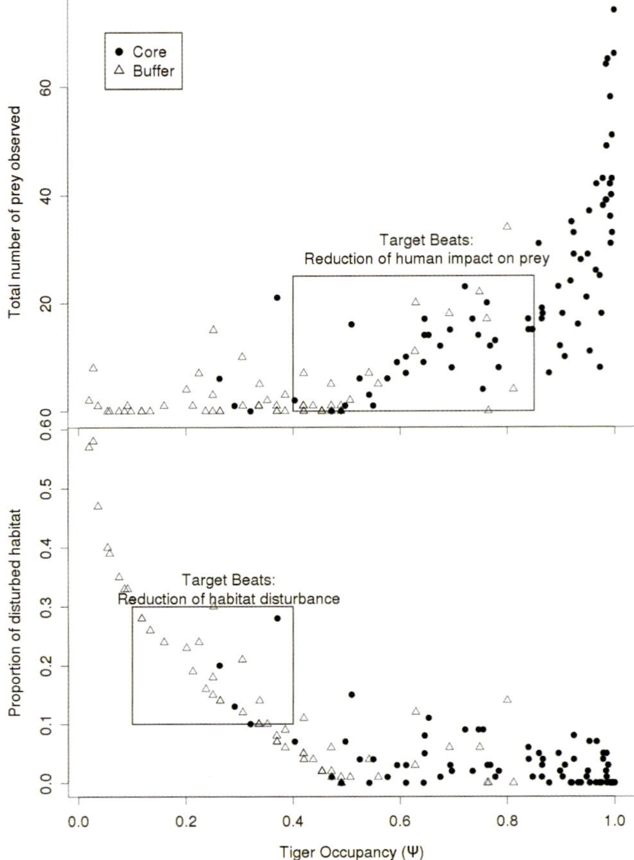

Figure 3.4. Management implications of tiger occupancy analysis in Similipal tiger reserve, Odisha, India. Solid circles are beats in the core area; open triangles are beats in the buffer area.

Suitability of Occupancy Analysis

In line with other workers (*e.g.*, MacKenzie *et al.*, 2005), we find occupancy analysis of cryptic carnivore sign survey data highly suitable for complex forested terrain such as that of Similipal. The sign survey method depends on substantial numbers of trained and experienced staff. It has a caveat that sampling protocols and procedures need to be well designed. Sign surveys have a very great advantage of low cost and low requirements for technical infrastructure, allowing them to be repeated frequently to build up robust time series of monitoring data for better understanding the impacts of management actions on the monitored populations. Sign surveys have important side consequences and correlates, including capacity building of staff, improving their day to day monitoring abilities and their abilities to recognize short term changes in their field area. Sign surveys are also highly suitable for engagement of local populations in monitoring and conservation efforts.

Spatio-Temporal Distribution of Tiger and Sympatric Co-predators

Tigers are in the apex of the ecological pyramid in the forests occupied by them. Other sympatric co-predators are leopard, dhole or wild dog, leopard cat, jungle cat, hyena, wolf, Indian jackal and sloth bear. Interaction among these carnivores in a wilderness area is fascinating. There are evidences of tigers killing leopard and devouring it (McDougal, 1988). Evidence for a tiger eating a leopard cub has been reported by Johnsingh (1979). Hyena usually wins one-to-one encounters over carcasses with leopards and tiger cubs, but may be dominated by adult tigers (Pocock, 1941; Mills and Hofer, 1998). Heptner and Sludskij (1980) report that in central Asia, tiger and leopard were, until recently, natural enemies of hyena. Tigers usually give sloth bears a wide berth, though some specimens may become habitual bear killers (Stephen, 2004) and it is not uncommon to find sloth bear fur in tiger scats (Richard, 1965). It is reported that Asian black bears (*Ursus thibetanus*) and Ussuri brown bears (*Ursus arctos lasiotus*) fall prey to tigers, and they constitute up to 40.7 per cent of the diet of Siberian tigers depending on local conditions and the bear populations (Mazak, 1996; Seryodkin *et al.*, 2005).

The sign survey data for tigers and the co-predators, *e.g.*, leopard, hyena, jackal, and sloth bear that were collected during the surveys of 2006 in Similipal were analysed. Other species, such as dhole were not analysed as too few signs were observed in this survey.

Examination of the relationships amongst tiger and co-predators occupancies, proportion of each beat with disturbed habitat, the number of total prey, and the distance of the beat from human habitation leads to the conclusions that evidence of leopards co-varies with tigers (**Figure 3.5**). But hyena and jackal were in higher numbers closer to humans and in more disturbed habitats. Bear occupancy could not be explained by any of the included covariates, as feeding ecology of sloth bear (Seidensticker *et al.*, 2011; Palei *et al.*, 2014) is quite different from the other carnivores.

McDougal (1988) and Seidensticker *et al.* (1990) report that leopards are more prevalent on the peripheries of Royal Chitwan National Park, Nepal sandwiched between prime tiger habitat, on the one side, and cultivated village land on the other, dependant on both natural prey and domestic livestock. But it is noticed in this study that evidence of leopards co-varies with tigers, strongly supporting the finding of coexistence between the two in Nagarhole national park, India by Karanth and Sunquist (2000).

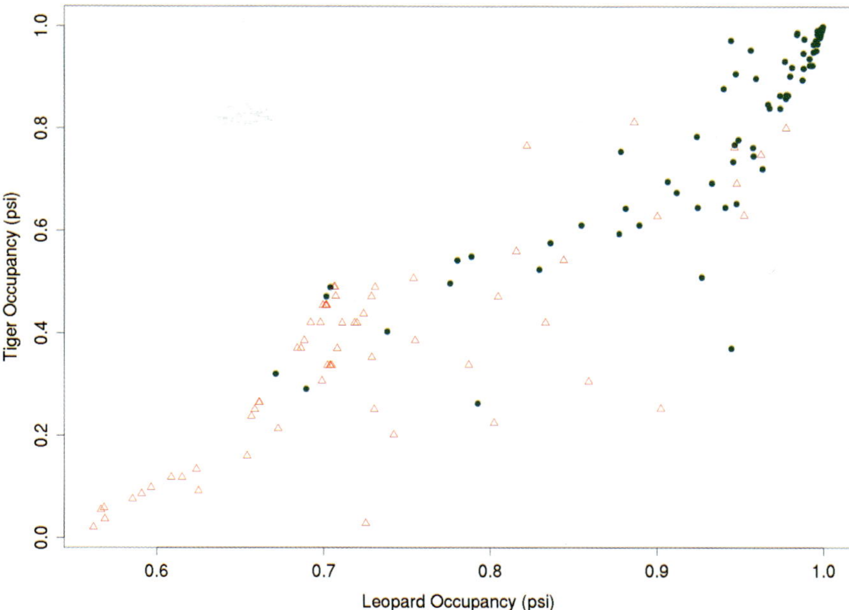

Figure 3.5. Correlations between tiger and leopard occupancy. Solid circles (green) are beats in the core area; open triangles (red) are beats in the buffer area.

Prey selection is different for tiger and sympatric co-predators in a reserve. Sambar and chital were the principal prey species for tiger and leopard, respectively, as inferred from the relative biomass consumption of prey remains in tiger and leopard scats in Mudumalai tiger reserve, Tamil Nadu, India (Ramesh *et al.*, 2009). Availability of prey in the appropriate size classes being not a limiting resource in Nagarahole, southern India, selective predation may facilitate large carnivore coexistence (Karanth and Sunquist, 1995). Abundance of ungulate prey species, as well as their availability in different size classes, are both critical factors that facilitate sympatry among the tiger, leopard and dhole (Andheria *et al.*, 2007). The considerably smaller leopard avoids competition from tigers by hunting at different times of the day and hunting different prey (Haemig, 2012).

Camera Trapping Exercises in Similipal and its Tiger Numbers

Based on beat wise tiger sign encounter rate data, all the beats of the reserve were classified into 'high', 'medium' and 'low' categories. Accordingly, 120 km^2 area of Upper Barhakamuda Range of STR belonging to the 'high' category was selected for camera trapping exercise of tigers. The area was divided into two blocks, each comprising 60 km^2. In each block of 60 km^2, there were 30 trap stations. The distance between two consecutive stations was kept between 1.5 km and 3 km. Reconnaissance survey with the local field staff was carried out to choose locations having maximum probability of getting a capture. Tigers are known to use animal trails, nullahs, river beds *etc.* The trap sites were selected based on clues of the use of the sites such as scats, scrapes, scent marking, claw marks, pugmarks *etc.* Two cameras facing each other were kept at a camera trap location for six consecutive days. The cameras were moved to the next 60 km^2 block and were set there again for six days. Once one round of fixing cameras was completed in 12 days, the cycle was repeated from first block. After selecting the best sites for locating camera traps the grass blades, leaves or any object that could be of potential interference to the infrared beam were removed. The transmitter and receiver were aligned properly, so that the receiver gets the strongest part of the infrared beam. The units (transmitter and receiver) were placed at 45 cm above the ground level (so as to target flanks of a tiger). Each camera was kept at a minimum of 4 - 4.5 meters from the centre of the trail to get a full frame picture of a tiger (**Plate 3.25**).

For the first cycle (2006) of all India tiger estimation, actual camera trapping exercises started from 26 March 2007 and ended on 26 May 2007. During these 60 days, seven tigers were captured photographically in the camera, of which three were found to be melanistic tigers (**Plate 3.26**). For the second cycle (2010) of all India tiger estimation, no camera trapping exercise was carried out in Similipal. But the number of tigers were projected based on sign survey.However, during camera trapping exercise in 2013 in Barhakamuda range of the reserve, photograph of a tigress with its three cubs, one of which was melanistic was captured (**Plate 3.27**).

Plate 3.25. Left and right flank of a normal tigress camera trapped during 2007 in Similipal tiger reserve of Odisha, India. Photo: Similipal TR archive.

Plate 3.26. One of the three melanistic tigers camera trapped during 2007 at Debasthali inside Similipal tiger reserve of Odisha, India. Photo: Similipal TR archive.

Plate 3.27. A tigress with its three cubs camera trapped in Barhakamuda range of Similipal tiger reserve, Odisha, India. One of the three cubs photo captured is melanistic. Photo: Similipal TR archive.

During the third cycle of tiger estimation in 2014, camera trapping session was carried out from 24 November 2014 to 27 January 2015 covering an area of 369 km^2. Five individual tigers were captured photographically of which one was melanistic tiger **(Plate 3.28A).** Similarly during all India tiger estimation of 2018 (fourth cycle of tiger estimation) 10 individual tigers were photo captured and three of them were melanistic **(Plate 3.28B).**

Plate 3.28A. Only melanistic tiger captured in the camera trap exercise conducted during 2014 in Similipal, Odisha, India. Photo: Similipal TR archive

Tigers have unique camera trapping patterns and this makes possible to differentiate between individuals from their photographs. Once camera trap survey was completed and the photographs developed, the task was to identify unique individual captures and build capture histories. The capture history of an individual tiger provides valuable information about when and at which place a particular tiger was captured. Using mark-recapture approach, Jhalla *et al.* (2008) estimated that there are 20 (17 to 23) tigers (adults only) in Similipal.

Plate 3.28B. Photo album of tigers captured in the camera trap exercise conducted during 2018 in Similipal, Odisha, India. SIMI_1, SIMI_4 and SIM_8 are melanistic. Photo: Similipal TR archive.

The camera trap survey has been widely used to estimate cryptic and endangered felids such as tigers, leopards and jaguars (Griffiths and van Schaik, 1993; Karanth, 1995; Karanth and Nichols, 1998; O'Brien *et al.*, 2003; Kawanishi and Sunquist, 2004). Carbone *et al.* (2001) used photographic rate to estimate density of tigers and other cryptic animals.

Accordingly tiger numbers estimated for Similipal by camera trapping exercise during 2006, 2010, 2014 and 2018 are as follows (Table 3.3):

Table 3.3. Tiger numbers estimated for Similipal tiger reserve, Odisha, India by camera trapping exercise during 2006, 2010, 2014 and 2018.

Year	*Tiger Numbers*		
	Population Number	*Lower Limit*	*Upper Limit*
2006	20	17	23
2010	23	12	34
2014	17	14	19
2018	12	11	13

Melanistic Tiger

Melanistic tigers are camera trapped very often since camera trapping exercise started in Similipal **(Plates 3.26, 3.27, 3.28A and 3.28B)**. Singh (1999) reports from analysis of the records of tigers with aberrant colours- the stripeless, white, melanistic and black *etc.*- that the body colour of tiger can vary over a wide range of aberrant colours with 'no stripes' to 'completely black' tigers. The intermediary stages include various shades of white tigers, the pallid or golden tiger, various shades of normal yellow tiger, the brown tigers, the melanistic tigers and the blue tigers. These entire possible colours occur according to a normal distribution curve in the wild gene pool of *Panthera tigris*. The middle of the dome is occupied by different shades of 'normal colour' tigers, while the aberrants occupy extreme regions of the dome of the curve. The aberrants reappear in a population in normal course of time as throwbacks and not because of identical repetitions of mutations.

On 21 July 1993, a tribal youth Salku Besra of Podagada village killed a young melanistic tiger in self defense that strayed from Similipal forest and entered his crop (maize) field. The skin measuring 195 cm is now with the Divisional Forest Officer, Karanjia Division. This was the first evidence of the presence of melanistic tiger in Similipal. But in 1970s, Bitanath Nayak, then Assistant Field Director, Similipal tiger reserve had spotted two full grown black (melanistic) tigers on the road to Matughar situated in the south Similipal and the same was not reported in the absence of any evidence to that effect. In 1991, Niranjan Mohanta, Forest Guard stationed at Debasthali claimed to have sighted a whole family of black (melanisitc) tigers- two full grown and two cubs. But this was dismissed as a case of 'mistaken identity' due to poor light condition in the dense forest.

During 1992, 265 cm long skin of a melanistic tiger was seized from the poachers by officers of wildlife protection cell at Tis Hazari, New Delhi. The rare skin was handed over to the National Museum of Natural History following the order of a Delhi Court in February 1993. This was probably the first physical evidence of melanistic tiger in the world. This melanistic tiger is distinguished by yellow stripes on its back and white ventral stripes. During January 2004, Priyakanta Mohanty, Assistant Field Director, Simipilal tiger reserve (STR), and his driver saw a melanistic tiger in the dense forest on the Debasthali-Nuagaon road inside the reserve. An adult black male tiger was also sighted by Debendra Nayak, Jeep Driver of STR along with Solem Deogam, Sabuja Bahini (Green Brigade) Volunteer of Jodapal beat on the Jenabil-Dhudruchampa road under Nawana South Range in the core area of the reserve on 5 September 2007. The black tiger was noticed with a normal coloured tigress. With the sighting of this new colour variety, Similipal became the home to three different colour variants of tiger that can be distinguished by their skin colour. The normal tiger has yellowish brown coat with black stripes, melanistic tiger has a black coat with yellowish brown stripes, and black tiger has deep black coat with a white abdomen, but no stripes. Similipal is the only tiger reserve in the tiger reserve net-work of the country to be inhabited by three different colour variants of tigers.

Unexpectedly on 28 July 2014 two melanistic male cubs (Krishna and Subhranshu) along with a normal female cub were born to Sneha, a white coloured female and Manish, a normal coloured male in the Nandan Kanan zoo at Bhubaneswar, Odisha. While Krishna is normal melanistic **(Plate 3.29)**, Subhranshu is whitish melanistic **(Plate 3.30).** Two years later in May 2016, another couple of Nandan Kanan (father-Samrat and mother-Renuka) delivered a still born melanistic cub. In August of the same year (2016), Sneha and Manish in their second litter produced one melanistic male and two white female.

Plate 3.29. Melanistic tiger (Krishna) born in Nandankanan zoo (28.07.2014), Odisha, India. Photo: Nandankanan archive

Plate 3.30. Whitish melanistic tiger (Subhransu) born in Nandankanan zoo, Odisha, India. Photo: Nandankanan archive

Interestingly, on pedigree analysis of these tigers, it comes to light that they are related to Similipal lineage. A seven week old normal coloured female cub was rescued from Similipal forests and transferred to Nanadan Kanan zoo on 21 November 1967 and named as Rani. Mishra *et al.* (1982) report that Rani was mated with captive bred normal coloured tiger, Deepak and gave birth to several cubs in different litters. One of their daughter of fourth litter, Ganga mated with father Deepak and gave a set of three white tigers, one male and two female. The breeding programme of tigers in Nanadan Kanan from aforesaid Rani and Deepak lineage produced Sneha, Manish, Renuka and Samrat, parents of the melanistic tigers born in the zoo.

Genetic Basis of Melanism

The question arises about the reason for such melanism in the tiger population of Similipal. Melanism in tigers is believed to be due to inbreeding. Singh (1999) predicts that normal colour tigers are due to dominant genes whereas white and melanistic tigers are manifestations of recessive genes. But the chemistry of pigmentation in animals is not that simple. Anderson *et al.* (2009) based on their study of North American gray wolves (*canis lapus*) reported that though mutations in the *Melanocortin 1receptor* (Mc1r) gene contributes to pigmentary diversity in natural populations of fish, birds, and many mammals, melanism in the gray wolf is caused by a different malanocortin pathway component, *the k locus*, that encodes a beta-defensin protein that acts as an alternative ligand for Mc1r.

Pigmentation in animals has been studied by several authors (*e.g.*, Castle and Allen, 1903; Cuenot, 1904; Haldane *et al.*, 1915; Wright, 1917a,b,c,d; Summer, 1921, 1929a,b; Benson, 1933; Dice and Blossom, 1937; Haldane, 1948; Russel, 1951; Russel and Major, 1957; Schlager and Dickie, 1966,1967; Morse, 1978; Shibahara *et al.*,1986; Barsh, 1996; Majerus and Mundy, 2003; Beermann *et al.*, 2004; Boswell and Takeuchi, 2005). Hoekstra (2006) reports that in mammal one aspect of pigmentation that has been a frequent substrate for natural variation is the regulation and distribution of pigment types produced by melanocytes. There are two types of mammalian pigments: eumelanin, which is responsible for black to brown colour, and pheomelanin, which is responsible for red to yellow colour. In melanocytes, several genes are involved in the coordination of 'pigment type-switching' between the synthesis of eumelanin and pheomelanin. This switch is controlled by the interaction of two primary genes: the *Melanocortin-1 receptor* (Mc1r), which encodes a seven-transmembrane receptor expressed in melanocytes, and its ligand, *agouti*, whose protein product is secreted from nearby dermal papilla cells and acts to inhibit Mc1r signaling (**Figure 3.6a**). In the absence of Agouti protein, basal levels of Mc1r activity keep levels of intracellular cyclic AMP (cAMP) sufficiently high to activate the eumelanin synthetic pathway. However, in the presence of Agouti protein, Mc1r activity is inhibited, cAMP levels are reduced, and melanocytes stop producing eumelanin and start producing pheomelanin. The interaction of these two proteins therefore plays a critical role in determining which pigment type is deposited along individual hairs.

Hoekstra (2006) further reports that additional genes are known to alter the density and distribution of

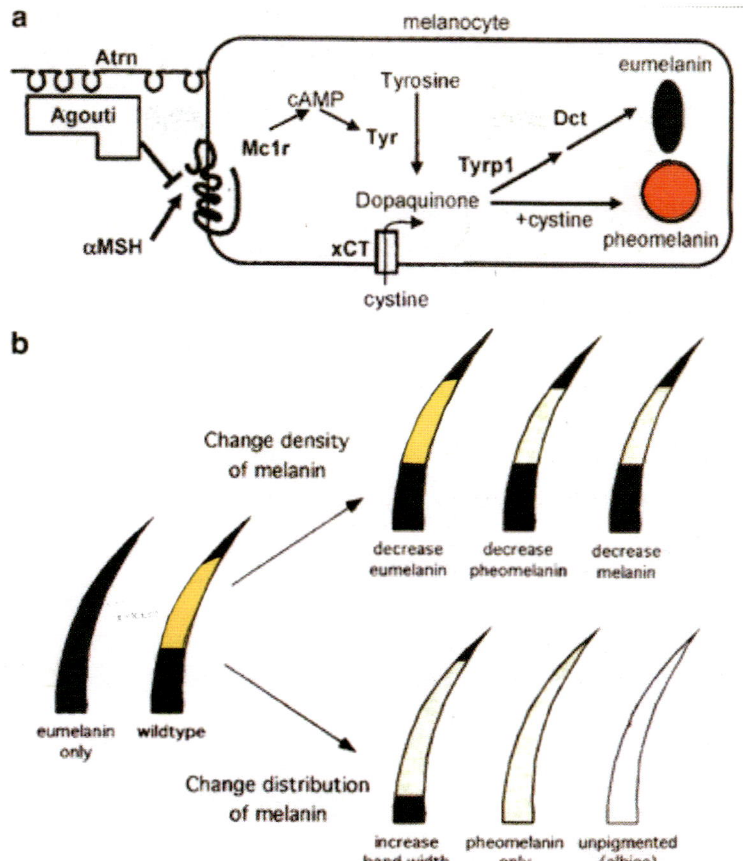

Figure 3.6. Genetic pathway regulating mammalian melanogenesis and phenotypic effect on individual hair pigment and pattern (Hoekstra, 2006).

(a): Circulating alphaMSH (a derivative of POMC) activates Mc1r, a G-protein-coupled transmembrane receptor, and signals via cAMP. Intracellularly, tyrosine is oxidized to dopaquinone, a reaction catalyzed by the enzyme tyrosinase (Tyr). Cyclic AMP is thought to affect the enzymatic activity of tyrosinase as well as eumelanic-specific enzymes, tyrosinase-related protein 1(Tyrp1) and dopachrome tautomerase (Dct). When all three of these enzymes function properly, eumelanin (brown to black pigment) is deposited in melanosomes. Agouti, the inverse agonist of Mc1r, binds to Mc1r with the aid of the extracellular protein Atrn to repress intracellular cAMP levels, resulting in the 'switch' to the production of pheomelanin (yellow to red pigment). The production of pheomelanin is dependent on the incorporation of cystine, whose uptake is at least partially regulated by xCT (the Slc7a11 locus).

(b): Overall coat colour in mammals is determined by the density of melanin and the distribution of melanin (or melanin types) on individual hairs. Pigment on individual hairs ranges from fully pigmented with dark eumelanin to complete absence of pigment resulting in albino hairs. Typical wild-type hairs in mammals have a subterminal band of light-coloured pheomelanin flanked by darker eumelanin, providing an overall brushed appearance.

melanosomes (pigment granules) found in melanocytes. *Tyrosinase*, for example, is the rate-limiting enzyme in melanogenesis. There are over 100 alleles of *tyrosinase* that have been characterized, ranging from null alleles, resulting in the complete absence of pigmentation (albino), to alleles with reduced function that limit the production of melanin. Other genes such as *tyrosinase-related protein-1* (Tyrp1) and *dopachrome tautomerase* (Dct or Tyrp2), primarily regulate the eumelanin pathway. *Gamma-glutamyl transpeptidase-encoding protein* (Ggt) affects the production of pheomelanin, and a second gene specific to pheomelanogenesis (*Slc7a11*) has also been described by Chintala *et al.* (2005). In addition, several genes, *Rab27a*, *Myo5a* and *Mlph*, are well studied as models for organelle transport because they coordinate the transport and distribution of melanosomes, both eumelanosomes, and pheomelanosomes, in melanocytes (Nascimento *et al.*, 2003). Mutations in these genes disrupt melanosome organization and can dilute overall coloration.

Therefore, many different molecular and developmental changes can affect the type, density and distribution of melanin on individual hairs and result in variation in overall pelage coloration (**Figure 3.6b**). Close examination of pigment and pattern on individual hairs can yield insight into the developmental changes and possible genes responsible for overall coloration. However, Hoekstra (2006) emphasizes that these candidates provide no guarantees as often changes in different genes can produce similar phenotypic effects.

Thus in mammals, melanocytes produce two types of pigment (eumelanin and pheomelanin), and the ratio of melanin types is largely responsible for variation in hair colour. The studies of linking genotype to phenotype (*e.g.*, Ritland *et al.,* 2001; Theron *et al.,* 2001; Nachman *et al.,* 2003; Mundy *et al.,* 2004) have contributed to the understanding of several questions about the process of evolutionary change:

(i) Are changes in coding or regulatory regions differentially responsible for adaptive morphology?

(ii) Are adaptive mutations generally dominant or recessive? and

(iii) Are the same genes responsible for similar adaptive phenotypes?

Hoekstra (2006) suggests that colour adaptation does not follow a simple evolutionary route. Mutations in both coding and regulatory regions, which are often dominant as in jaguar (*Panthera onca*) reported by Eizirik *et al.* (2003), pocket mice (*Chaetodipus intermedius*) reported by Nachman *et al.* (2003), bananaquit (*Coereba flaveola*) reported by Theron *et al.* (2001), deer mice (*Peromyscus maniculatus*) reported by Dodson (1982), and semi dominant as in jaguarundi (*Herpailurus yaguarondi*) reported by Eizirik *et al.* (2003), Arctic skua (*Stercorarius parasiticus*) and lesser snow geese (*Anser caerulescens caerulescens*) reported by Mundy *et al.* (2004), but other times recessive as in Japanese wild mice (*Mus musculus molossimus*) reported by Wada *et al.* (1999), black bears (*Ursus americanus*) reported by Ritland *et al.* (2001), cave fish (*Astyanax fasciatus*) reported by Protas *et al.* (2006) contribute to pigmentation diversity in nature, and furthermore, sometimes the same but sometimes different, genetic mechanisms are involved in parallel and convergent colour evolution in vertebrate populations. As mentioned above, melanism has been linked to dominant or semidominant mutations in the *Mc1r* locus, and in the exceptions of Japanese wild mice, black bears (recessive alleles at the Mc1r locus responsible for pale coloration of both the animals) and cave fish (recessive null alleles at the Oca2 locus responsible for loss of pigmentation), the adaptive significance of the colour variation remains unclear. Therefore it is observed that in pigmentation studies, recessive null alleles often (although not always) result in lighter coloration (*i.e.*, loss or reduction of pigmentation), whereas dominant mutations are often associated with darker colour (*i.e.* gain of pigmentation) (Hoekstra, 2006).

Under the above background discussion, understanding the melanism in tiger population of Similipal requires knowledge of multiple adaptive traits, and morphological, physiological and behavioural characters of the species. The function of tiger's stripe is to camouflage, helping to serve tigers to conceal themselves amongst the dappled shadows and long grass of their environment as they stalk their prey and therefore the stripped pattern of their habitat is found on the skin of the tiger. In Similipal open grasslands are not numerous except a few frostbitten pockets of South Similipal. Tigers in Similipal hunt their prey hiding inside dense forest vegetation. The principal prey animal of tigers in this forest is sambar which is common in such dense tropical moist deciduous sal forest. Whereas, chitals are common in grassland habitats, dense forest of Similipal, having canopy density more than 40 per cent over 93.32 per cent in the core area (Swain, 2014), is an ideal habitat of sambars; sambar is basically a browser, whereas chital is a grazer. High rainfall (over 2000mm), high temperature (mean annual temperature 19.1 to 22.2°C) and high humidity (over 90 per cent) may have resulted in the melanistic mutation of Similipal tiger which have helped it to camouflage against dense forest to hunt the prey animal, sambar. However, the use of molecular markers and complete genome sequencing of Similipal tigers is necessary to identify the genetic basis of melanism.

I was head of National Tiger Conservation Authority (NTCA) during 2017-18 and collaborated with Uma Ramakrishnan (Principal Investigator) of National Centre for Biological Sciences (NCBS) and researchers from NCBS, Wildlife Institute of India and Stanford University to study the genetic basis of melanisms in tigers. Our findings (Sagar *et al.,* 2021) reveal that the pseudomelanistic phenotype in tigers is inherited as an autosomal recessive trait. In this paper these tigers are called pseudomelanistic, as they display irregular stripe pattern and it is not completely black. The pseudomelanistic tigers carry a mutated copy of *Taqpep* gene on both the homologous chromosomes. In domestic cats and king cheetah, *Taqpep* protein (*Transmembrane aminopeptidase Q*), the product of LVRN gene, is reported to regulate the occurrence of dark patches on the coat by controlling the expression of *Endothelin3* (*Edn3*) gene (Kaelin *et al.,* 2012). *Edn3-Ednrb* signaling precedes Mc1r signaling where activation of *Ednrb* after *Edn3* binding causes activation of Mc1r, leading to production of eumelanin (Barsh, 1996). Pseudomelanistic tigers display disrupted pattern of dark patches like tabby blotched cats and king cheetah.

Therefore, LVRN presents a critical candidate gene for regulating pseudomelanism in tigers.

We identified a single base change in the exon nine of *Taqpep* from C to T (*Taqpep* c.1360C>T) that was present in all pseudomelanistic tigers on both the homologous chromosomes. Nonpseudomelanistic tigers either did not have this variant of *Taqpep* or carried it on only one of the homologous chromosomes. We genotyped twelve unique individual tigers and found two wildtype homozygotes, four mutant homozygotes, and six heterozygotes. These results provided an estimated *Taqpep* c.1360T allele frequency of 0.58 in the sampled population of Similipal tigers. We also genotyped 200 tigers at the *Taqpep* c.1360 locus from different tiger populations in India and observed that all of them were homozygous for the wildtype *Taqpep* c.1360C allele. Similarly we analysed the genotypes of 54 tigers including that of Amur, Malayan, Sumatran, and Bengal tigers outside India and confirmed that all of the 54 tigers were homozygous for wildtype allele at *Taqpep* c.1360 locus. Therefore, our results indicate that *Taqpep* c.1360T is likely an exclusive allele only available in the wild in Similipal. We also observed that the mean relatedness between Similipal tigers is 0.34 (median = 0.36). This indicates that Similipal tigers may be inbred, which is supported by the high F_{IT} value for Similipal (F_{IT} = 0.172). Population genetic theory suggests strong effects of genetic drift in such small and isolated population. Our study shows that Similipal tigers are genetically close to central India population and to some extent southern India population substantiating to the fact that there was a continuous patch of forest from Similipal to central India and to southern India as evident from the predicted forest cover map from 1800 of Ellis *et al.* (2010). Therefore, Similipal population started diverging from other populations in the past 200 years, implicating the increase in frequency of the pseudomelanistic allele from a low frequency to 50 per cent in the last 50 or less generations. However, our study emphasises the high frequency of pseudomelanistic tigers in Similipal and there might be a call for genetic rescue of this population, and the study advocates the need for detailed genomic studies in this globally unique and threatened population.

Tiger Habits

Sighting of tigers in dense forests of Similipal is not easy and that is the beauty of this landscape. The rolling tableland, with its dense vegetation cover and perennial flow of water, everywhere provides effective cover to tigers. The continuous sampling of sighting tigers by patrolling staff of the reserve from May, 2005 to April 2006 was possible only 18 times and during the same period next year was 32 times. In contrast sampling of tiger signs during the same period (May to April) of 2005-06 and 2006-07 reveal that there were 63 evidences of scats, 101 times pugmarks, 150 times scrape marks, 62 times vocalization and four kill evidences during 2005-06; and 119 evidences of scats, 117 times pugmarks, 229 times scrape marks, 136 times vocalization and eight kill evidences were recorded during 2006-07. This is in no way different from sightings recorded by Choudhury (1999). He observed in nearly two years, October 1974 through June 1976, of continuous sampling of tiger evidence in Similipal tiger reserve, faecal matter had been detected only seven times as against pugmarks 374 times, sighting 32 times, kill evidence 26 times and roars (vocalization) four times. In the following year, July 1976 to June 1977, the corresponding ratings were, faeces eight times, kill evidences 36, and vocalization 13 times. Therefore, in this book, the observation of tiger habits is based on indirect evidences of tiger signs, daily routines of a free-living tigress (named Khairi) recorded by the first Field Director of Similipal tiger reserve (S.R. Choudhury), and observing tiger activities in the open grasslands of Kanha and Bandavgarh, which are part of Central India landscape.

Khairi

Khairi **(Plate 3.31)** was a two-month-old female cub stolen from her mother by tribals (Kharia tribe) of Similipal and handed over to the then Field Director, S.R. Choudhury on October 5, 1974 (Choudhury, 1999). The tiger cub was hand-reared by S.R. Choudhury and allowed to move freely in his house and forests of Similipal. The tigress lived up to March 28, 1981 and

Plate 3.31. Tigress Khairi from Similipal tiger reserve, Odisha, India. Photo: Similipal TR archive.

died possibly because of rabies. Khairi's activity was recorded round the clock.

Kanha

Located in the Maikal ranges of the Satpuras, and falling in the Mandla and Balaghat districts of Madhya Pradesh, Kanha tiger reserve is known for its typical Central Indian floral and faunal attributes. Apart from supporting a viable population of the tiger, the National Park has also distinguished itself in saving the highly endangered hard ground barasingha (*Cervus duvauceli branderi*) (**Plate 3.32**) from extinction, and has the unique distinction of harbouring the last world population of this deer species. The reserve was one of the nine initially established tiger reserves of the country along with Similipal in 1973 and it covers an area of 1949 km² with a core area of 940 km².

Plate 3.32. Barasingha (*Cervus duvauceli branderi*) in Kanha tiger reserve, Madhya Pradesh, India. Photo: D Swain.

Bandhavgarh

The reserve named after the highest hill Bandhavgarh (807 m) in the centre of it, falls between the Vindhyan hill range and the eastern flank of Satpura hill range and is located in Shahdol and Jabalpur districts of Madhya Pradesh. It has been an excellent habitat of tiger and is known for the highest density of tigers in the world. Considering the importance and potentiality of Bandhavgarh National park, it was included in the Project Tiger Network in 1993. The adjoining Panpatha Sanctuary too was declared as a part of the reserve. The total area of the reserve is 1537 km² with core area of 717 km².

Similipal Tiger Size

Khairi weighed only 6.2 kg on the day following it was rescued (about two months old) and was 88 cm long from the tip of the nose to the tip of the tail along the spine. It weighed nearly 200 kg measuring 273 cm along the spine, and stood 91 cm at the shoulder when it was fully grown. Prater (1965) reports that an average Indian female tiger measures about 260 cm and weighs about 135-185 kg. With a slightly short tail and a long body, Khairi was a large female of the big Similipal tiger race (Choudhury, 1999).

Prey Animals of Similipal tiger

Experiments on Khairi suggested that it did not take any crow, cat or fox meat. She did not accept fish, crab or shrimp to eat. Langur and Rhesus rated topmost in her menu. Bush meat – cheetal, sambar, muntjac, four horned antelope and rufustailed hare was almost equally palatable. Porcupine ranked even lower than goat, and sheep was preferred to goat. Beef was least preferred (Choudhury, 1999).

Scat and kill samples of Similipal tigers were collected by patrolling staff from May, 2005 to October, 2007 during my stay as Field Director and analysed to find out animals eaten by tigers. 195 scat samples and 22 kill samples were collected in 883 days. All these samples were examined to identify the animals killed. The prey animals killed were Langur in 49 samples, wild boar in 59 samples, sambar in 60 samples, muntjac in 28 samples, Rhesus monkey in one sample, spotted deer in one sample and porcupine in four samples.

In Similipal besides the above animals, tigers are known to prey on elephant calves. Baby elephants are considered to be one of the favourite prey animals of tiger for which it has been noticed that whenever a tiger is found to be on prowl, elephants keep their calves in the middle of the herd **(Plate 3.33)**. In many cases, elephants give birth to their calves near to the human habitation just to avoid tigers.

Plate 3.33. An elephant calf between two adult female elephants and another standing close to its mother at Debasthali inside Similipal, Odisha, India when a tiger found to be roaring nearby. Photo: D Swain.

I saw a sick and old female elephant with a calf **(Plate 3.34)** during my visit to Jadapal Beat of Nawana (South) Range of the reserve and the next day I received a message that the mother I sighted was killed and her calf had been preyed upon by a tiger.

Plate 3.34. A sick and old female elephant with a calf sighted at Jadapal Beat of Nawana (South) range of Similipal reserve, Odisha, India were later killed by a tiger. Photo: D Swain.

It is observed that tigers follow elephant herds just to kill elephant calves as evident from distribution of tigers and elephants in Similipal (**Figure 3.7**). Gaur, which is present in Similipal is also a prey animal of the tiger.

However, an analysis of scat and kill data mentioned above shows that the mean weight of prey killed by a tiger varys considerably. The smallest animal species killed was Rhesus monkey and its mean weight is 6 kg. The largest animal species killed was sambar and its mean weight is 150 kg. The relative number of animals of different species killed, mean body mass of each animal species and mean mass of all prey killed derived from scat and kill data of Similipal are presented in Table 3.4.

Table 3.4. Relative number of animals of different species killed by tigers, mean body mass of each animal species and mean mass of all prey killed derived from scat and kill data from May, 2005 to October, 2007 of Similipal tiger reserve, Odisha, India

Species	*Mean Body Mass in kg*	*Relative Number Killed by Tigers*	*Total Mass in kg*	*Mean Mass of Prey Killed by Tigers in kg*
Sambar	150	30.6	4590.0	
Wild boar	38	30.1	1143.8	
Langur	10	25.0	250.0	
Muntjac	20	14.3	286.0	
Total		**100**	**6269.8**	62.7

The mean mass of prey animals killed by tigers, therefore, in Similipal is 62.7 kg (Table 3.4). The mean mass of prey animals killed by tigers elsewhere is presented in Table 3.5. The low value of the mean weight of prey killed in Huai Kha Khaeng of Thailand likely reflects the paucity of large prey at the site when the study was made.

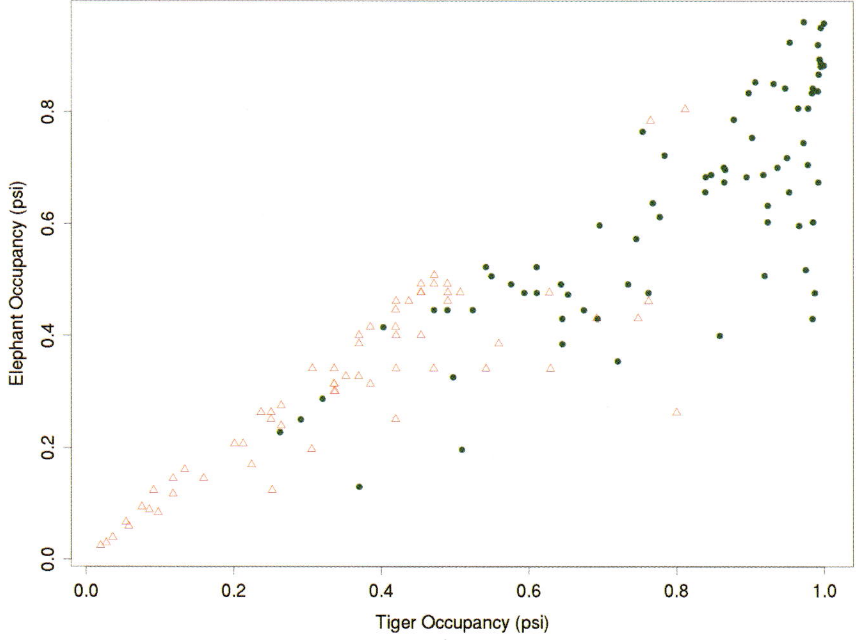

Figure 3.7. Distribution of tigers and elephants in Similipal tiger reserve, Odisha, India. Solid circles (green) are beats in the core area; open triangles (red) are beats in the buffer area.

Table 3.5: The mean mass of prey animals killed by tigers in some selected sites.

Site	*No of Scat/ Kill Samples Examined*	*Mean Mass of Prey Killed in kg*	*Reference*
Kanha, India	300	66.0	Schaller 1967
Nagarahole, India	455	65.5	Karanth and Sunquist 1995
Similipal, India	217	62.7	In the present study
Chitwan, Nepal	123	61.8	McDougal 1977
Huai Kha Khaeng, Thailand	38	14.7	Rabinowitz 1989

It is seen that sambar constitutes bulk of the biomass in the diet of tigers in Similipal. Sambar is also the main prey animal in Chitwan and Nagarahole (Sunquist *et al.*, 1999). Therefore, tiger follows the distribution of sambar in Similipal as evident from **Figure 3.8**. Chital is the bulk of the biomass in the diets of tigers in Kanha. But Langur constitutes 4 per cent of the biomass in the diets of tigers in Similipal which may be an exception in comparison to other Parks of Asia. The defecation of seven to eight weeks old captured tiger cub, Khairi contained Langur hair suggesting that her mother had ingested regurgitated partially digested skin and flesh of Langur (Choudhury, 1999).

The density estimation of prey species have been carried out in Similipal during 2012 using line transect survey (Nayak, not dated). Common langur was found to be the most abundant prey species in Similipal with a density of 10.2/km^2 in winter and 16/km^2 in summer followed by rhesus macaque (6/km^2 in winter and 7/km^2 in summer), chital (5/km^2 in winter and 3.8/km^2 in summer), wild pig (3.2/km^2 in winter and 4.8/km^2 in summer), sambar (2.8/km^2 in winter and 4/km^2 in summer), muntjac (1.6/km^2 in winter and 1.6/km^2 in summer) and mouse deer (0.6/km^2 in winter and 1/km^2 in summer). Palei *et al.* (2016) report that densities of ungulate species in the intensive study area of 421 km^2 of the southern part of Similipal from September 2012 to May 2013 were: sambar (6.65/km^2), wild pig (4.81/km^2), muntjac (3.82/km^2), chital (2.82/km^2) and mouse deer (1.88/km^2). Palei *et al.* (2016) further states that the estimated ungulate density and biomass of 19.98/km^2 and 1264 kg/km^2 respectively are relatively low compared to other tropical forests of India.

Tigers are sometimes seen chewing grass, may be needing some roughage or the chlorophyll in the grasses for easy digestion of meat they consume. On my visit to Bandhavgarh tiger reserve during October, 2015, I saw three cubs ingesting grass leisurely while following their mother at a distance (**Plate 3.35**).

Predatory Behaviour

Acute sensitivity, secretiveness and ability to surprise, untiring perseverance, agility in attack, tenacity to follow and hold, and strength to overpower are the qualities necessary in a successful predator; and the tiger possesses them all in extraordinary measure (Sankhala, 2005). The tiger has a set of powerful canine teeth combined with strong jaw muscles. Choudhury

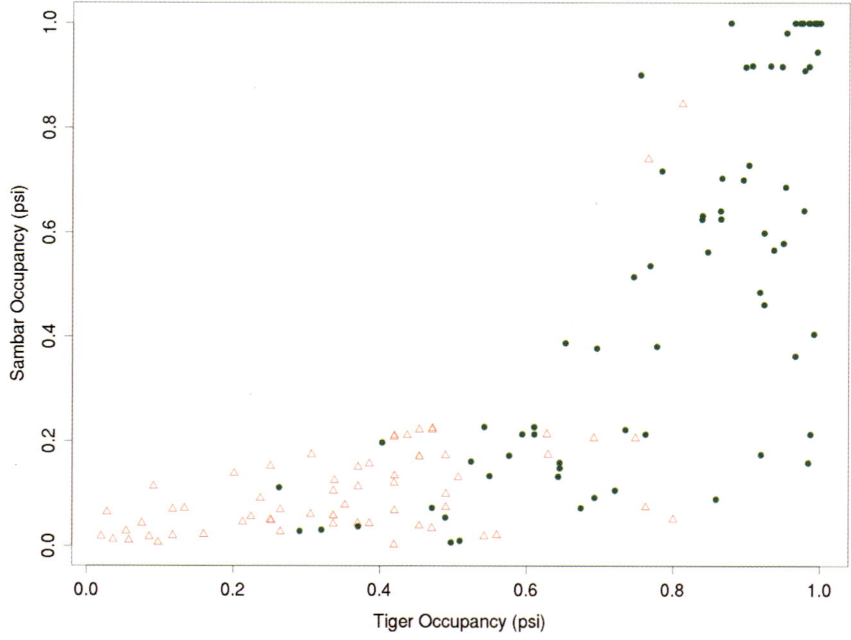

Figure 3.8. Distribution of tigers and sambars in Similipal tiger reserve, Odisha, India. Solid circles (green) are beats in the core area, open triangles (red) are beats in the buffer area.

Plate 3.35. Tiger cubs seen chewing grass leisurely while following their mother at a distance in Bandhavgarh tiger reserve, Madhya Pradesh, India. Photo: D. Swain.

(1999) reports that the milk teeth of Khairi were changed mediolaterally starting with her incisors, in dorso-ventral, ventro-dorsal succession: upper-lower, lower-upper and lower-upper. The medial (inner) right of her upper jaw was the first to be shed when she was eight months old. The lateral (outer) two teeth of her upper jaw were the last when she was 10 months old. The replacement was by coaxial push out, the edge of the permanent tooth peeping through the gum pit within two or three days. Khairi's permanent incisors in her upper jaw were notched in antero-posterior asymmetry, and those of her lower jaws were symmetrically sharp edged. The frontal edge of the upper notches overhang the top line of the lower incisors, the edges of which fitted into corresponding notches in the upper incisors. The lower incisors thus functioned as cleats for a non-slipping grip on anything that they wedged into the upper incisors. This enabled the tigress to gnaw, tear open and create dents for the incurved barbules on her tongue to hook and manipulate sideways in her cutting and crunching molars. The outer incisors in the upper jaw had in addition a pseudocanine look, which appeared to have a transitional function between the incisor and the true canine. The permanent canine teeth of Khairi started cutting her gum, all four within five days, when she was about one year old. These continued growing close to the inner side of the deciduous canines. The replacement was by lateral push out and milk canines peeled off sideways within a month. After changing her deciduous denture, Khairi killed its first natural prey, a chital fawn, at the age of about 18 months. Choudhury (*supra*) further reports that in her search, selection and capture of prey, playful or real, Khairi was used to apply her ears, eyes and nose.

Breeding

Sunquist *et al.* (1990) report that under favourable conditions tiger populations grow rapidly; gestation period is 103 days. Females breed relatively early, and they come into oestrus rapidly following loss or dispersal of young. Females first breed at about three years of age, after establishing residency. Litter size is commonly three, but may be upto five, and the interbirth interval may be as short as 20 months. I sighted a mother with four cubs (**Plate 3.36** shows one of the four cubs) in the Kanha tiger reserve during January, 2004.

Plate 3.36. One of the four tiger cubs sighted in Kanha tiger reserve, Madhya Pradesh, India. Photo: D Swain.

I had also the opportunity of observing mating of tigers in the Bandhavgarh tiger reserve on 15 January, 2004. I heard the growling of tigers at 6.30 AM while moving inside the reserve on elephant back. On a search, I noticed that there were three tigers - one adult female, one adult male and one subadult male, a few meter from the forest road on the uphill inside the forest. The Mahout (elephant driver) from his past observations in the reserve informed that the subadult male (**Plate 3.37**) was the offspring of the female and she was driving away this male child in order to mate with the adult male.

Plate 3.37. A subadult male tiger in Bandhavgarh tiger reserve, Madhya Pradesh, India. Photo: D Swain.

At about 7.30 AM, the adult male started mounting the tigress. Each mounting (**Plates 3.38A and B**) lasted for about 20 to 30 seconds and such mating continued throughout the day. There was an intervening period of about 10 to 20 minutes between successive mounting.

Plate 3.38A. Mating of tigers in Bandhavgarh tiger reserve, Madhya Pradesh, India. Photo: D Swain.

Plate 3.38B. Each mounting of the male tiger lasted for about 20 to 30 seconds and such mating continued throughout the day. There was an intervening period of about 10 to 20 minutes between successive mounting. Photo: D Swain.

In Similipal a pregnant tigress was captured in the camera trap during the camera trap exercise of 2007 (**Plates 3.39A and B**). **Plate 3.39A** was taken on 6 April, 2007 and **Photo 3.39B** was that of the same tigress on 14 May, 2007. Pregrancy of the tigress is clearly visible from its bulging belly which had increased in size during the period from 6 April to 14 May, 2007.

Plate 3.39A. A pregnant tigress captured in the camera trap during the camera trap exercise on 6 April, 2007 in Similipal tiger reserve, Odisha, India. Photo: Similipal TR archive.

Plate 3.39B. The same pregnant tigress of Photo 3.39A on 14 May, 2007 showing clearly visible bulging belly which increased in size during the period from 6 April to 14 May, 2007. Photo: Similipal TR archive.

Defecation

Choudhury (1999) reports that the place where Khairi eliminated her excreta was inherently selective. She had a strong inclination to drop her faeces at such places where its smell would be completely sealed (in water) or muffled (in pits, crevices, boulder gaps and thick leaf litter). She also defecated on dung heaps. The fresh smell of her faeces got mixed up with that of the dung heap and thus lost its identity. During her free living in the Similipal forests in summer of 1976, it was observed that from 11 April to 30 April, 1976, defecation of Khairi was recorded 45 times in 20 days. Out of defecations, 23 were in water, nine each under leaf litter and inside trenches/pits, and two each in the open ground and human toilet. When the summer became intense from 14 May to 31 May, 1976, defecation in water increased. Out of 38 defecations in 18 days,

26 were in water, four each under leaf litter and inside trenches/pits, two inside boulder crevices, and one each in open ground and on road/foot path. In contrast during the winter season from 5 January to 23 January 1977 within 19 days, there were 27 times defecation and only three were noticed in water; there were nine defecations inside trench/pits, six in human toilet, four in open ground, two each inside boulder crevices and on road/foot path, one on elephant dung pile and none under leaf litter. Concealment of faecal matter by wild tigers in Similipal is confirmed by the rarity of getting their scat samples. It is, therefore, not surprising that tiger scats are not noticed in Similipal during summer and even scrape marks on ground are hard to find because of hard and stony surface. It is therefore advisable to carry out sign survey of tigers and co-predators during the winter season only. And that population estimates and measurement of tiger density in Similipal would, therefore, be far more difficult than in drier open forests. The intensity of tiger signs would be comparatively low than that of Kanha and Bandhavgarh in the same central Indian landscape.

Tiger excreta vary in shape, size, colour and texture based on age of the animal, meat consumed and more importantly time lapsed after last meal. When the stomach gets empty, there is excretion of white materials consisting of digested and undigested bones **(Plate 3.22C)**, and after this a tiger is ready for another kill (Anand Singh Negi, personal communication, 26 December 2004).

Scent-marking of Territory

When Khairi was over seven months old, lipids (fat) appeared as white colloids in her hitherto clear urine and she made her first property marking at the age of about one year by lifting her tail to squirt a jet through her urino-genital opening. Irregular and low frequency markings were continued over a month and regular marking of territory, den and properties started thereafter. Choudhury (1999) states that the frequency of marking is directly influenced by territory and the necessity for renewal of individual scent flags. The territory is basically sized to food potential and predation success. Entrance into new territory and re-entrance into old ones are invariably followed by heavy markings. The marking fluid is ejected postero-medially upward to 45° according to the object marked – flat trajectory on upright flat face and sharply arched up for the underside of foliage. The placement of marking is dually purposeful, in its proximity to the nose **(Plate 3.40)** and for protection from obliteration by the impact of natural showers. The distribution of territorial marking is generally along the most frequented trails used by the marker and likely to be used by its conspecifics. It is intended to increase the probability of redetection by self and act as a prominent sign-post for others. Marking stopped during seasonal oestrus of the tigress Khairi.

Plate 3.40. A tigress smelling the scent mark in Bandhavgarh tiger reserve, Madhya Pradesh, India. Photo: D Swain.

Sarkar *et al.* (1994) based on their experiment with nine female and three male tigers of Nandankanan zoo of Odisha, report that the frequency of marking may exceed ordinary urination by 10 to 600 times. Their study indicated that removal of a neighbour from beyond a boundary (*i.e.*, keeping it vacant) sharply reduces the frequency of marking. They further report that high rates of marking by proestrous females markedly decrease during mating. The rates of marking before and after litter dropping were almost the same. Brahmachry *et al.* (1990) report that signaling molecules in the urine and marking fluids of tigers include a substance having an aroma like that of Basmati rice. The urine and marking fluid are found to contain 14 major saturated short chain free fatty acids: acetic (73 per cent), propionic (3.6 per cent), isobutyric (1.6 per cent), butyric (0.7 per cent), isovaleric (0.4 per cent), valeric (1.2 per cent), isohexanoic (0.4 per cent), hexanoic (7.7 per cent), isoheptanoic (1.0 per cent), heptanoic (1.6 per cent), isooctanoic (1.3 per cent), octanoic (2.0 per cent), isonanoic (0.6 per cent) and nonanoic (4.5 per cent) acids (Sarkar *et al.*, 1991).

Man Eating Tigers

In Similipal man eating by tigers is rare. It has been said that Similipal has been depopulated several times because of man-eaters (Das and Das, 1968). In early parts of nineteen forties, tiger havoc was widespread in the village Budhabalang and continued for five years. Villagers abandoned their homes and took shelter elsewhere. The headman of the village Mana Ho left for Garh Similipal. In 1948, he came across a number of persons belonging to Munda tribe, who showed interest to settle down permanently in Similipal. They brought a hunter named Chaitanya Gunju who shot several tigers and performed rituals to avert tiger menace in future and Mana with others resettled in the village. But there is no such high incidence of man-eating by tigers in recent times in Similipal. I collected information about human killing from records of the reserve, interviewing the old and retired forest staff and village headmen inside and around Similipal reserve.

CASE ONE

On 3 December 1981, on a chilly morning, 10 Kharia tribals of Upper Barhakamuda went out as usual to collect honey from the forests and this time in the Balidar Beat area of Similipal. They were walking in a file at a distance of 40 metres from one another looking to the crown of trees in search of nest of honey-bees, where honey would be available. Daitari Dehuri positioned at the end of the file was attacked by a tigress probably when he came very close to its den. Daitari called his fellowmen loudly and Gora Dehuri who was near to him rushed towards Daitari and threw his axe at the tigress. The tigress did not leave Daitari, rather roared at Gora. Nobody went near the tigress out of fear. Gora came running to Upper Barhakamuda forest office and reported the matter in the evening. As it was already night, nobody dared to visit the site of the incidence. The next day in the morning, a Forester alongwith 10 to 15 other people located the partly eaten dead body of Daitari at about 11 AM. For further investigation they did not lift the dead body from the site. To dissuade the tigress, they collected dry wood and litter and put up a fire around the corpse. They have hardly covered 10 km back from the incidence site; they met the then Field Director, S.R. Choudhury who was on his way to this place after being informed about the incidence over VHF. When all of them returned to the place where the corpse was last left, they found that the tigress had dragged the dead body to a nearby stream side and had consumed the legs and lower portion of the body. They noticed deep canine injury on the throat of the corpse. The pugmarks of the tigress were visible on the bank of the stream. In the evening all of them returned to Barhakamuda. Nextday when they came back to inspect the deadbody, it was noticed that the tigress had again dragged away the corpse about 200 metres from the site and had consumed the remaining portion of the body leaving the skull and a few bones. On search of the place where the corpse was first located, the plastic belt of Daitari Dehury was recovered. There were canine marks of the tigress and its cub on the belt.

During September 1982, the above tigress attacked a wood cutter at Ranibhol near Baniabasa of Udala Range, about 30 km North-East of Balidar and ate away the entire body of the wood cutter. The same tigress again in the first week of October, 1982 attacked a girl of about 20 years, when she was returning with a group of women after collecting firewood, leaves and tubers from the forest. The girl was in the middle of the group. But the tiger jumped at the girl and dragged her away into the forest. The incident occurred at the site named Gadei in the Chandanchaturi forest Beat of Pithabata Range, about 30 km North-East of Baniabasa where the tiger had killed the wood cutter. It was afternoon and the message reached the forest staff at Chandanchaturi. The staff rushed to the spot. They saw that the tigress had already killed the girl and had started to eat the body. The staff made a machan on a nearby tree to observe the tigress. They noticed that the tigress ate part of the flesh and sat at a distance away from the corpse. The process was repeated until the entire body was consumed leaving a few bones and the skull. Again after about 10 days, the tigress attacked a father-son duo of village Digdiga at the core line adjoining their village, about 10 km North of Chandanchaturi. The tigress tried to drag the father and the son threw his axe at the tigress hitting her body hard. The tigress had to leave the father and ran away. The son brought the grieviously injured father to the district headquarters hospital at Baripada. But the father succumbed to his injury. The tigress was declared a man-eater. After about 15 days, in the first week of November 1982, the tigress came to Bahalda village near Chandanchaturi and killed a cow. A machan was made on a tree near the carcass and when the tigress came to the kill, it was shot dead.

CASE TWO

On 3 July 1987 Jaga Dehuri (of Kharia tribe) with his wife was on his way back home at Upper Barhakamuda after collecting wild mangoes from the forest. He faced a tiger on Tarinibilla road near a place called Mandadar. The tiger caught hold of Jaga Dehuri just below his neck, but Jaga could be able to escape from the grasp of the tiger. The entire body of Jaga Dehuri was wet due to dribbling of saliva from the mouth of the tiger. Jaga Dehuri reported about the incidence to the forest office of Barhakamuda. There were signs of four canine marks below the neck of Jaga Dehury. While injury due to one canine tooth was deep, the other two canine teeth had not pierced his body. There was very little injury due to the fourth canine tooth indicating that the tiger had become old and one canine tooth only was intact, due to which it was not able to catch hold of Jaga Dehuri properly as a result of which he could escape. On 4 July 1987, Forest staff of Barhakamuda together with 12 other Kharias went to the site of incidence. While the staff was inspecting pugmarks of the tiger, the tiger rushed and attacked them with a thunderous roar. The tiger could be able to catch hold of Gora Dehuri; but the tiger left Gora when beaten by the other people present there with axe and wooden club. Gora Dehuri got injury on his palm due to piercing of nail and teeth bite of the tiger. Now they did not like to stay there anymore after this incidence and they decided to return to Barhakamuda. They had hardly gone 50 metres from the place of incident, when the tiger came rushing towards them and grabbed Asadhu Dehuri just below his waist. Asadhu Dehuri slipped on the road face down and remained firm. The tiger again got a beating from the people. The tiger left Asadhu and went into the forest. Asadhu Dehuri was badly injured and he was first taken to Barhakamuda and then to the hospital at Baripada. All the injured persons were treated including Asadhu and there was no casualty.

CASE THREE

Kharias of Jenabil village went out to collect sal resin in the Compartment, Sanjo 17 (P) near Core line on 8 September 1988. Suddenly one of the Kharias named Sriram Naik got close to a tiger and the tiger attacked him. All other Kharias left the place out of fear without taking any step to rescue Sriram Naik and reported the matter to the forest staff at Jenabil. As it was evening, no body dared to search for poor Sriram Naik. The next day, forest staff and other people reached the site and found that the body of Sriram Naik had been dragged about 500 meters from where he was first attacked and the tiger had already consumed three-fourths of the corpse. Clothes of Sriram Naik, sal resin basket and his sandals were found scattered. But strangely, the tiger did not resort to human kill thereafter. No further human kill of any other tiger in Similipal tiger reserve has been reported.

Reason for Man-eating Behaviour

Prater (1965) reports that the larger carnivores are not usually dangerous to man. A tigress or panther with cubs may make a demonstration of frightfulness and charge home. Again, taken unawares, these animals may attack in self-defence, or turn on their pursuers, as when wounded or harried in a beat, or the attack may be made in error a man being mistaken for more legitimate prey.

Sankhala (2005) reports that during the last 30 years in the States of Kerala, Tamil Nadu, Rajasthan, Assam and Bihar, there has been no report of man-eating tigers. But cases of human kill by tigers in recent times have been reported from Corbett in Uttarakhand (Sankhala, 2005) and Sunderbans in West Bengal and Bangladesh (Montgomery and Briggs, 2001; Neumann-Denzau and Denzau, 2010). Based on his study in the Sunderbans of Bangladesh, Hendrichs (1972, 1975) reports that the salinity of the water has something to do with the man-eating habits of tigers. But Choudhary and Chakrabarti (1972) do not support the view of salinity of water being responsible for man-eating behaviour of Sunderban tigers, as they have recorded more man-eaters in areas where the concentration of salt in water is less. Neumann-Denzau and Denzau (2010) examining the extent of human-tiger conflict in the Sundarbans forest of Bangladesh predict that the high level of disturbance by forest resource users seems to provoke man-eating behaviour among local tigers.

The human kills that occurred between December 1981 and November 1982 in Similipal reported here started accidentally by a tigress with cub(s) and the tigress became man-eater killing four persons in one year. It was ultimately shot dead. The second one was old and infirm having three out of four canine teeth damaged. The human killing of the third tiger was never repeated. Therefore, human killings in Similipal could be either accidental or freakish, or due to the tiger becoming old and infirm.

THE WILD DOG OR DHOLE

Population of wild dog or dhole (*Cuon alpinus*) in Similipal seems to have fallen steeply over the years, so much so that there have been no recent sighting records for the past several years. This is not surprising as sudden fluctuations in wild dog populations with a cyclical rise and fall is a phenomenon, which has been known from other parts of our country, especially South India. Srivastava and Singh (2003) mention that in Similipal, wild dog was found in small groups during 1970s, and the status was low in 1980s. An estimate in 1990 puts the number at 400-428 dogs in 104 packs. In the late 1990s dhole was rarely sighted. Competition faced from village dogs is considered as one of the factors responsible for the changing status of dhole in Similipal. The gradual reduction in sighting of dhole in Similipal between 1970s and 1990s could be a positive sign for the growth of population of tiger, but the possible reason for their gradual reduction is a matter of concern for biodiversity conservation. However in recent times dholes are caught in camera traps occasionally in Similipal (**Plate 3.41**).

THE ELEPHANT

The Asian elephant (**Plate 3.42**), *Elephas maximus*, once roamed over a vast area of Asia extending from the Tigris-Euphrates in West Asia to China in South East Asia through Iraq, Iran, Afghanistan, Pakistan, Nepal, India, Bangladesh, Sri Lanka, Myanmar, Thailand, Indo-China, Malay Peninsula, Java, Borneo and Sumatra. It has since disappeared from Mesopotamia, Syria (East

Plate 3.41. A dhole caught in camera trap in Similipal, Odisha, India. Photo: Similipal TR archive.

Plate 3.42. A charging male Asian elephant (*Elephas maximus*) in musth at Patbil inside Similipal, Odisha, India. Photo: M V Nair.

Mediterranean), Iraq, Iran, Afghanistan, Pakistan, Java and most of China. In India, elephants were distributed over a very large area extending from the Himalayan foot hills (1524m MSL) down South through Western Ghats, Karnataka, Kerala; on the East to Assam through Madhya Pradesh, Bihar, Bengal and Odisha. During recent years, they have become much depleted in numbers due to destruction of their habitats because of uncontrolled growth of human population, excessive demand for cultivable land to raise crops, unplanned industrialisation, mining and lust for ivory (Behura, 2004).

At the turn of the 19th century, there were five to ten million elephants both Asian and African put together. By 1979 they had been reduced to only about a million. By 2001 their numbers became further reduced to 400,000 African and 40,000 Asian elephants. In the last 100 years we have lost more than five million elephants due to hunting and habitat destruction (Swain, 2004). The process of annihilation is continuing unabated and the future of elephants remains bleak. Several measures were initiated both globally and locally for the conservation of this magnificent animal of our time. Ivory imports were banned; the elephant was placed as an Appendix-I animal in CITES' (the Convention on International Trade in Endangered Species of Wild Fauna and Flora) regulation; African and Asian elephant specialist groups were formed to suggest conservation measures; Elephant Projects were initiated in elephant holding countries; and national and international funds have been setup for funding conservation activities.

In India, Project Elephant was launched in February, 1992 and elephant reserves (ERs) were named to protect elephants with specific action plans. There are 30 reserves across India extending over 65,450 km^2. Mandated by Conference of Parties' (COP) resolution of CITES, Monitoring of Illegal Killing of Elephants (MIKE) programme started in South Asia in 2003 to measure levels and trends in the illegal hunting of elephants, to determine changes in these trends over time and to determine the factors causing or associated with such changes, and to try and assess in particular to what extent observed trends are a result of any decision taken by the COP to CITES. MIKE is being implemented in 10 ERs in India and they are: Chirang Ripu (Assam), Dhang Patki (Assam), Eastern Doars (West Bengal), Deomali (Arunachal Pradesh), Garo Hills (Meghalaya), Mayurbhanj (Odisha), Mysore (Karnataka), Nilgiri (Tamil Nadu), Shivalik (Uttarakhand) and Wayanad (Kerala).

Dung Density Count of Elephants in Similipal

As part of the MIKE/CITES programme, elephant census based on dung density count was taken up in the above 10 ERs. The methodology adopted and findings of such a census in Mayurbhanj ER is presented here.

Mayurbhanj ER

The Mayurbhanj ER (**Figure 3.9**) was constituted in September, 2001 and is located between 21°10' to 22°35' North latitude and 85°45' to 87°05' East longitude, spreading over 7043.74 km² in Mayurbhanj, Balasore, Bhadrak and Keonjhar districts of Odisha state. It

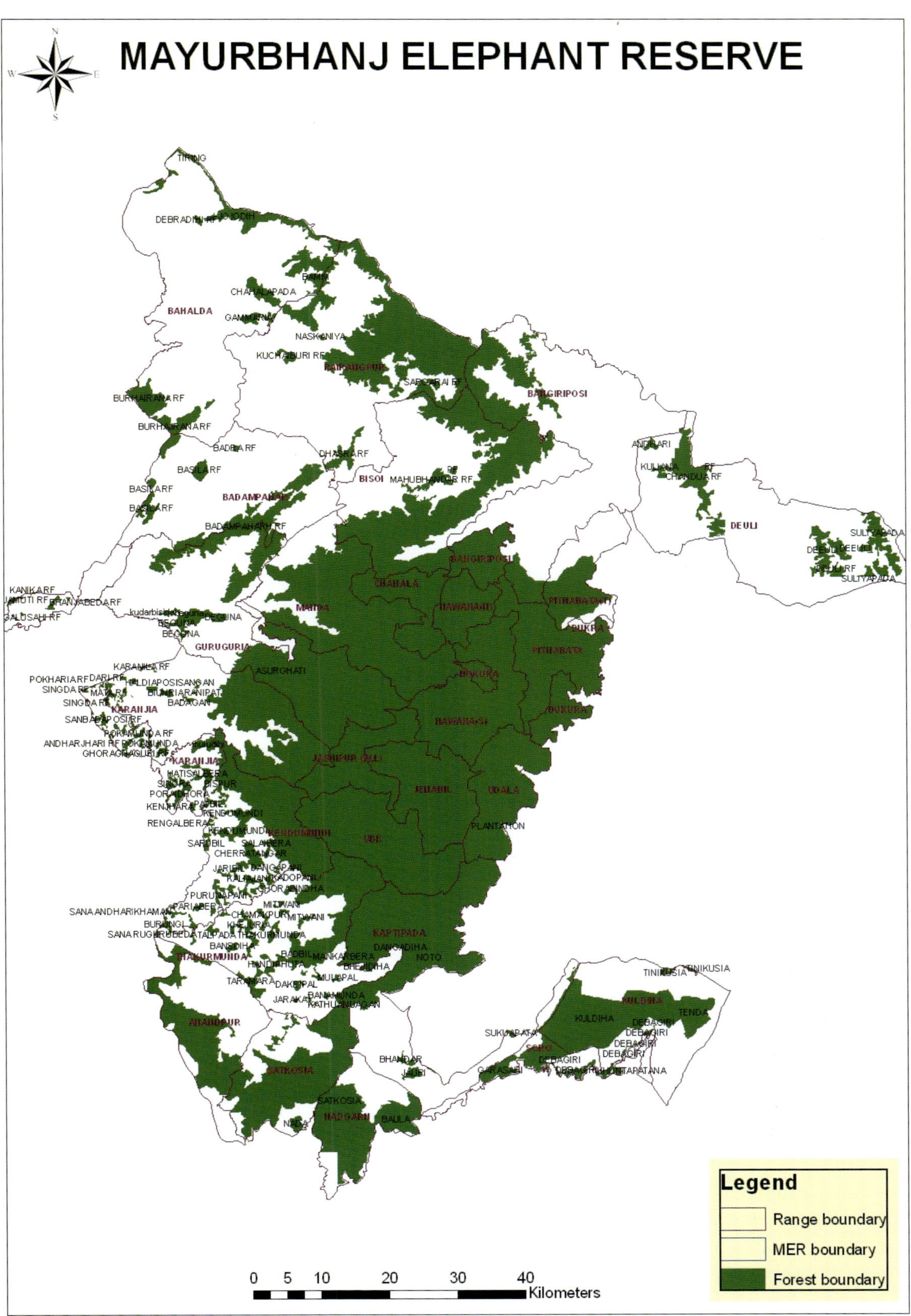

Figure 3.9. Mayurbhanj elephant reserve, Odisha, India.

comprises of a continuous tract covering Similipal tiger reserve (2750 km²), Kuldiha sanctuary (272.75 km²), Hadgarh sanctuary (191.06 km²) and other reserved forest, protected forest, village forest, revenue and homestead land (3213.81 km²).

CENSUS DESIGN

Since the early 1980s biologists have used dung counts for surveying elephants in the forests of both West and Central Africa (*e.g.* Short, 1983; Merz, 1986; Barnes *et al.*, 1995). The same techniques have also been used for Asian elephants (Dawson, 1990; Santosh and Sukumar, 1995; Varman, Ramakrishnan and Sukumar, 1995). An estimate based on dung counts usually requires estimates of three variables – dung-pile abundance on the ground (*i.e.* dung-piles per km²), defecation rate, and dung decay rate – which are later combined to give the final estimate of elephant numbers (Barnes, 2001). For this purpose, the following steps were followed in Mayurbhanj ER.

A. *Stratification of Elephant Zone*

1. Each forest division (FD)/protected area (PA) constituting the reserve was stratified as follows on the basis of the elephant densities known from the previous (*viz.*, 2002) census:
   a. High density/high usage strata (H): > 1 elephant/km²
   b. Medium density/medium usage strata (M): 0.5-1 elephant/km²
   c. Low density/low usage strata (L): Below 0.5 elephants/km²
   d. No usage-elephant free zone (not included in the calculation of 'effective elephant habitat' while extrapolating sample data).
2 For quick stratification, a 'forest beat' was taken as the unit of stratification and it was allotted to H, M or L on the basis of the majority of its area falling under a particular stratum. A stratified map of the FD/PA was prepared on the scale 1:50,000.

B. *Measuring Dung Decay Rate (also called Dung Disappearance Rate) or DDR*

1. Calculation of DDR needed an experiment on a sample of dung piles scattered over various strata. A number of visits to the field were made, adding fresh dung piles to the sample and recording the state of the previously marked dung piles.

   A fresh dung pile was one which was 0-24 hours old.

   State of a dung pile was recorded as Present (=1) or Absent (=0).

   Present was any stage where some dung material was left.

   Absent was a stage where only traces (*e.g.* plant fibre remains, termite moulds, mud, *etc.*) were left and no dung material was present. Absent also included 'total disappearance' of dung pile (*e.g.* washing away in heavy rains).
2. Number of Dung piles to be marked for calculating DDR:

   An ideal target was to have about 120 dung piles for each stratum within the ER. As there were three strata, the total target for the ER = 360 and in fact 840 dung piles were marked (Table 3.6).
3. Initiating the Experiment:

   It takes about 105 days to complete the experiment. The experiment began about 14 weeks (98 days) before the proposed date of census by dung count method.
4. Selecting dung piles:

   It required to visit the field every fortnight as per the schedule in Table 3.6, searching for fresh elephant dung (less than 24 hours old) in each stratum and marking the same for future reference.
5. Marking dung piles:

   Each dung pile was marked and numbered uniquely using wooden pegs with number written on top with indelible paint and also written on an aluminium foil tagged to the peg or nearby tree. GPS reading of the location was recorded.
6. Observations:

   During each visit, the previously marked dung piles were visited and their state (Presence/Absence) noted. During the last visit (*i.e.* 6.5.2005), the state (Presence/Absence) of all previously marked dung piles were noted. No fresh dung pile was marked on this visit.
7. Recording Observations:

   A specially designed DDR observation form was filled up for each Range/Division. However, a consolidated form was prepared for the entire ER based on the observations made on the last day (*i.e.* 6.5.2005).

C. *Estimating Dung Densities*

Dung densities were estimated using the line transect method where the perpendicular distance from the center of each dung pile to the transect line (**Figure**

3.10) was measured as per distance sampling (*e.g.*, Buckland *et al.*, 2002; Thomas and Karanth, 2002).

Results and Inferences

Dung decay rates of high (Upper Barhakamuda and Jenabil ranges of Similipal tiger reseve), medium (Nawana North, Nawana South, Pithabata, Chahala and National Park ranges of Similipal tiger reserve), and low (buffer of Similipal tiger reserve coming under Baripada, Karanjia and Rairangpur divisions, and whole of Hadgarh and Kuldiha sanctuaries) strata are furnished in Table 3.7. Elephant density and elephant numbers based on dung density counts are set in Table 3.8. As per the said dung density count of elephants there are at least 788 elephants in Mayurbhanj elephant reserve (Table 3.8). In contrast, it was reported to be only 465 elephants by total count method during that period (2005). The elephant numbers are therefore under represented for this reserve in total count method. The various census methods used to estimate elephant numbers in India are total counts, sample counts, waterhole counts, sight and resight counts and dung counts. In total counts the entire elephant area is traversed and sighting of elephants is recorded. This method of counting elephants has been practised in different parts of India from the late seventies (Krishnamurthy, 1991; Lahiri-Choudhury, 1991; Singh, 1978; Swain, 1996). A sample count is one where a predetermined portion of the entire elephant area is searched, usually in a number of small distinct sample areas. The details of the methods and analysis have been reported by Dawson and Dekker (1992), Karanth (1991), Rodgers (1991), Sagar and Singh (2001) and Sukumar *et al.* (1991). In the waterhole count technique elephants are counted as they visit waterholes. This technique was employed to count elephants of Odisha in 1999 (Swain, 2004; Swain and Patnaik, 2002). Sight and resight method (Rodgers, 1991) has been adopted to a sight, recognize and resight methodology which works on simple proportions in accordance to the Petersen or Lincoln Index (Overton, 1971). Sukumar (1985) has reported 'registration and sex ratio method' and has estimated the elephant population of Biligirirangan in South India using this methodology. Elephants in Mayurbhanj elephant reserve have always been counted as per total count method and the populations so arrived at in 2005 and 2007 are 465 and 517 respectively.

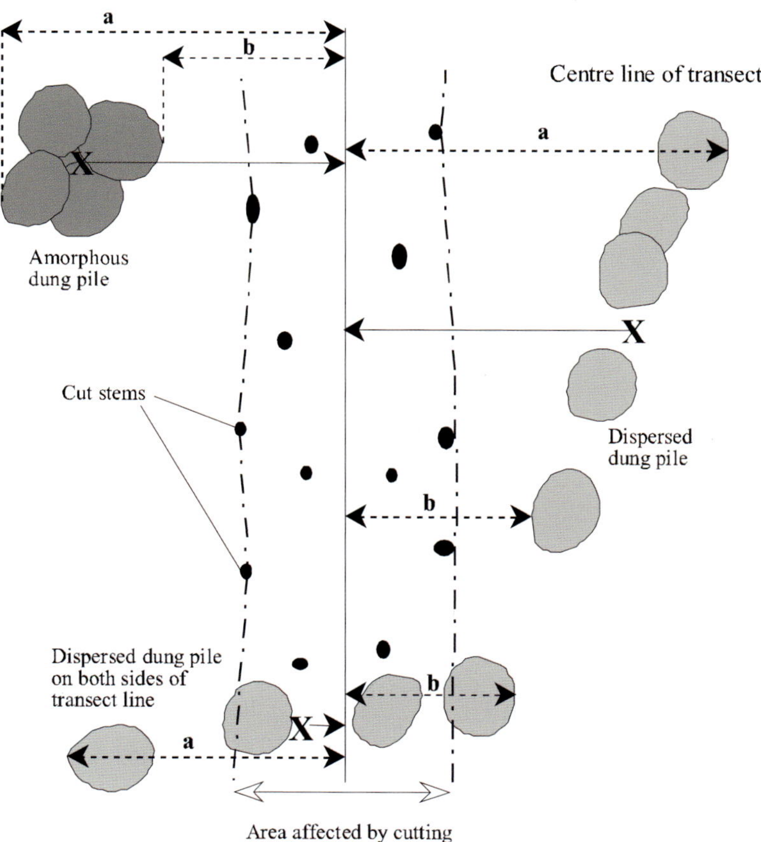

Figure 3.10. Elephant census based on dung density count: Measuring perpendicular distances from line transects (White and Edwards, 2000). The perpendicular distance of the dispersed dung pile (X) is calculated by measuring the distance to the furthest edge of the furthest bolus (a), and the nearest edge (b) of the nearest bolus to the centreline, adding the two measurements and dividing by 2. Where a dung pile traverses the centre line, the perpendicular distance (X) is found by subtracting the smaller distance (b) from the larger (a), and dividing by 2.

Table 3.6. Marking of dung piles for calculation of dung decay rate in Mayurbhanj elephant reserve, Odisha, India during 2005

No. of Visit	Date of Visit	No. of fresh dung piles to be selected and marked							Total
		Stratum H (Jenabil and UBK)	Stratum M (Chahala, Nawana (N), Nawana (S), Pithabata and National Park	Balasore (WL) Division (L1)	Anandpur (WL) Division (L2)	Karanjia Division (L3)	Rairan-gapur Division (L4)	Baripada Division (L5)	
1	21.01.05	17	17	17	17	17	17	17	11
2	05.02.05	17	17	17	17	17	17	17	11
3.	20.02.05	17	17	17	17	17	17	17	11
4	07.03.05	17	17	17	17	17	17	17	11
5	22.03.05	17	17	17	17	17	17	17	11
6	06.04.05	17	17	17	17	17	17	17	11
7	21.04.05	18	18	18	18	18	18	18	12
8	06.05.05	Final recording of observations (Present/Absent)							
		120	120	120	120	120	120	120	840

Note: High density strata (H) comprise of Upper Barhakarmuda (UBK) and Jenabil Ranges of the core area. Medium density strata (M) comprise Nawana North, Nawana South, Pithabata, Chahala and National Park ranges of the core area. Low density strata (L) comprise of Rairangpur (R), Baripada (B) and Karanjia (K) Divisions.

Table 3.7. Dung Decay Rate calculation in Mayurbhanj elephant reserve, Odisha, India during 2005

Strata	No. Dung Piles Monitored	Mean Days for Decay (Days)	CV
Karanjia (L3)	38	72.6	0.0548
Rairangpur (L4)	67	77.02	0.0504
Baripada (L5)	67	77.02	0.0504
L (R, B K Div.)	142	72.88	0.0642
M (Similipal)	155	153.93	0.3443
H (Similipal)	125	89.54	0.0634

Note: High density strata (H) comprise of Upper Barhakamuda (UBK) and Jenabil Ranges of the core area. Medium density strata (M) comprise of Nawana North, Nawana South, Pithabata, Chahala and National Park ranges of the core area. Low density strata (L) comprise of Rairangapura (R), Baripada (B) and Karanjia (K) Divisions.

Table 3.8: Elephant density and elephant numbers in Mayurbhanj elephant reserve, Odisha, India during 2005 based on dung density count.

Strata	No of Transects	Elephant Density/ km^2	Strata Area in km^2	Elephant Numbers	CV of Elephant Density	UCL (ED)	LCL (ED)	UCL (EN)	LCL (EN)
Baripada	18	0.430	455.00	196	0.273	0.695	0.251	316	114
Karanjia	12	0.335	507.34	170	0.256	0.572	0.196	290	99
Rairangpur	09	0.401	342.17	137	0.300	0.779	0.207	267	70
L (R,B,K, Ku,H))	39	0.429	1766.49	758	0.165	0.584	0.302	1032	534
M (Similipal)	20	0.188	572.20	108	0.407	0.420	0.084	240	48
H (Similipal)	14	0.981	331.77	325	0.211	1.546	0.623	513	207
Total of all strata (HML)				**1191**				**1785**	**788**

Note: High Density strata (H) comprise of Upper Barhakamuda (UBK) and Jenabil Ranges of the core area. Medium density strata (M) comprise of the Nawana North, Nawana South, Pithabata, Chahala and National Park ranges of the core area. Low density strata (L) comprise of Rairangpur (R), Baripada (B) and Karanjia (K) Divisions, Kuldiha(Ku) and Hadgarh(H) sanctuaries.

ED=Elephant density; EN=Elephant numbers; UCL= Upper limit of 95 per cent confidence interval; LCL=Lower limit of 95 per cent confidence interval.

Barnes (2001) opines that there are few more options for counting elephants in forests. One of them is the individual recognition of animals at salticks or water sources (Turkalo and Fay, 1995; Turkalo, 1996). But it is not possible to adopt as it requires particular conditions that are rare (Barnes, 2001). Another method is thermal imaging which has been successfully practised for deer (Gill, Thomas and Stocker, 1997) and the same is impractical for dense tropical forests (Barnes, 2001; Prinzivalli, 1992). Barnes (2001) emphasizes that the only method superior to dung density count is DNA sequencing of elephant dung identifying individual elephants in a forest and application of mark capture-recapture framework in the calculation of elephant density. However, it requires advanced laboratory skills and expensive materials (Barnes, 2001). But this can be used to validate the dung count or new development introduced to dung count. Dung counts therefore remains as the only practical and scientific method of counting elephants in tropical/subtropical forests of India. The methodology is also easy to learn by wildlife officers and researchers. Furthermore, dung counts have the advantage that they give data not only on numbers but also on the distribution and deferential habitat use. Dung counts are likely to remain the most practical option for most wildlife managers in the forest zone, at least for the foreseeable future (Barnes, 2001).

Elephant Habits and Habitats in Similipal

There are 129 plant species belonging to 44 families under 23 orders constituting the food plants of elephants in Odisha (Swain, 2004). Elephants are generalised feeders and they consume at least small amounts of most of the plants available in their habitats. Dependence on plant species varies from season to season. From the onset of monsoon grasses and sedges of different species are abundantly available. During the rainy season and prior to winter, grasses and sedges constitute the bulk of elephant diet. Important species grazed by elephants are *Cyperus* sp., *Eragrostis* sp., *Digitaria* sp., *Ischaemum* sp., *Themeda* sp., *Panicum* sp., *Cynodon dactylon* and *Brachiaria* sp. I observe that elephants spend most of their time in grazing on these grasses and sedges in the grasslands at Chahala, Debasthali and Upper Barhakamuda of Similipal forests (**Plate 3.43A**).

Elephants also graze on tall coarse elephant grass (*Typha angustata*). I observed that in Similipal near Deokund, an old tusker faced an accidental fall into a 150 m gorge and died while feeding on this grass species (Swain, 1993). From October onwards and until the crops are harvested there are incidences of crop raiding by elephants. Along with grass and sedges, elephants consume a good quantity of leaves, branches, roots and

Plate 3.43A. Elephants grazing in the grassland at Upper Barhakamuda inside Similipal tiger reserve, Odisha, India. Photo: M V Nair.

tubers which are plentifully available during and after the rainy season until the breaking of forest fires. They depend on leaves and branches of a large number of plants *viz*., *Mallotus philippensis* (**Plate 3.43B**), *Kydia calycina, Grewia tiliifolia, Helicteres isora, Mimosa pudica, Boswellia serrata, Lannea coromandelica, Acacia pennata etc*. Tubers of *Asparagus* and *Dioscorea* and roots of young plants of *Calamus* and *Diospyros* are eaten after uprooting the plants with the help of trunk and front feet.

Plate 3.43B. *Mallotus philippensis*, an important food plant of elephants in Similipal, Odisha, India. Photo: K K Mandal

Elephants also consume bark and small stems of a number of plants. At the end of winter there is leaf shedding and grasses and sedges dry out which make the forest floor a perfect stage for catching fire. Fire burns out many of the elephant food plants. At this time elephants depend on bark and small stems. Important among these are *Bauhinia vahlii* (**Plate 3.43C**), *Trewia nudiflora, Boswellia serrata, Largerstroemia parviflora, Sterculia urens* and *S. villosa*. On analysis of 27 boluses of elephant excrement collected from the Similipal forests, it was noticed that 90 per cent of all these boluses contained long fibres and the rest remnants of grasses and leaves. Elephants like to eat the bark of sal (*Shorea robusta*). The sal trees which abound around an artificial salt-lick at Chahala (Similipal) are debarked by elephants as a result of which 90 per cent trees wither out. It was observed that generally tuskers with the help of their tusks resort to debarking. Female elephants twist the tree and then remove the bark for feeding.

Plate 3.43C. Barks and small stems of *Bauhinia vahlii* are consumed by elephants when fire burns out many of the elephant food plants during summer in Similipal, Odisha, India. Photo: D Swain

Elephants have penchant love for *Mahua* (*Madhuca longifolia*) flowers. Flowers appear in the tree during the period of leaf shedding (February and March). Elephants in herds perk themselves under the *Mahua* trees during the season and feed on flowers that have fallen on the ground. They also use their trunks for plucking flowers. As a trick to attract wild animals including elephants for close-observation, the park staff throw dried *Mahua* flowers near the salt-licks and the adjoining meadow. Wild animals are so fond of this flower that they come out from hiding and feed on them. On the night of 12 August 1992, a herd of three female elephants while grazing on the ground near Chahala Forest Rest House managed to cross the trench and take away a gunny bag full of dried *Mahua* flowers stored in the verandah of the staff quarters. A country liquor is brewed from the dried *Mahua* flowers by the local tribals. Sometimes elephants raid huts of tribals to drink country liquor (Swain 2004). Many species of other animals like bear, gaur and even cattle congregate under *Mahua* trees to feed on the flowers as it contains a good quantity of alcohol.

During the summer season when leaves and grasses are scarce in the forest, elephants live on fruits of different species. Important among them are mango (*Mangifera indica*), 'Bela' (*Aegle marmelos*) (**Plate 3.43D**) and 'Kumbhi' (*Careya arborea*). A large amount of undigested seeds voided out in the faeces of elephants germinate at the onset of the monsoon (**Plate 3.43E**). Elephants are fond of mango so much so that in the mango season they congregate under mango trees to eat ripe mangoes in the forest. This facilitates elephant census in any forest habitat. Previously, elephant census was carried out by counting them under mango trees for which elephant census was otherwise known as Mango Census of elephants.

Plate 3.43D. Bela (*Aegle marmelos*) plant with unripe fruits in Similipal, Odisha, India. Ripe fruits of *Aegle marmelos* are cherished by elephants during summer. Photo: H Hazebrock

Plate 3.43E. A large amount of undigested seeds of Bela (*Aegle marmelos*) voided out in the faeces of elephants germinate at the onset of the monsoon in Similipal, Odisha, India. Photo: D Swain

As mentioned above, consumption of food plants varies from season to season according to their availability, palatability and nutritional value. In any season the browsing and grazing intensities vary. Sukumar (1989) reports that during the dry season there is 69 per cent browsing and 31 per cent grazing in a South Indian population studied by him. Feeding on grass picks up (up to 54 per cent) during the first wet season. Once again after the second heavy rains there is increased feeding by browsing (56 per cent). Field and Ross (1976) report a similar observation for African elephants in the Kidepo valley National Park, Uganda. McKay (1990) reports that in the Ruhuna National Park, Sri Lanka elephants spend about 86 per cent of their feeding time on short grasses. Of course, Vancuylenberg (1977) has strong reservation of this observation of McKay (1990) as the observations were made in open grassy areas without considering occupancy of different habitats. Olivier (1978) reports that in the rain forests of Malayasia grasses are also preferred food plants, but their overall contribution is much less than that of browsing owing to their low availability.

The plant species which serve as the staple food of elephants in Similipal are *Kydia calycina* (leaves), *Grewia tiliifolia* (leaves and small stems), *Helicteres isora* (leaves and branches), *Mallotus philippensis* (leaves and branches), *Trewia nudiflora* (bark), *Aegle marmelos* (fruit), *Boswellia serrata* (leaves, branches and bark), *Garuga pinnata* (leaves), *Lannea coromandelica* (leaves and branches), *Mangifera indica* (fruits), *Ziziphus mauritiana* (leaves), *Bauhinia racemosa* (leaves), *Bauhinia vahlii* (leaves and small stems), *Cassia fistula* (leaves and small stems), *Acacia pennata* (leaves and small stems), *Mimosa pudica* (leaves and inflorescence), *Careya arborea* (leaves, small stems and fruits), *Madhuca longifolia* (flower), *Dioscorea* sp. (tubers), *Asparagus racemosus* (tubers) and *Smilax zeylanica* (leaves and small stems).

Two species of bamboos, *e.g.*, *Bambusa arundinacea* (Dhaba bamboo) and *Dendrocalamus strictus* (Salia bamboo) play an important role in the survival of elephants. Natural bamboo is rare in Similipal. But abundance of bamboo has been mentioned in Hatt's Working Plan written in 1896-97. At present while Salia bamboo occurs rarely, Dhaba bamboo is conspicuously absent in Similipal. Leaves of Salia bamboo contain 14.4 per cent crude protein, 0.25 per cent calcium and 0.01 per cent sodium; whereas that of Dhaba bamboo contains 14.5 per cent crude protein, 0.29 per cent calcium and 0.03 per cent sodium (Sukumar, 1989). The high nutritional values of the aforesiad two species of bamboos suggest that they should be sown/planted in large scale inside Similipal for the benefits of elephants. There are natural and artificial salt-licks inside Similipal and elephants frequently visit the salt-licks to quench their salt (sodium) requirement. By eating the salt mixed with soil (**Plate 3.44**) the elephants are seen to be infected by pathogens affecting their health. Sometimes natural salt-licks are poisoned by poachers for which there are large scale mortality of the pachyderms in Similipal (**Plate 3.45**).

Plate 3.44. A herd of elephants in the salt-lick at Debasthali, Similipal tiger reserve, Odisha, India. Photo: D Swain.

Plate 3.45. Death of a tusker due to poisoning in Similipal tiger reserve, Odisha, India. Photo: D Swain.

Elephants are also observed to dust their bodies to keep off flies and mosquitoes. Villagers and tourists often litter forest floors with various wastes including polythenes which enter into elephant body **(Plate 3.46).**

Plate 3.46. Polythene in elephant dungs at Pithabata inside Similipal tiger reserve, Odisha, India. Photo: D Swain.

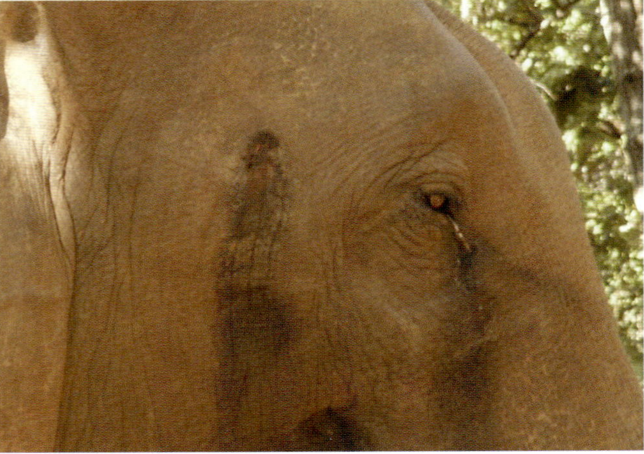

Plate 3.47B. Secretion of musth fluid from temporal region of the elephant. Photo: D Swain

Musth in Similipal Elephants

Males over 25 years of age are susceptible to come into annual musth **(Plates 3.42 and 3.47A and B)**. It is believed that adult male Asian elephants and both sexes of African elephants manifest annually a physiological and behavioural phenomenon known as musth (Chandrasekharan and Cheeran, 1996).

Plate 3.47A. Mahendra, a captive male elephant in Similipal tiger reserve, Odisha, India,shows the manifestation of musth, charecteristic of male Asian elephant. Photo: D Swain

On 28 July, 2002 I observed a female elephant at Devasthali in southern Similipal when she had a clear mark of fluid-secretion from the temporal region **(Plate 3.48A)**. The secretion was similar to the condition of a male elephant in musth. The animal, about 40 years old, was quite normal and mixed freely with the other elephants in a group of 14, and had a calf of one year old. The fact was reported and discussed by Swain and Singh (2003).

Plate 3.48A. A female elephant at Debasthali in Similipal, Odisha, India had a clear mark of fluid-secretion from the temporal region. Photo: D Swain.

Elephant and Forest Interaction in Similipal

My study on biomass of wild herbivores in Similipal (Swain, 2004) reveals that the elephant constitutes 28 per cent of the total wild herbivore biomass in Similipal, while the highest is that of sambar which is 36 per cent. The biomass of gaur and wild boar constitute 19 per cent and 10 per cent respectively. In case of other wild herbivores, the percentage of biomass is negligible compared to the total biomass. It is, therefore, understood that elephants have to face competition from sambar with regard to food material. But in fact sambar is basically a browser whereas elephant is a generalised feeder and feeds on both grass and woody vegetation. An elephant feeds on the branches and leaves of a tree up to which its trunk reaches which may stretch up to a height of 3 to 5 meters. While sambar is a browser, gaur is basically a grazer although it does undertake some browsing. Wild boar feeds mostly on roots and tubers.

Similarly, langur and rhesus macaque feed on leaves and fruits of trees and shrubs almost from top to bottom branches. They extensively feed on the inflorescence of *Mimosa pudica* which is also liked by the elephant. Thus, an individual species of wild herbivore has very little effect on the food materials of the elephant. But the sum total effect of all the wild herbivores combined with the large numbers of domestic animals from enclaved villages is quite heavy.

I further found out that elephant dungs in Similipal on dry weight basis contain 1.2 per cent nitrogen, 1.17 per cent phosphorus, 0.098 per cent potassium, 1.7 per cent calcium, 1.0 per cent magnesium, 1520 (mg/kg) iron, 261 (mg/kg) manganese, 99 (mg/kg) zinc and 8.4 (mg/kg) copper (Swain *supra*). Therefore, elephants in Similipal supply enormous humus and nutrients to the forest ecosystem. A variety of insects feed upon the faeces of elephants. During the rainy season and in the dry period along the banks of streams close to water, dung beetles, *Heliocopris* sp. (Scarabaeidae) are extremely common. These beetles constitute the most important agency to remove, utilise and simplify elephant dung on the forest floor. Another insect which feeds on elephant dung is the termite. During dry periods the dung is mostly decomposed by the termites **(Plate 3.48B)**.

Plate 3.48B. Decomposition of elephant dung by termites in Similipal, Odisha, India. Photo: D Swain.

Elephants play a vital role in the dispersal of seeds of several forest species. The dispersal of seeds by elephants have the following advantages : (a) Elephants feed on the fruits and disperse the seeds to long distances through faeces, thus reducing probable chances of density of seedlings and mortality near the parent plants. (b) While the elephant moves through the forest it creates gaps by pushing down some of the understorey trees or it may remove climbers thus creating a light gap, an ideal condition for establishment of plant offsprings. (c) Further, light gaps as opened up by elephants have been shown to be inimical to fungal pathogens that cause damping off of seedlings. (d) The fruits dispersed by elephants are characterised by juiciness, dull colour, unfresh odour, larger size, undetachability from the tree even after maturation with tough pericarp, dangling on long stalks or attached to the main stem or branch as in the case of *Aegle marmalos*, *Careya arborea* and *Mangifera indica*. Only a dispersal agent like the elephant can carry out the job more thoroughly. Thus, elephants help in the survival of many plant species in the forest and they reduce dominance of single species in the forest community.

THE CROCODILE

Of the 21 species of Crocodilians that are found in the warm subtropical and tropical regions of the world, three occur in the Indian subcontinent (Singh 1999b).These are the Gharial (*Gavialis gangeticus*), the Saltwater or Estuarine crocodile (*Crocodylus porosus*) and the Mugger (*Crocodylus palustris*). All occur in the state of Odisha.Gharial is found in the Mahanadi, the Saltwater crocodile is found in Bhitarkanika Sanctuary and the Mugger is found in the inland river systems of the State.

Results of a survey for mugger conducted in November-December, 1979 indicate that the mugger had virtually become extinct in Similipal (Anonymous, 1999). Although the reasons for the extinction of mugger from Similipal were not clear, Anonymous (1999) cites adverse factors like (1) fishing by using explosives and nylon nets, (2) use of DDT and other insecticides with an intention to control malaria, (3) fire in the river banks forming nesting sites, and (4) natural affects of activities such as predation and or exhuming of a nest by wild boars and monitor lizards.

Mugger management project in Similipal was started thereafter at Ramatirtha near Jashipur to maintain a viable population of the mugger in Similipal, conduct management-oriented research, and provide muggers for re-stocking elsewhere under the financial support from UNDP, FAO, Government of India and Government of Odisha. An initial stock of 150 young muggers were brought from Tamil Nadu and six of them were retained for captive breeding at Ramatirtha. Over the years, 788 crocodiles have been released into the river systems of Similipal between 1981 and 2002 **(Plate 3.49A)**.

Survey of Muggers in Similipal

Survey of muggers in Similipal is conducted mostly during winter. The direct sighting method is used to

Plate 3.49A. Release of mugger crocodiles in Similipal, Odisha, India. Photo: D Swain.

count their number. The survey teams survey the river side on foot. During the survey, direct sighting **(Plate 3.14)** as well as indirect evidence such as, tunnel, basking place, scute mark on the ground are taken into account to count. The different routes surveyed are indicated in Table 3.9. The body length is measured by occular estimation during direct sighting. From indirect evidence the same is measured from the tracks of hind paw mark and also from the tail mark. The body length is equal to approximately 14 x paw length and approximately 65 times the maximum distance between the two lines created because of movement of the tail (Singh, 2000).

Distribution

The numbers counted in the rivers West Deo, Khadkei, Budhabalanga, Khairi, East Deo and Palpala are furnished in Table 3.10. The size/length wise distribution of crocodiles as per 2004 and 2011 censuses is presented in Table 3.11.

The Mugger is the most adaptable of the three crocodile species and has been encountered up to 400 m above mean sea level in clear hill streams, in sewage treatment ponds, and in cold deep rivers in the Himalayan foot hills (Whitaker and Andrews 2003). In

Table 3.9. Different routes surveyed for muggers in the river systems of Similipal, Odisha, India

Sl.No.	River System	Route
1	BUDHABALANGA	1) Budhabalanga village to Jambu bridge
		2) Jambu Bridge to Barehipani Fall Head
		3) Barehipani fall head to Bhatunia
		4) Bhatunia to Majhigaon
2	EAST DEO	5) Sarua pool
3	KHAIRI	6) Jenabil to Ransa
		7) Ransa to Kabataghai
4	WEST DEO	8) Debasthali to Kandadhenu
		9) Kandadhenu to Jalchinda
		10) Patabil to Upper Baraha Kamuda (UBK)
		11) UBK to Manda Darah
5	KHADKEI	12) Khadkei to Baunskhal
		13) Baunskhal to Sorispal
		14) Haldia down stream
		15) Panasia Reservior
6	PALPALA	16) Entire river route

Table 3.10. Number of Muggers sighted in different river systems of Similipal Odisha, India during 1999, 2000, 2004 and 2011

River System	*1999 (Sahu and Swain, 2008)*	*2000 (Sahu and Swain, 2008)*	*2004 (Sahu and Swain, 2008)*	*2011 (Mishra et al., 2013)*
WEST DEO	34	58	42	64
EAST DEO	5	1	5	3
KHAIRI	17	26	20	6
BUDHABALANGA	8	10	9	1
KHADKEI	7	2	7	9
PALPALA	0	0	0	2
TOTAL	**71**	**97**	**83**	**85**

Table 3.11. Distribution of Muggers in different river systems of Similipal during 2004

River Systems	***Number of Crocodiles***									
	<1m		***1-1.5m***		***1.5-2.0m***		***>2.0m***		***Without Size***	
YEAR	2004	2011	2004	2011	2004	2011	2004	2011	2004	2011
WEST DEO	14	2	8	27	16	16	4	14	0	5
EAST DEO	1	2	4	1	0	0	0	0	0	0
BUDHABALANGA	0	0	2	1	5	0	2	0	0	0
KHAIRI	5	2	3	3	10	1	2	0	0	0
KHADKEI	2	0	3	3	2	1	0	0	0	5
PALPALA	0	0	0	2	0	0	0	0	0	0
TOTAL	**22**	**6**	**20**	**37**	**33**	**18**	**8**	**14**	**0**	**10**

the Similipal reserve, they have been seen at an elevation of 850m.

It was found that in river West Deo more muggers were sighted in comparison to other river systems. Most part of this river system is confined to the core area of Similipal. There is reduced anthropogenic disturbance in this area, presence of *darahs* (deep water area) and adequate basking sites **(Plate 3.49B)**. Though East Deo is also situated in the core area, there is less *darahs* and less basking places for the muggers. In Budhabalanga, the river bank is steep and rocky. The availability of basking places is less and biotic interference is more. In Khairi, the population of mugger is more from Jenabil to Jadi *darah* in the Jenabil to Ransa route. Here, the number of muggers is lower in comparison to the West Deo as the river bed is rocky which provides less area for the basking of the muggers. A detailed systematic study has to be undertaken during different seasons in order to study the seasonal behaviour and their ecology as the crocodile species plays a vital ecological role as a master predator in the aquatic habitat in which it lives (Whitaker and Andrews 2003).

Plate 3.49B. One of the several *darahs* (deep water areas) and basking sites for crocodiles in river West Deo inside core zone of Similipal tiger reserve, Odisha, India. Photo: D Swain.

From Table 3.11 it is seen that during the 2004 census, 33 crocodiles within body length of 1.5 to 2.0 m and eight crocodiles of more than 2.0 m body length were sighted, which indicate that 47 per cent of crocodilians were adult. In 2011, there are 32 crocodiles within the body length of 1.5 to 2.9 m (18 crocodiles) and more than 2.0 m body length (14 crocodiles) which is 38 per cent of toal population. In Tamil Nadu, the wild mugger population was 465 with 52 per cent adult (Andrews 1999) and in Gujarat, the population was 492, with 88 per cent adults (Vijaya Kumar *et al.*1990). However, the censuses show that the mugger population is stable in the Similipal reserve.

THE GIANT SQUIRREL

The genus *Ratufa* is an Indo-Malayan taxon of four species of tree squirrels (order Rodentia, family Sciuridae, subfamily Sciurinae). There are about 121 species of tree squirrels including the giant and pigmy squirrels (Moore, 1959). Ellerman (1940) reports that about 56 species under 16 genera of tree and ground squirrels occur in the oriental region. Tropical Asia contains the most diverse squirrel fauna than any other region in the world. The Malabar or Indian giant squirrel (*Ratufa indica*) occurs in Peninsular India. The earliest classification of Sciuridae was based on the skeletal and dental morphology (Ellerman, 1940; Moore, 1959). Based on immunological studies, Hight *et al.* (1974) showed that *Ratufa* is different from genera such as *Calliosciurus, Sundasciurus, Tamiops, Funambulus* and *Mentes* which are closely related. The pelage colour in squirrels exhibits the greatest intra specific variation and the squirrels are found in diverse habitats ranging from deciduous to evergreen forests (Ramachandran, 1988).

The giant squirrel (*Ratufa indica*) is exclusively a forest animal and is the most brightly coloured of the different races of Indian giant squirrels. This species is distributed from evergreen to moist deciduous forests of Western Ghats in Peninsular India. *Ratufa indica* is a top canopy dwelling species, rarely travels on the ground (Ramachandran, 1988; Borges 1989; Datta, 1993), feeds in the upper canopy levels (Payne, 1979, 1980; Ramachandran, 1988) and requires tall profusely branched trees for construction of nests or dreys. In the light of studies conducted by Ramachandran (1988), Borges (1989) and Joshua (1992) it is expected that structural attributes of forests play a major role in habitat use by the giant squirrel.

The distribution pattern of the giant squirrels indicates the quality of forest canopy of Similipal reserve (Singh, 1997). Canopy continuity near nesting trees is essential to facilitate moving along the arboreal pathway, as tree squirrels have been observed to build nests on trees near neighbouring trees with interlinking crowns (Patton, 1975; Hall, 1981; Ramachandran, 1988). Thus, the observations of the abundance of *Ratufa* in different compartments of Similipal indicate that it is one of the luxuriant and healthy forests in Odisha.

Food Plants

The squirrels are diurnal and are active during the early hours of the day and late hours of the evening. The squirrels rest during mid-day and this activity forms 25 per cent of the total activity (Johnsingh and Joshua, 1991). A study was made on the feeding habit of giant squirrel in Similipal (Rout and Swain, 2006). Twenty-four species of food plants were recorded as mentioned in Table 3.12. The Giant squirrel fed principally on the seeds of *Careya arborea, Terminalia bellirica, Terminalia chebula, Tamarindus indica* and *Terminalia tomentosa*. Parts of plants that were found at the bottom of the trees include bud, flower, fruit, leaf and petioles. The squirrel travels to the edge of the branch and cut the peduncle of the pod or fruit with its incisors. It holds the pod in the mouth. The tail is used as a balancing organ during the feeding activity. The squirrel manipulates the pod or fruit with its fore arms while gnawing. Mango (*Mangifera indica*) and Remeli (*Protium serratum*) appeared to be used frequently for gnawing.

Nesting Habits

The location of the drey is at the interlinking of the crown of neighbouring trees so as to allow easy access for jumping and moving away from the drey. This method of drey construction has been observed in Kaibab squirrel (Hall, 1981). Data on nesting habits were randomly collected during the field trips in Similipal. Fourteen species of plants have been recorded to hold *Ratufa* nests. Drey construction is mostly of a peculiar type. It is constructed by depositing a large number of twigs with leaves and found mainly on the forked twig. The mouth of one old nest on a Kusum tree was one cm in diameter. The entry into nest was placed horizontal to the ground. The depth of the nest was 35cm and the inner diameter 40cm. The nest occupied an area of 123x54cm. Inside the nest it was smooth. The leaves were arranged in three to four layers between twigs. Only a few hairs of the squirrel were found and no excreta or food particles were found inside the nest chamber.

These nests of squirrels are distinct from bird nests in having leaves of the nesting tree interwoven in the middle of twigs. Freshly constructed nests were sighted mostly during the months of September-November. The nests are amidst small branches at a mean height of about 17.3m above the ground (range=12-30m for n=53nests). When a fresh nest is detected on a tree the squirrel is found within about 100m from it. One of the criteria for locating nests at higher reaches of the trees appeared to be for escape from damage by langur and Rhesus macaques. When a langur is on the same tree as the squirrel is, they maintain distance. At times the *Ratufa* remains clinged against a branch but keeps watching at a distance. Usually a single squirrel is seen on a tree. Occasionally when two are seen, they may be a breeding pair.

Maximum species of trees like Champa (*Magnolia champaca*), Sal (*Shorea robusta*) and Kusuma (*Schleichera oleosa*) were noticed to be used for nesting (Table 3.12). Plants where nests were found but there were no signs of feeding included Baranga (*Kydia calycina*), Chadheigudi (*Vitex peduncularis*), Sidha (*Lagerstroemia parviflora*) and Paja (*Litsea monopetala*). Similarly, food plants where no nests were detected included Anla (*Phyllanthus emblica*), Bahada (*Terminala bellirica*), Chara (*Buchanania lanzan*), Gambhari (*Gmelina arborea*), Harida (*Terminalia chebula*), Kadamba (*Anthocephalus chinensis*), Kasi (*Bridelia retusa*), Kumbhi (*Careya arborea*), Kuruma (*Adina cordifolia*), Poijamu (*Eugenia operculata*), Rai (*Dillenia pentagyna*), Rasna (Orchid spp.) (*Bulbophyllum triste*), Sunari (*Cassia fistula*) and Tentuli (*Tamarindus indica*).

A giant squirrel will often construct more than one drey during a season. This is more pronounced in the case of adult lactating squirrels. Within the home range there are several dreys (Ramachandran, 1988). Individual squirrel used more than four nests within a territory at a given time. The multiple nest phenomenon is meant either to escape from nest predators like langur and Rhesus macaque or to provide protection against climatic factors like high temperature, cold or rain. Single squirrels have been seen visiting up to three nests in any season. These nests may be in the same

Table 3.12. Food plants recorded from feeding signs and the maximum number of nests located in the study area along Kabatghai-Ransa-Ganapati (compartment nos: WD15, KH22, KH18, KH17, KH19, KH14, BLW19, BLW18, BLW17) of Similipal reserve, of the Indian Giant squirrel

Sl.No.	Scientific Name	Local Name in Odia	Plant Part(s) Eaten	Number of Trees with Nest
1.	*Adina cordifolia*	Kuruma	Petiole of leaf	0
2.	*Anogeissus latifolia*	Dhaura	Petiole of leaf	1
3.	*Anthocephalus chinensis*	Kadamba	Fruits, Petiole of leaf	0
4.	*Bombax ceiba*	Simuli	Flower	3
5.	*Bridelia retusa*	Kasi	Petiole of leaf	0
6.	*Buchanania lanzan*	Chara	Fruits, Petiole of leaf	0
7.	*Bulbophyllum triste*	Rasna	Fleshy pseudobulbs	0
8.	*Careya arborea*	Kumbhi	Fruits, Petiole of leaf	0
9.	*Cassia fistula*	Sunari	Fruits, Petiole of leaf	0
10.	*Dillenia pentagyna*	Rai	Petiole of leaf	0
11.	*Phyllanthus emblica*	Anla	Fruits	0
12.	*Eugenia operculata*	Poijamu	Petiole of leaf	0
13.	*Gmelina arborea*	Gambhari	Petiole of leaf	0
14.	*Kydia calycina*	Baranga	—	3
15.	*Lagerstroemia parviflora*	Sidha	—	1
16.	*Litsea monopetala*	Poja	—	1
17.	*Mangifera indica*	Amba	Petiole of tender leaf	3
18.	*Magnolia champaca*	Champa	Flower	7
19.	*Protium serratum*	Rimli	Fruits, Petiole of leaf	4
20.	*Schleichera oleosa*	Kusuma	Fruits, Petiole of leaf	5
21.	*Shorea robusta*	Sal	Petiole of leaf	6
22.	*Syzygium cumini*	Jamu	Fruits, Petiole of leaf	2
23.	*Tamarindus indica*	Tentuli	Fruits	0
24.	*Terminalia bellirica*	Bahada	Fruits, Petiole of leaf	0
25.	*Terminalia chebula*	Harida	Fruits, Petiole of leaf	0
26.	*Terminalia tomentosa*	Asan	Fruits, Petiole of leaf	4
27.	*Vitex peduncularis*	Chadheigudi	-	2
28.	*Xantolis tomentosa*	Jastimadhu	Fruits and bark	2

or different trees located within about a 20 m radius. The distance between two dreys was only less than 10m (Ramachandran, 1988). The squirrels visited the nests only a few times and spent very little time in them during the day. The purpose of visiting so many nests requires further study.

Conservation

The Indian Giant Squirrel inhabits the deciduous, mixed deciduous, and moist evergreen forests of peninsular India, south of the Ganges (Prater, 1980) and listed in Appendix II of CITES (Favre, 1989) and Schedule–II of the Wildlife (Protection) Act, 1972 of Govt. of India. The distribution of the species has been reduced to isolated forest patches (Borges *et al.*, 1992) due to the forestry operations and tree felling even in protected areas. Hence, for effective management and conservation of *Ratufa indica,* it is important to document the distribution, food plants and nesting behaviour of the species. We are still far from understanding the ecology of this species given that no serious study has ever been undertaken in Similipal to assess the status and biology of the animal. In the absence of such a study, it would be difficult to suggest sensible management policies except to recommend the protection of already known squirrel habitats.

Another threat to the long- term survival of *Ratufa* comes from the activities of nomadic tribes of Kharias, Mankidias and Shabaras. They make illicit felling of old trees when they collect honey, wax and resin. Besides, the animal is also killed by some group of tribes in Similipal for use in medicinal preparations and for its attractive fur. There is a thriving trade in smuggled infant *Ratufa* to keep as pets. Therefore, immediate conservation measure is warranted or else we run the risk of losing our only living Giant squirrel. This would not mean just the loss of a single species but a body blow to the entire ecosystem. Some of the urgent conservation measures that have been initiated are:

(i) Strengthening of their habitats in terms of forest protection efforts so as to provide adequate food, stress free living and breeding.

(ii) Monitoring of giant squirrel populations, their birth and death rates, and predator-prey ratio. Their ecology needs to be studied so that causes which hamper their population growth could be identified and eliminated.

(iii) Establishment of long-term ecological monitoring programme.

(iv) The development of an integrated conservation strategy.

THE NON-HUMAN PRIMATES

Three species of non-human primates occur in Odisha. They are: The Hanuman langur, *Semnopithecus entellus entellus* (Dufresne 1797), the Rhesus macaque, *Macaca mulatta* (Zimmermann 1780), and the Bonnet macaque, *Macaca radiata* (E. Geoffroy 1812) (Cercopithecidae) (Mishra *et al.*, 1996). Of these, *M. radiata* is confined to the Malkangiri district in South Odisha. The other two species of non-human primates occur in the forests of Similipal hills.

Sixteen sub-species of *Semnopithecus entellus* are distinguished based on minor morphological variations, habits and the area inhabited, of which fourteen occur in India. *Semnopithecus entellus entellus* (Dufresne 1797) inhabits southern West Bengal, southern Bihar, southern Chhatisgarh, Jharkhand, northeastern Maharashtra and Odisha in eastern India, with a probably introduced population in western Bangladesh (Brandon-Jones, 2004). *Semnopithecus entellus* and *Macaca mulatta* occur in quite large numbers in the Similipal tiger reserve.

Hanuman Langur

In our study on Hanuman langur (Swain *et al.*, 2008), we observed 144 Hanuman langur troops **(Plate 3.50A)**, the size ranged between one and 50. We could not ascertain number of males in a troop in 106 groups. Of the 144 Hanuman langur troops sighted, four troops consisted of single individuals. Of them, only in one, the sex could be noted to be male, whereas, in the remaining three of single individual troops, the sex could not be identified. In a single troop of the two individuals, both were found to be males. We noticed a single male in twelve bisexual groups composed of three to 13 individuals. In seven bisexual groups of three to 20, there were two males; in six troops of seven to 34, males numbered three; in two troops of six and 12 individuals, males numbered four; in two troops of seven and 20, there were five males; whereas in two troops of eight and two troops of 15 individuals, males numbered six in each of the four troops.

Plate 3.50A. Langurs (*Presbytis entellus*) browsing sal seedlings in Similipal, Odisha, India. Photo: H Hazebrock

Hanuman langurs live in fairly stable groups consisting of juveniles of all ages and males and females. Behura (1987) reported that in a social bisexual Hanuman langur group, the number of individuals varies between 13 and 35, and in all male groups the number varies between four and 15. Prater (1965) states that in North India, a Hanuman langur group consisted of 18 to 25 individuals and in the south around 15. Roonwal and Mohnot (1977) report that in northern and central India, most bisexual groups had two to six adult males; one bisexual group had only one adult male; there were a number of all male groups and a few solitary females. In 39 groups of almost 1000 Hanuman langurs, the group size varied from 2 to 120; most groups in forested areas had 18-25 members. In central India, Sugiyama (1964) found the bisexual group size varying between 15 and 61, with an average of 27.8. In Gujarat, bisexual groups were of 16-48, with occasional all male groups of two or three. In peninsular India, Sugiyama and his co-workers found a significant difference in group size between relatively open and forested areas. The mean group size (both bisexual and all male groups) in forested areas was 14.4 individuals and in the open area 17.1. The average group size for both areas was 15.1 (Roonwal and Mohnot, 1977). In the Similipal reserve, in 144 groups of 1505 Hanuman langurs, the mean troop size works out to 9.6 ± 1.19. It thus appears that observations here tally fairly well with the findings of workers in other areas of India. The observations further corroborate that the Hanuman langurs live in social groups of various sizes and compositions in well forested hilly areas.

In Similipal only a single all female group of six was sighted. This appears somewhat extra-ordinary. The highest and lowest density of Hanuman langur in the Similipal reserve was 1.625 (Kuanribil) and 0.017 (Kalkam) per hectare respectively. Thus, higest density of langur population (162.5 per km^2) in Kuanribil appears to be quite high compared to 100 langurs per km^2 at Dharwar as reported by Sugiyama (1964). This may be due to the higher concentration of preferred food plants in the concentrated area.

Rhesus Macaque

In our study on Rhesus macaques (Swain *et al., supra*), we noticed that compared to the sightings of the Hanuman langur in Similipal, the number of Rhesus macaques **(Plate 3.50B)** met with was far less in the number of troops and in the number of individuals. A total of twelve macaque troops with a total population of 138 individuals were observed. The troops were made up of 4 to 25 individuals, with a mean of 11.5 ± 1.82. Unfortunately, in none of the troops, the male could be distinguished. The population density was 10-98.3 per km^2, with a mean of 26.86.

Plate 3.50B. Rhesus macaque (*Macaca mulatta*) in Similipal, Odisha, India. Photo: Similipal TR archive.

Neville (1968) found the population density of about 5-15 per km^2 in the elevated chirpine forests of Uttar Pradesh and about 57 per km^2 in the moist deciduous areas at lower elevations, as against 753 per km^2 in the town (Roonwal and Mohnot, 1977). It appears that the high population density of 10-98.3 per km^2 in true dense forest dwelling Rhesus macaques in the Similipal reserve, is probably due to preferred host-plant concentration in the census units surveyed.

THE CHAUSINGHA

Prater (1965) mentioned the distribution of Chausingha or Four-horned Antelope (*Tetracerus quadricornis*), as peninsular India, south of the Himalayas, where the country is wooded and hilly, but not too densely forested, and that it was not found on the Malabar Coast. Ranjitsinh (1987) reported that the range of Chausingha extends from the Terai forests and Jammu to Mudumalai Sanctuary in Tamil Nadu, though in the past it was reported in the Palani Hills, and that it occurs in the Barda and Gir forests of Gujarat, extending eastwards to Bihar. It was never reported to occur in Odisha. Rice (1991) conducted a questionnaire survey on the status of Chausingha in 1986-1987 and cited the information gathered from Odisha through the questionnaire.

The state of Odisha is known to have populations of Chausingha in Badrama sanctuary, Chandaka-Dampada sanctuary, Kotgad sanctuary, Satkosia George sanctuary and Similipal sanctuary. Chausingha is locally extinct in Karlapat sanctuary (Mishra *et al.*, 1996). Singh and Swain (2003) report distribution of Chousingha in Similipal.

During early 1970s Chausingha occurred in Similipal in open wood lands along the edge of large valleys to the south (Choudhary, 1974). During a reconnaissance survey in Bachurichara Valley in May 1972, two Chausinghas were seen and in May 1973 five dung-heaps were found.

From November 1987 to January 2002 information on Chausingha was collected through different sources. These are from direct observations of Chausingha, observations of their dung heaps, and second-hand information made available by staff through their 'diary'. Chausingha was sighted near Nigirda-Panasia stretch in central Similipal and near Bachhurichara Valley in the south. Other information include their availability near Panasia, Dhudruchampa, western part of Sunpokhri, Tarinivilla, Nuagaon, Devasthali, Patbil and Bhanjabasa. Dung heaps are best seen in Bachhurichara and Debasthali valleys. All the location represent the south-central and southern parts of the core zone of Similipal tiger reserve.

The Chausingha population of Similipal assumes significance from two respects. First, it is within the eastern-most limit for the species in the state of Odisha that remained unnoticed for a long time, and second, the population is very small, perhaps about a hundred heads, in the reserve **(Plate 3.51)**.

Plate 3.51. Chausingha or Four-horned Antelope (*Tetracerus quadricornis*) camera trapped inside Similipal, Odisha, India. Photo: Similipal TR archive.

Plate 3.52. Mouse deer (*Moschiola indica*) camera trapped inside Similipal, Odisha, India. Photo: Similipal TR archive.

Chausingha is included under Schedule I of the Wildlife (Protection) Act, 1972. Rice (1991) reported that since Chausingha occurs in low densities and small population sizes, and has attracted little scientific attention, it was necessary to have trial efforts for techniques in genetic monitoring and conservation management.

The population of Chausingha in Similipal reserve form an example to the concern expressed by Rice (1991) that with intensification and spread of human impacts Chausingha populations are becoming increasingly isolated. The rate of the sighting of Chausingha in Similipal has become occasional, and the population is subjected to adverse human impacts.

THE MOUSE DEER

On 8 May 2002 at about 03.30PM, I noticed a mouse deer (*Moschiola indica*) running in front of the vehicle and then stopping inside the grass patches by the side of the forest road. The mouse deer remained standstill at the spot feigning death. Immediately there were flies all around its face, on the eyes, nostrils and mouth (Swain and Singh, 2004). A few photographs taken from the vehicle with a standard lens were not clear. When I came out of the vehicle and approached close within a touching distance, the deer jumped and ran away. The observation was made along the Nuagaon-Kabatghai forest road of National Park Range in southeastern part inside Similipal tiger reserve. Mouse deer **(Plate 3.52)** in Similipal are generally nocturnal. Very seldom these are seen during the day. Considered as a timid and gentle creature, feigning death during the daytime as observed here, may be a defensive behaviour to escape further attention from a potential natural predator, but the behaviour appears to be unsuccessful against human-intruders. In a similar encounter the deer could have been easily killed by any of the many people who use Similipal during the day. The population of mouse deer in Similipal was estimated to be about 1500-1800 (Anonymous, 1991), but believed to be more than that. Although sighting of a mouse deer is very occasional, their preponderance in certain sections of the reserve is considered high from the numerous spoor seen on pug impression pads that are specially laid to obtain pugmarks of large carnivores during tiger and leopard census.

CONCLUSION

Similipal is a repository of large fauna. The flagship species are tiger, elephant and marsh crocodile. The tiger is a highly endangered species. The threats to the surviving populations produce a very strong need to obtain tightly and temporally spaced monitoring data in order to rapidly react to declines in populations. Monitoring needs to be simple and effective in the field with clear implications for management. Occupancy analysis of sign survey data is a simple and low cost monitoring method; it can be easily repeated to obtain time series and to identify changes in estimates of the population. With repeated surveys building up a time series, it is able to identify both declines and potential causes for the declines.

Similipal tiger reserve, by its size and qualities, is one of the most important tiger reserves in the Indian subcontinent, with an identified minimum population target of 219 tigers (Ranganathan *et al.*, 2008). In a single season occupancy analysis of tiger sign survey data from Similipal reserve, it has identified habitat disturbance as an important correlate of low tiger occupancy, and tiger prey species abundance as an important correlate of occupancy in areas where habitat disturbance is low. This analysis has direct management implications, particularly reducing the proportion of habitat disturbance and human impact on the reserve in areas with more than 10 per cent habitat disturbance, and increasing prey abundance in areas of lower habitat disturbance. Analysis will help identify which villages need to be prioritized for relocation by identifying those with greatest impact on habitat degradation and prey populations. It is recommended that regular repeated sign surveys in the future need to be carried out along with cross correlation with other monitoring data for standardization, particularly camera trap capture/recapture studies and pug mark sampling.

Elephant is the largest herbivore in the reserve. It is also a charismatic species; Similipal constitutes a large part of Mayurbhanj elepant reserve. Monitoring illegal killing of elephants (MIKE), a programme under CITES is implemented here. There is the first time census of elephants taken up in Mayurbhanj ER by dung density count. The dung density count method, a statistically robust methodology of censusing elephants, indicated elephant density from 0.188 per km^2 to 0.981 per km^2 in the reserve. As per the dung density count of elephants there are at least 788 elephants in Mayurbhanj ER. In contrast, it was reported to be only 465 elephants by total count method during that period. The elephant numbers are therefore under represented for this reserve in total count method. To arrive at a precise population figure of elephants in the reserve, repeated counting of dung density is necessary in different seasons of a year.

Elephants are generalised feeders and they consume at least small amounts of most of the plants available in their habitats. One hundred and twenty nine (129) plant species belonging to 44 families under 23 orders could be identified as food plants of elephants of Odisha. Consumption of food plants varies from season to season according to their availability, palatability and nutritional value.

Mugger crocodile (also known as snub-nosed, marsh, or fresh water crocodile) though present in Similipal earlier, was extinct from 1979. Since then, the programme of 'rear and release' of this species has been undertaken and so far over 800 crocodiles have been released in the various streams of Similipal. Monitoring of these released crocodiles is necessary to know the success of the programme.

Indian giant squirrel makes nests on the top branches of a tree. For commuting from one tree to another, closeness in canopy cover is necessary, for which the species is a good indicator of closed canopy. Twenty-four (24) species of food plants have been recorded as food plants of giant squirrel in Similipal.

Ecosystems of a biosphere reserve are dynamic and there are constant interactions among its components. The ecological integrity refers to the ecosystem of a pristine natural habitat of a locality where the interactions among its components are in a steady state of equilibrium with ideal species composition and diversities. Under the influence of anthropogenic pressure, a major ecosystem stressor, as in the case of Similipal which is surrounded by 1200 villages within the transitional zone, besides 62 villages inside the buffer as well as core zone of the reserve, a steady state of equilibrium is just not existing and there are retrogressions in the ecosystem structure and functions. Early detection of retrogressive changes is *sine qua non* for better management prescriptions to arrest degradation in ecosystem components and processes. Therefore, regular monitoring of keystone species like tiger, elephant, crocodile and giant squirrels, is needed.

Simlipal

Sketch Ink on Paper
by Dr. ***J.P. DAS*** *Cardiologist*

Chapter 4

The People

"It is forbidden to decry other sects; the true believer gives honour to whatever in them is worthy of honour."

– Ashoka the Great [c. 304 BCE – 232 BCE]

INTRODUCTION

I was once trekking from Kaliani to Gudgudia inside Similipal. On the way at the bank of river Khairi, I saw a group of people belonging to Kharia community (**Plate 4.1**), a primitive tribe residing in and around Similipal. They apprised me that they were on their way to a place named Mahubhandar, about 10 km from that location deep inside the forest. They would stay there few days and collect honey, arrowroot, resin and other such jungle products. They further told me that entire Similipal had been divided amongst their community for collection of the various forest produce and they used to enjoy this right generation after generation without encroaching on the area of other people from the same community. Therefore, even though they resided at Kapand near Jashipur at the fringe of Similipal, they travelled such a long distance to collect the forest produce. Incidentally, these are the people shifted from Jamunagarh and Kabatghai inside core area of Similipal in 1994. I explained to them that collection of NTFPs (non-timber forest products) from the reserve was illegal and they would be prosecuted. It was not convincing to them as they believed it was their right since time immemorial to collect NTFPs from the assigned area and if they did not obey the law of the community, the forest deity would be annoyed inviting misfortune on them.

There is another group of nomadic tribe called Mankidias (Birhor tribe) who catch monkeys and even eat them. They move inside Similipal collecting Siali fibres and sell them in local market.

Plate 4.1. Kharia tribes moving inside Similipal forest, Odisha, India in search of honey and other non-timber forest products. Photo: D Swain.

According to a local folklore, a group of aboriginal people used to take cattle of the king of Mayurbhanj everyday inside Similipal for grazing. Years passed by and when the king disengaged these people from the practice of grazing his livestock, some of them resorted to settled agriculture outside Similipal and they are now called Bathudi tribe. The people who did not take up agriculture and roamed inside Similipal and depended on roots, tubers, barks and leaves are called Kharia tribe. Another set of people who remained in Similipal and resorted to kill monkey and other small creatures for their consumption are called Mankidia tribe.

HISTORY OF HUMAN SETTLEMENTS IN SIMILIPAL

The earliest evidence of man in Mayurbhanj district, where Similipal is located, is believed to belong to the Lower Palaeolithic or Early Stone Age culture (Senapati and Sahu, 1967). The Early Stone Age man lived here during the Middle Pleistocene time. The sites of this culture are concentrated along the banks of river Budhabalanga and its tributaries, the Sone and the Gangahar. Senapati and Sahu (*supra*) report that Middle Stone Age Culture in Mayurbhanj flourished on the western side of the Similipal hill range along the river Kharkei and its tributaries. After the Middle Stone Age the Microlithic culture of the Late Stone Age also flourished in this region. In comparison with the vast collections of the Early and Middle Stone Age tools, the microlithic collection of this district looks very insignificant and gives indication that this culture did not live long in this district. But the other Late Stone Age culture, *viz.*, the culture of the polished stone celts had a very flourishing time in this district. The long list of sites of this culture in Mayurbhanj, gives this district a unique place of the Neolithic civilization in eastern India. For the first time, evidence of the prehistoric pottery is noticed here. Besides pottery, grinding and polishing of the stone tools and domestication of animals might have been practiced during this period. The people lived in well-built houses and settled as agricultural communities. The shouldered celt culture is a later development of the Neolithic. The people of this culture knew the use of metal. It came to India with the people speaking Austro-asiatic language. The tribal people of the district, namely, the Santals, the Mundas, the Kolhas and the Savaras, now speak this language. There is also evidence of primitive copper celts in this district. Senapati and Sahu (*supra*) further state that the earliest type of celts, the shouldered is dated to the Yin dynasty (about 1300-1028 BC).

The tribes residing in the district are Bathudi, Bhumija, Kharia, Kolha, Mankidia, Munda, Santal, Bindhani and others. Not much study has been made about the tribes of Similipal. One of the earliest studies about the economic condition of the inhabitants of Similipal was reported by Das and Das (1968). Dash (1998) gives an account of the life and culture of the Kharia Savara, Ujia Savara and Birhor (Mankidia Savara) in and around Similipal hills. Barik (2003) gives an account of tribal population and settlement inside Similipal. According to a legend, Bathudis are the earliest settlers of Similipal (Das and Das, 1968; Dash, 1998). The exact time of settlement is not known. Dash (1998) citing Vasu (1911), the time period of early settlement is predicted to be that of King Ashoka's period as the Bathudis of Similipal are said to be descendants of Madan Nayak and his three sons Nanda, Benu and Beni. Madan Nayak, as believed by Bathudis, was either a Sardar in Ashoka's army or simply belonged to Ashoka's army or just belonged to Ashoka's period in Odisha.

Kharias were originally living in Panchpir and were eking out their livelihood by collecting forest produce from Similipal. Their Chief was at Jashipur and Bathudis under the leadership of Nand Das defeated the Kharia Chief and captured Jashipur fort. Kharias after the defeat of their chief migrated into the deep dense forests of Similipal. Similarly, Bamanghaty which was then under the control of a Gond king was captured by Bathudis and Gonds migrated and settled in certain parts of Similipal. But due to depredation of wild animals, diseases like malaria and gastroenteritis, and constitution of Similipal as a reserved forest and Game Reserve for the royals, large scale settlement was never possible. It is stated that depopulation happened as many as seven times in Similipal and fresh migrations took place (Das and Das, 1968).

After accession of Maharaja ShriRama Chandra Bhanja Deo to the throne of Mayurbhanj state in 1890, he initiated steps to improve the living conditions inside Similipal. Temporary leases for cultivation were granted to a few settlers. One Peter Dubraj, a Kolha from Singbhum who was working as a Postmaster at Chaibasa was granted a lease of 920,642 acres at an annual rent of Rs. 254 and Annas 9 only. He was declared intermediary Sardar for the area and was allowed to establish villages inside Similipal. Similarly, a lease was granted to Durjadhan Mahakud for Kasirapirh, where seven revenue villages were formed in due course. When Dubraj died in 1934, it was found that he was in heavy arrear of rent. His successors were forced by circumstances to sell their lands. Ganesh Ram Ho of Nawana purchased some lands in 1949. In 1955, the Sardari system was abolished. Lands of Peter Dubraj were auctioned towards arrear revenue in 1965 and were purchased by a group of persons belonging to Mahato caste. The Mahatos of Bhanjkia set up a Doharaghar (farm house) in Similipal.

After integration of Mayurbhanj state into the Union of India on 6 November 1948, it became part of Odisha as a district on 1 January 1949 and forced labour and monopoly system of timber extraction were abolished. Forest coupes were auctioned to the highest bidders. People from inside Mayurbhanj district and from outside the district, even far off places of the then Bihar (now Jharkhand) settled inside Similipal for forest operations and collection of minor forest produce. People cleared forests in the valley near the perennial water sources to settle down in protected areas for which they were penalized several times by the Forest Department. Finally in 1955 those lands were declared as 'Nayabadi' and were allotted to the occupants. The availability of Nayabadi lands stimulated others to encroach upon forest land and people from far off places like Ranchi and Singbhum rushed to Similipal to establish villages encroaching upon forest lands and destroying forest wealth. This infiltration of outsiders has considerably increased the population, particularly at Jajdihi, Kukurbhuka, Budhabalang, Saruda and Bunduriabasa. They mostly belong to Munda tribe converted to Christianity (Das and Das, 1968).

Das and Das (*supra*) further state that other processes through which Similipal became populated are matrimonial relations and kinship bond. On many occasions, the son-in-laws are made to live in Similipal after marriage. Many did immigrate on request of their relatives who had settled in Similipal earlier. Many came down to this place to remain as '*barmasia*' or annual agricultural labourer. Subsequently, they acquired patches of land and settled down permanently.

Gudugudia Panchayat is found to be populated by immigrants from other parts of the district of Mayurbhanj, whereas Astakunhar is mainly populated by outsiders. Bamanghati kolhas are found in older settlements like Sanakasiria, Barheipani, Budhabalang, *etc.*, whereas, Kolhan kolhas are noticed in Nawana, Balarampur, and Garh Similipal.

Settlements in the interior of Similipal did not take place during any particular period. It took place over a long period of time by successive flow of immigrants from within Mayurbhanj and from outside. During 1911 to 1931, immigration was casual, but it increased from 1931. While Bathudis and Kharias migrated from within Mayurbhanj, the majority of Kolhas came from the adjoining Kolhan Government Estate of Singbhum which formed an integral part of the territory of Mayurbhanj till 1937, when their administration went over to the British Government and which formed a compact block in the Eastern frontier of Kolhan estate (Das and Das, 1968).

The virgin soil, abundant perennial flow of water, easy access to forest resources and availability of engagement in forest operations, encouraged the people from adjoining places particularly from the then Bihar state (and present Jharkhand) to settle inside Similipal. The old inhabitants were also not hostile to the new comers. Initially, the forest department did not take much notice of it as land resources were extensive and there was easy availability of migrant labourers for forest department work of various types. But when the population started swelling enormously and encroachment of forest land increased the forest department started a drive to evict the unauthorized emigrants.

Those who could produce Amalnamma signed by Dubraj, Mahakud or their successors are considered as bonafide settlers and others are treated as encroachers. There are two types of immigrants, permanent and temporary. Permanent immigrants came down from within and outside the State, encroached on forest lands, destroyed forests, and settled in regular villages. The temporary immigrants were persons who migrated from outside Similipal for some specific purpose and returned after that. The herdsmen who used to come with their cattle for grazing belonged to the temporary category.

SOCIO-ECONOMIC CONDITION

There are 62 villages inside the Similipal sanctuary – four in the core (**Plate 4.2**) and 58 in the buffer zone of the reserve. In 1971, 7,700 people lived in these villages, but according to 2001 census, the population swelled to 12,449 in those thirty years. In 2011, the population became 17,472 inside Similipal. Similarly, about 1200 villages are located within 10 km of the reserve (transitional zone), with a total population of approximately 450,000.

Plate 4.2. A core area village inside Similipal, Odisha, India. Photo: H Hazebrock.

My study (Swain, 2009) shows that the people in and around Similipal reserve lead a subsistence living. The annual per capita income for people in the core, buffer, and transitional zones are respectively 203 US$ (~Rs. 10,150), 272 US$ (~Rs. 13,595) and 139 US$ (~Rs. 6,950), whereas that of Odisha and India are 346 US$ (~Rs. 17300) and 543 US$ (~Rs. 27150) respectively in 2006. The situation is similar for the other socio-economic profiles mentioned in Table 4.1. The literacy rate is abysmally low in the core and buffer areas (16.51

per cent in the core, with a female literacy of 3.53 per cent, and 12.31 per cent in the buffer, with a female literacy of 6.19 per cent), in comparison to 51.91 per cent of the district of Mayurbhanj (female literacy 37.84 per cent) and 63.08 per cent of the state of Odisha (female literacy 50.51 per cent).

Plate 4.3A. Collection and processing of palua (arrawroot) by the core area people inside Similipal, Odisha, India. Photo: D Swain.

Plate 4.3B. Selling of mushrooms in a local market by tribal people collected from Similipal forest, Odisha, India. Photo: H N Thatoi.

The people's livelihoods depend upon the collection of NTFPs (non-timber forest products) (**Plate 4.3A and B**), agriculture (**Plate 4.4**) and rearing cattle, goat, and poultry (**Plate 4.5**). Almost all the households in the core, buffer, and transitional zones depend upon the Similipal forests for fuel wood and small timber requirements. Some of the landless, marginal and small farmers resort to illegal felling of trees for fuel wood and timber to eke out their living.

Table 4.1. Comparison of demography, socio-economy, and health and development in the core, buffer and transitional area of Similipal biosphere reserve (Swain, 2009)

Sl. No.	Criteria	Core	Buffer	Transitional
1.	Sex ratio (F: 1000M)	1076	994	974
2.	Average household (Number of members)	7	5	6
3.	Social category	Tribe 100 per cent	Tribe 98 per cent	Tribe 90 per cent
4.	Housing pattern	Mud 100 per cent	Mud 100 per cent	Mud 92 per cent
5.	Literacy per cent (overall)	16.51	12.31	42.77
	(Male)	34.47	18.39	44.69
	(Female)	3.53	6.19	40.80
6.	Education per cent			
	Primary	77.77	47.03	53.52
	Secondary	5.50	30.81	29.82
	High School	11.00	16.76	12.62
	College	5.50	5.41	4.04
7.	Average landholding (Ac)	8.40	3.0	2.7
8.	Productivity (paddy in qtl/Ac)	3.18	3.01	4.70
9.	Landholding pattern (per cent)			
	Big (>10Ac)	11.11	13.89	4.74
	Medium (>5 and <10Ac)	55.55	13.89	9.62
	Small (>2 and <5Ac)	33.34	61.11	67.04
	Marginal (<2Ac)	0	11.11	9.96
10.	Land less (per cent)	0	0	8.64
11.	Avg.cattle/household	27.87	16.19	11.19
12.	Annual household income (in Indian Rupees) (1US$=50 INR in 2006)			
	Agriculture	7430	5030	1325
	Animal husbandry	1120	1350	695
	NTFP	800	1775	4515
	Wage earning	800	5440	415
	Total	10150	13595	6950
13.	Chief NTFP (per cent)			
	Sal leaves	10	40	70
	Kusum seeds	10	25	10
	Resin	40	05	01
	Honey	40	10	02
	Mahua flower and seeds	0	20	17
14.	Irrigation facility (per cent)	90	50	63

Abbreviations used: F (Female), M (Male), Ac (Acre, 1Acre=0.4 Hectare), qtl (quintal, 1qtl=100kg), NTFP (Non-timber forest product).

Plate 4.4. Cultivation of rice inside Similipal, Odisha, India. Photo: H Hazebrock.

Plate 4.5. Cattle and goats of people residing inside Similipal, Odisha, India. Photo: H Hazebrock.

Therefore there is extreme poverty, unavailability of schools and remoteness of tribal habitations in the core area of Similipal reserve. Mainly the villagers are small and marginal farmers, and the land is rain-fed. The people are involved in collection of various NTFPs, but these are at subsistence level, and they are facing problems of transporting and marketing of the material. The population of livestock is increasing in the core area, which causes degradation of grassland for herbivorous wildlife. The problem of unemployment is also prevailing. Most of the villagers are involved in wage earning, mainly provided by the forest department. The people in the core area face a number of health hazards due to poor drinking water facilities and unhygienic living conditions. In the buffer zone the pre-dominant tribal communities are in higher sex ratio and enjoy better educational facility. Many features in relation with the socio-economic condition, village institutions and development in the buffer zone are same as in the core area. In comparison with the core area the villagers in the buffer area are in an advantageous position in the collection of NTFPs and medicinal plants which support the income generation by selling in the local market. In the transitional zone, the villagers belong to the Scheduled Tribes, Scheduled Castes, other Backward Classes and General Castes. The literacy rate in the transitional zone is better than the other two zones. The average income of the households in the transitional zone from all sources is comparatively less, but they earn more by selling of NTFPs. Due to implementation of watershed activities, the irrigation facilities in the transitional zone are better. In the buffer and transitional areas, some Civil Society Organizations are involved in watershed management, herbal cultivation, and rural employment through SHG (Self Help Group) formation. In the transitional zone, in some villages, the electricity and telephone facilities are available.

VILLAGE RELOCATION

In search of inviolate space for tiger and associated fauna and flora, families living years together in deep forest are displaced out of the forest. This is a painful exercise for people being relocated and people involved in relocation process. Displacement resulting from the establishment and enforcement of protected areas has troubled relationships between conservationists and rural groups in many parts of the world (Brockington and Igoe, 2006). Colcester (2004) describes that conservation through the establishment of 'national parks', pioneered in USA and applied worldwide, has violated the rights of indigenous people causing impoverishment and social problems. But conservation related displacement, revealing the forms of deprivation, social and equity issues have not been studied properly in south Asia as revealed by Agrawal and Redford (2009), Ferraro (2008), and Rangarajan and Shahabuddin (2006). Lasgorceix and Kothari (2009) opine that local communities usually have unclear or unrecognized rights to park resources and, being located away from mainstream society, are already disadvantaged. This contributes towards their poverty, which further worsens when non-consultative relocation occurs and causes conflict with protected area managers. Panwar (2003) citing the case of Kanha tiger reserve and Kuno national park says that when displacement is consultative and participatory has a beneficial impact on tribals and other forest dwellers.

In 1994, 11 Kharia families from Jamunagrh and 30 Kharia families from Kabatghai were shifted to Kapand near Jashipur at the fringe of Similipal. They were provided only residential accommodation and without any provision for earning their livelihood. Kharias are basically nomadic tribes and they were relocated outside

Similipal under Kharia and Mankdia Development programme of Odisha. The resettled tribes depended on Similipal for collection of NTFPs like honey and resins. Four years later 23 tribal families from Jenabil were relocated to Ambadiha near Udala township **(Plate 4.6)**. This time each family was provided with residential accommodation along with two acres of irrigated land as per the policy of Project Tiger, Government of India. The families were given eight decimal each of homestead land and during shifting the families were provided free rations and transportation cost of their household belongings. They were even provided temporary accommodation at the rehabilitation site, so that staying there, they constructed their own house as per their requirements within the fund provided to them. Such a scheme became attractive to others and in 2003, eight families from Kabatghai opted to come out to Ambadiha rehabilitation colony. Remaining families of Jenabil was also willing to be shifted to Ambadiha. But it could not be possible due to Kalinganagar mishap where 13 tribals were killed in 2006 when they protested against acquisition of their land for Tata Steel Limited. In the mean time, when National Tiger Conservation Authority was constituted in 2006 and village relocation policy from inviolate area of tiger reserves was revisited in 2007, two options were formulated.

Plate 4.6. A relocated tribal family from Jenabil village inside Similipal forest to Ambadiha, near Udala township just outside of Similipal forest, Odisha, India. Photo: DFO, Baripada.

Option I- Payment of one million rupees (about US$15,000) per family as one time package in case the family opts so, without involving any rehabilitation/ relocation process by the forest department.

Option II- Carrying out rehabilitation/relocation of village by the forest department at the above rate of one million rupees per family in the following manner:

(a) Agriculture land procurement (two hectares) and development - 35 per cent of the total package

(b) Settlement of rights- 30 per cent of the total package

(c) Homestead land and house construction – 20 per cent of the total package

(d) Incentive – 5 per cent of the total package

(e) Community facilities commuted by the family (access road, irrigation, drinking water, sanitation, electricity, telecommunication, community center, religious places of worship, burial/cremation ground) – 10 per cent of the total package

With the revised package of one million rupees, rest 61 tribal families from Jenabil came out of Similipal and settled at Ambadiha in 2010. There were few Kharia families kept at Upper Barhakamuda and Bahaghar **(Plate 4.7)** in Upper Barhakamuda Range since tiger reserve was constituted for getting labour to carry out departmental work. With the passage of time, these families grew in size for which they were required to be shifted out of Similipal. In 2013, there were 22 families in Barhakamuda and 10 families in Bahaghar camps of Similipal. They were shifted to Asankudar village near Karanjia.

Plate 4.7. Kharia families residing at Upper Barhakamuda inside Similipal,Odisha, India. They are now relocated to outside Similipal. Photo: D Swain.

Further in 2015, 35 families from Jamunagarh were shifted to Nabra near Podadiha. Only three families were left behind at Jamunagarh, who did not give their willingness as per Scheduled Tribes and Other Traditional Forest Dwellers (Recognition of Forest Rights) Act, 2006 to come out of Similipal. They are still continuing at Jamunagarh. In 2016, rest 47 families of village Kabatghai, resettled at Manda near Jashipur **(Plate 4.8).** The name of villages and number of families relocated from the core area of Similipal tiger reserve are furnished in Table 4.2.

Table 4.2. Name of villages and number of families relocated from the Core Area of Similipal tiger reserve

Village/ Hamet	Family	Year	Year-wise Break Up of Families Relocated and Rehabilitated in the Rehabilitation Colony						Balance Family (as per survey)
			Kapand	Ambadiha	Asankudar	Nabra	Manada	Total	
Jamunagarh	49	1994	11	-	-			46	03 (2014- Survey)
		2015	-	-	-	35			
Jenabil	84	1998	-	23	-			84	0
		2010		61	-				
Kabatghai	85	1994	30	-	-			85	0
		2003	-	8	-				
		2016	-	-	-	-	47		
Bakua	61	-	0	0				0	61 (1998-Survey)
Barhakamuda	22	2013	0	0	22			22	0
Bahaghara	10	2013	0	0	10			10	0
Total	**311**		**41**	**92**	**32**	**35**	**47**	**247**	**64**

Plate 4.8. Relocated village at Manda, near Jashipur from Kabatghai inside core area of Similipal, Odisha, India. Photo: A. Satpathy

Plate 4.9. Construction of houses under progress by the villagers relocated from Kiajhari inside Similipal, Odisha, India. Photo: DFO, Karanjia.

Besides shifting of villages from core area, a village remotely located in the buffer, namely Kiajhari, was relocated out of Similipal. Entire 79 families of that village were shifted and settled at Khandiadar, near Karanjia **(Plate 4.9)** in 2018.

Mahapatra *et al.* (2015) examined in detail the income from different sources and livelihood diversification of displaced dwellers from similipal and rehabilitated at Ambadiha and found an increase in per capita income for poorer segments with an overall 8 per cent increase in absolute household income and corresponding improvement in the poverty ratio and the poverty gap index for the relocated community. Mahapatra *et al.* (*supra*) further state that the tribes at Ambadiha relocation colony received sizable compensation packages and had closer market access. The rehabilitated families find themselves in a new world with little forest resource but better infrastructure. Educational facilities and access to transport and health care are some of the major benefits. Damage to property by the elephants, to crops by herbivores, and fear of Left Wing Extremists attack are the risk factors with which they had lived inside Similipal. Pecuniary benefit has come in the form of compensation to construct houses and create agriculture land, which has helped some households to come out of below poverty line status. The social integration with locals, complications in observing tribal rituals-cum-festivals and ethnic discrimination post-displacement considered major social setbacks in conservation-induced displacement are not evident at Ambadiha.

Rehabilitation of families from Similipal is a win-win situation as evident from the case study in **Box 4.1.**

BOX-4.1 REHABILITATION OF FAMILIES FROM SIMILIPAL AT AMBADIHA – A CASE STUDY

Sundar Mohan Nayak belongs to Bathudi community and he was resident of Jenabil village inside core area of Similipal tiger reserve. He was the first person opted to receive rehabilitation package and settled at Ambadiha in 1998. He received two acres of irrigated land and eight decimal of homestead land. His father simultaneously had also received two acres of irrigated land and eight decimal of homestead land. Sundar Mohan has two sons and two daughters **(Plate 4.6)**. They are now reading in nearby school and college. Every year the family gets about 25 quintals of paddy from their two acres of land. Further 25 quintals of paddy are received from two acres of land assigned to Sundar Mohan's father who is now no more and therefore Sundar Mohan enjoys father's property. Sundar Mohan cultivates NTFPs (like arrowroot), vegetables and pulses in his land and earns handsomely. He has a pisciculture pond (**Plate 4.10**), and on the bank of it he cultivates mango and banana. He also raises goats and poultry. Udala township being nearby he gets better price for his crops and animals. His wife works as a cook in the Anganwadi centre of the village. There are modern amenities like television, smart phones and cooking gas in their home. There are also better health and transport facilities. With four acres of land, cost of each acre being about 10 million of Indian Rupees (1 INR= 70 US$), Sundar Mohan is a millionaire. He is happy here at Ambadiha and never wants to go back to Similipal.

Plate 4.10. Rehabilitated people at Ambadiha earns livelihood from fishery and horticultural cultivation besides rice cultivation. Photo: DFO, Baripada.

Similar is the case with Rabi Nayak, a neighbour of Sunder Mohan Nayak. He has no child. But both Rabi Nayak and his wife are happy at Ambadiha. They have about 200,000 INR in their Bank Account. Sushil Dehury (**Photo 4.11**) belonging to Kharia community has a son and a daughter. While he married off his daughter, his son after marriage works at Chennai. Sushil lives happily at Ambadiha with his wife and daughter-in-law.

Plate 4.11. Sushil Dehury, a Kharia settled at Ambadiha after relocated from Jenabil, a core area village inside Similipal, Odisha, India. Photo: DFO, Baripada

Relocation of Jenabil village from core area of Similipal to Ambadiha is a win-win situation for communities and wild animals of Similipal. Forest dwellers got a satisfactory life as that of mainstream population. There is now a large inviolate space available for wild animals to roam around inside Similipal. The village Jenabil after shifting of inhabitants has been developed into an expansive meadow **(Plate 4.12)**.

Plate 4.12. Part of an expansive meadow at Jenabil developed inside Similipal after relocation of inhabitants to Ambadiha near Udala of Odisha, India. Photo: D Swain.

IMPLEMENTATION OF FOREST RIGHT ACT

The Scheduled Tribes and Other Traditional Forest Dwellers (Recognition of Forest Rights) Act, in short 'Forest Right Act' was promulgated in December 2006. The preamble of the Act *inter alia* states that it is an Act to recognise and vest the forest rights and occupation in forest land in forest dwelling Scheduled Tribes and other traditional forest dwellers who have been residing in such forests for generations but whose rights could not be recorded; to provide for a framework for recording the forest rights so vested and the nature of evidence required for such recognition and vesting in respect of forest land.

There are 62 villages and hamlets inside Similipal, besides three uninhabited villages. Four villages are situated in the core area of the tiger reserve and 58 villages are in the buffer zone of the reserve and all these villages lie in the centre of the Similipal sanctuary spreading over an area of 111.18 km^2. Besides these villages inside Similipal, most of the Rayati lands of over 250 villages lie close to the boundary of the sanctuary/ reserve. However, so far 2.21 km^2 individual rights have been settled under Forest Right Act, 2006 (Table 4.3).

Besides individual rights, community rights of 1751.73 ha settled in the buffer area of Similipal tiger reserve for collection of non-timber forest produce.

CONCLUSION

There are 62 villages inside Similipal inhabited mostly by tribals like Bathudi, Bhumija, Kharia, Kolha, Santal, and others. Out of four core area villages of Similipal tiger reserve, namely Jenabil, Jamunagarh, Kabatighai and Bakua, only Jenabil and Kabatghai villages have been relocated outside Similipal successfully. Except three families of Jamunagarh, rest 35 families have also been rehabilitated. The study reveals that there is extreme poverty, unavailability of schools and remoteness of tribal habitations in the core area. In comparison with the core area the villagers in the buffer area are in an advantageous position in the collection of NTFPs that support the income generation by selling in the local market. In the transitional zone, the people belong to the Scheduled Tribes, Scheduled Castes, other Backward Classes and General Castes. The literacy rate in the transitional zone is better than the other two zones. The average income of the households in the transitional zone from all sources is comparatively less, but they earn more by selling of NTFPs.

However, relocation and successful rehabilitation of core area families at Ambadiha, close to Udala municipality, is a win-win situation for management of Similipal tiger reserve. There is an increase in per capita income and a decent living for the families of relocated community. Simultaneously, wild animals found extensive inviolate area for movement and reproduction. The Government of India have promulgated the

Table 4.3. Individual rights for homestead and agricultural purposes settled under Forest Right Act 2006 in Similipal tiger reserve, Odisha, India. Abbreviations: STR (Similipal tiger reserve), FRC (Forest Right Committee), DLC (District Level Committee)

Sl. No.	Name of the Forest Division	Villages in STR	No. of FRC Constituted	Individual Rights - Titles Issued after Approval from DLC		Area in Acres	
				Homestead	Agriculture	Homestead	Agriculure
1	Baripada	15	1	9	26	7.58	110.85
2	Karanjia	21	4	0	126	0	161.93
3	Rairangapur	29	26	14	116	7.14	261.46
	Total	**65**	**31**	**23**	**268**	**14.72 Acre or 5.93 ha**	**534.24 Acre or 215.42 ha**

Scheduled Tribes and Other Traditional Forest Dwellers (Recognition of Forest Rights) Act, 2006. The Act grants legal recognition to the rights of traditional forest dwelling communities. Granting of individual and community rights is in progress. It is therefore necessary to involve local communities in the protection and management of Similipal reserve failing which conservation of tiger and its habitat will remain a distant dream.

Chapter 5

The Management

"Before you start some work, always ask yourself three questions - Why am I doing it, What the results might be and Will I be successful. Only when you think deeply and find satisfactory answers to these questions, go ahead."

– Kautilya [370BCE - 283BCE]

INTRODUCTION

(Late) Amarendra Lal Bose, a veteran journalist of yesteryears from Baripada (Headquarters of Similipal tiger reserve), was known for his covering stories on Similipal, tigers and environment in national newspapers. He was the Honourary Wildlife Warden for Mayurbhanj district and awarded the 'Biju Patnaik award for wildlife conservation' for his life time work for preservation of wildlife. He used to tell me, "Tigers are burning bright in Similipal as the dense and undulating forest of Similipal is the safe haven for tigers. When forest staff do not see the tigers often in Similipal, it will not be easy on the part of poachers to find them." He was right.

The tiger (*Panthera tigris*) is the largest obligate terrestrial carnivore in all of the mammalian assemblages in which it occurs in Asia (Seidensticker *et al.,* 1999). It once roamed from the land adjacent to the Caspian Sea to the Russian Far East in the north, to Java and Bali in the south, through the greater part of Indochina and the Indian Subcontinent (Maza'k, 1996; Nowell and Jackson, 1996). There are now about 160 distinct and fragmented populations of tigers in a broad variety of forest types, including dry deciduous, moist deciduous, semi- evergreen, wet evergreen, riverine, swamp, and mangrove. They are also found in the coniferous-deciduous forests of the Russian Far East, in the tall-grass habitats south of the Himalayas, and in the tropical rain forests of Sumatra and Malaysia (Kitchener, 1999; Sunquist *et al.,* 1999). Tigers have gone extinct at the extreme ends of their distribution -notably, the Caspian region and the islands of Java and Bali-and are probably extirpated in southern China (Dinerstein *et al.,* 2007).

Tigers were found throughout India from the Himalayas to Kanya Kumari, except the deserts of Rajasthan, Punjab, Kutch and Sindh and the tiger left its track in the winter snows of the Himalayas at heights of 3050m MSL (Prater, 1965). In the early twentieth century, the tiger had a far denser and more dynamic population in those habitats across India, and it was perceptibly plentiful during the 1930s and 1940s. Due to the wanton killing of tigers and loss of habitats, there were merely 1,827 tigers in India in 1972, as per the first all-India census of tigers (Srivastava, 1979). After this census, concern for the tiger's future led to the establishment in 1973 of Indian tiger reserves under the aegis of Project Tiger. Initially, there were nine tiger reserves and from 1974 to 2017, 41 more reserves were added; thus, in total, 50 tiger reserves are currently spread over 18 states in India.

Using tiger density estimates and vegetation data, Ranganathan *et al.* (2008) set tiger population targets that could be attained through reasonable management of the Indian subcontinent's nature reserves. Depending on the quality of the surrounding landscape matrix, their model indicated that the subcontinent could potentially hold about 3500–6500 tigers, in up to 150 reserves. Similipal landscape has been predicted as the 4th reserve of the subcontinent where more than 200 tigers can persist for the future through reasonable management inputs.

PAST AND PRESENT SYSTEM OF FOREST MANAGEMENT

Forest management in Mayurbhanj dates back to later part of nineteenth century. A forest policy was declared before 1885 by the then Maharaja of Mayurbhanj. Similipal forest and 300 km² of other forests of the then Mayurbhanj state were notified as reserve forest. The practice of shifting cultivation in Similipal was stopped during that time (Patnaik, 1997). The reserve forests of Mayurbhanj were under the management and control of the forest department where as other protected forests were under the charge of the revenue department. The reserve forests were more or less stable and permanent in nature, but protected forests were maintained to meet the requirement of the royals and residents and also subject to clearance for cultivation. The forest area being given under 'Amalnama' lease by the revenue authorities and leases for reclamation of reserve forests were given under the special sanction of the ruling chief. Thus the extent of reserve forests and protected forests decreased. In 1904 the Mayurbhanj narrow gauge railway line was laid from Rupsa to Baripada. This line was of immense use in transportation of timber in huge quantity extracted from plain forest mainly reserves. In 1906 a survey party demarcated the boundary line from Talabandha to Similipalgarh to form a working circle for giving lease to M/S Bholanath Borooah and Co. In 1907 a forest state department was created with J.A.Martin, state engineer as head of the department. The 30 year lease of Bholanath Borooah and Co expired during1946. There was no systematic working of the forests for which Similipal reserve during the 30 years lease, was worked twice and north Similipal three times. East Similipal, which was withdrawn from the lease in the year 1922, was worked like the rest of the Similipal forests through several contractors.

The first working plan was prepared by M.C. Hatt in 1896-97.This was revised by B.M. Dasgupta in 1946. Dasgupta prepared the first working plan for the whole of Similipal reserve forests for working under a selection cum-improvement system (Dasgupta 1946). But after about 6 years, Dasgupta's plan was replaced by the working plan of reserve forests of Mayurbhanj state by Sripal Jee during 1953-54 after the integration of Mayurbhanj state into the Union of India on 6th November 1948 and became part of Odisha as a district on 1st January 1949. This plan was revised separately for Karanjia and Baripada divisions by R. Mishra and S. Bose respectively in 1973-74 which expired in 1992-93. A.O.F. Bakhla revised the Working Plans of both Karanjia and Baripada together in 1996-97 which is followed at present, even after its expiry during 2015-16. The said Working Plan is under revision. However, in the mean time in 1973 Similipal was constituted as one of the nine tiger reserves of the country and the first management plan was prepared by S. R. Choudhury.

Despite practicing commercial forestry, supplying railway sleepers and other utility timber to outside Mayurbhanj, the rulers were very rigid in their forest protection measures and employed a large number of forest staff, much higher in number in comparison with other princely states and even the directly British administered areas with good network of forest roads and communication facilities.

Protection of forests suffered a lot after independence in 1947 when forested areas were drastically reduced. There was a proposal to declare Similipal as a national park in 1952 and a national park division was created in 1957. But the actual notification was not issued due to fear of losing revenue and Similipal was notified as a Game sanctuary by the Conservator of Forests, Angul Circle on 17.02.1970. On 04.12.1973, 2750 km² of Similipal was declared as one of the nine tiger reserves of India and Saroj Raj Choudhury, an eminent wildlife conservationist of the country took charge as first field director of Similipal tiger reserve. The notification to declare Similipal as a sanctuary under the Wildlife (Protection) Act, 1972 was issued on 03.12.1979 by the State Government. The core area of the sanctuary is a proposed national park as per notification issued on 06.08.1980 and 11.06.1986. Similipal Forest Development Corporation (SFDC) was formed in 1979 to work out the timber and non-timber forest produce (NTFP) in Similipal besides taking up other developmental work within the forest. After a complete moratorium on the tree felling was imposed in 1988, SFDC stopped working in Similipal. On 22.06.1994, Similipal RF and 10 km periphery of this forest came under Similipal biosphere reserve as notified by the Ministry of Environment and Forests, Government of India. The area of the biosphere reserve is 5569 km². It came into the international network of biosphere reserve on 22 May, 2009 by a notification of UNESCO. About 7000 km² of area encompassing Similipal and spreading over the districts of Mayurbhanj, Balasore and Keonjhar was constituted as Mayurbhanj elephant reserve in February, 1992.

Silvicultural System

Nothing relevant is known regarding silvicultural system implemented in Similipal forest prior to 1885. Departmental exploitation had been practiced in a small scale upto 1904. Then forests were leased out to traders at concessional rates for working under selection system. Subsequently, large chunks of forest areas were leased out to contractors. During 1906, work in western Similipal were conducted by eight contractors, those in east Similipal by four contractors and north-western Similipal by a corporate contractor registered as Ms. B. Barooah and Company. In the rest of Similipal, several smaller leases were operative. From 1916, the whole of Similipal reserve was handed over to Ms. B. Barooah and Company on a long term lease of 30 years. In 1919 the Barooah Timber Company was formed which took over all the assets and liabilities of the lessee B. Barooah and Company with the consent of the lessor. East Similipal reserve was excluded from the lease under mutual consent and the maximum permissible conversion was raised from 5,00,000cft to 8,00,000cft. After expiry of Hart's Plan, the forests were worked as per the prescriptions of Dasgupta's Plan (1947-1967) and Jee's Plan (1953-1972). The system first followed was Selection-cum-Improvement as prescribed in Hart's Working Plan which prescribed to remove trees of six feet (1.8m) girth and above to avoid development of hollowness during 1904-1905. Subsequently, during the period of 10 years lease (1906-1916), the lessee was entitled to all trees of six feet (1.8m) above girth except *Schleichera oleosa*, *Terminalia chebula*, *Strychnos nuxvomica*, *Terminalia bellirica*, *Bombax ceiba*, *Ficus bengalensis*, *Terminalia tomentosa;* all edible fruit bearing species, *viz.*, *Madhuca longifolia*, *Buchanania lanzan*, *Diospyros melanoxylon etc.*; all bamboo clumps; all trees on or along the bank of streams and nullahs. During the 30 years lease period, the above mentioned "Selection System" was in practice only for one species of tree *viz.*, "Sal" with five feet (1.5m) and above girth. Annual target of converted Sal was kept between 3,00,000 cft to 5,00,000 cft. The maximum permissible conversion was raised from 5,00,000 cft to 8,00,000 cft. By 1939, exploitation was as per schedule in north and west Similipal whereas in south Similipal inaccessibility compelled the lessee to exploit less than the stipulated amount. To compensate the lessee, the exploitable girth of sal was brought down to 4'-6" (1.35m) from earlier 5' (1.5m). There was no marking rule based on silvicultural considerations except the observance of girth limit and excluding the trees on the banks of streams/nullahs.

After the expiry of 30 years lease in 1946, Dasgupta's Plan was strictly adhered to. The forests were put under a selection-cum- improvement system of working. Exploitable girth for Sal was fixed at 5'-6" (1.65m) at breast height (Dasgupta, 1946). Prescriptions of Dasgupta's Plan was abandoned midway during 1953-54 and Jee's Plan was followed. The high forests of Similipal were worked in Selection System. The felling cycle was fixed at 20 years with a provision to retain one-third of selection trees (Sal) (Jee, 1968). From 1972-73, separate working plans were written for Baripada and Karanjia divisions (Mishra, 1978; Bose, 1980) in which the areas covered by Similipal tiger reserve (now, north Similipal) were excluded *i.e.*, 25 compartments of Baripada Division and seven compartments of Karanjia Division. Hence the remaining portion of the core area *i.e.* 542 km^2 proposed to be declared as a National Park vide Government's Notification No.8F(T) 8/85- 19525/FFAH dated 11-06-1986 was worked as per the prescriptions of Mishra's Working Plan for Karanjia Division and Bose's Working Plan for Baripada Division. However, all commercial forestry operation was discontinued after May, 1982 as a result of Government's notification dated 03-12-1979 declaring the tract as a sanctuary.

The practice of selection system followed by natural regeneration resulted in an uneven age crop. Certain patches of forest in South Similipal (in frost prone valleys) look like an even aged crop due to repeated dying back effect. Similarly, forest areas on both the sides of Jenabil-Nuagaon Forest road seemed to be even-aged, the cause may be over exploitation (a near total clear felling by area type) in the past.

INDICATORS OF ECOLOGICAL INTEGRITY AND CONSERVATION ISSUES

Ecological indicators are the instruments to connote a small suite of basic environmental measures including biologic (*e.g.*, species abundance and distribution), physical (*e.g.*, atmospheric temperature and rainfall), and chemical (*e.g.*, soil pH) components and processes, that present the entire array of such features in an ecosystem, and spatial changes in the said ecosystem structure and function. Indicators of ecological integrity are also known as focal, flagship, umbrella, keystone or simply indicator components of biodiversity (Simberloff, 1998; Zacharias and Roff, 2001). Davis (2005) termed them as 'vital signs' to indicate that a small selection of a very large array of potential ecosystem features was made in an initial attempt to discover a tractable number of critical ecosystem features that would represent system conditions, corresponding to the medical selection of a few critical measures used to assess basic human health condition. The stressors or conservation issues for the ecosystems of Similipal are:

(i) Encroachment
(ii) Fire
(iii) Illegal felling of timber and firewood
(iv) Illegal collection of NTFPs including medicinal plants
(v) Grazing of cattle
(vi) Poaching and tradition of *Akhand* shikar (mass hunting of animals)
(vii) Tourism
(viii) Increasing human population inside Similipal
(ix) Invasive species

Encroachment

Clandestine encroachment (**Plate 5.1**) in and around the enclaved villages inside Similipal has been noticed. Even people from nearby Jharkhand State are illegally settling in the revenue land clearing the forest. In the transition zone between forests and villages, girdled trees manifest the severity of encroachment.

Plate 5.1 Clandestine encroachment in and around the enclaved villages inside Similipal, Odisha, India. Photo: H Hazebrock.

Fire

Fires (**Plate 5.2**) are common and frequent due to high deposition of combustible materials like dried grass, a mat of leaf litter and other debris like fallen trees.

Barring certain moist valleys, entire Similipal forests come under conflagration although fires are generally confined to ground only (**Plate 5.3**). The causes of fire are purely biotic mainly by NTFP collectors, hunters, poachers and callous tourists. The fire season commences in mid-February and continues up to mid-June normally. However, sporadic rains, pre-monsoon showers and summer storms curtail the intensity of fires. Early control burning of the meadows helps a lot in reducing fire hazard; core lines also serve as fire lines.

Plate 5.2: Fires are common and frequent in Similipal, Odisha, India. Photo: D Swain.

Plate 5.3 Though most of the fires are ground fire, but they cause irreparable damage to the biodiversity, especially ground flora and fauna in Similipal, Odisha, India. Photo: D Swain.

Though most of the fires are ground fire, but they cause irreparable damage to the biodiversity, especially ground flora and fauna in Similipal. Saranya *et al.* (2014) state that from 2003 to 2013, as it is revealed from satellite, an area of 2175.9 km^2 (59.6% of total vegetation cover) has been affected by varied rate of frequency of forest fires in Similipal biosphere reserve.

The year 2018 has been the worst year in recent times as there were as many as 1731 incidents of fire occurred in Similipal (**Figure 5.1**).

Rampant Illicit Felling

Because of illicit felling of trees, *Dalbergia latifolia* (Rose wood) which was once abounded across the reserve, is rarely seen even in the most protected, the core area of the reserve. Now the most sought after timber species for illicit removal is *Pterocarpus marsupium*. This has been selectively felled by timber smugglers. If

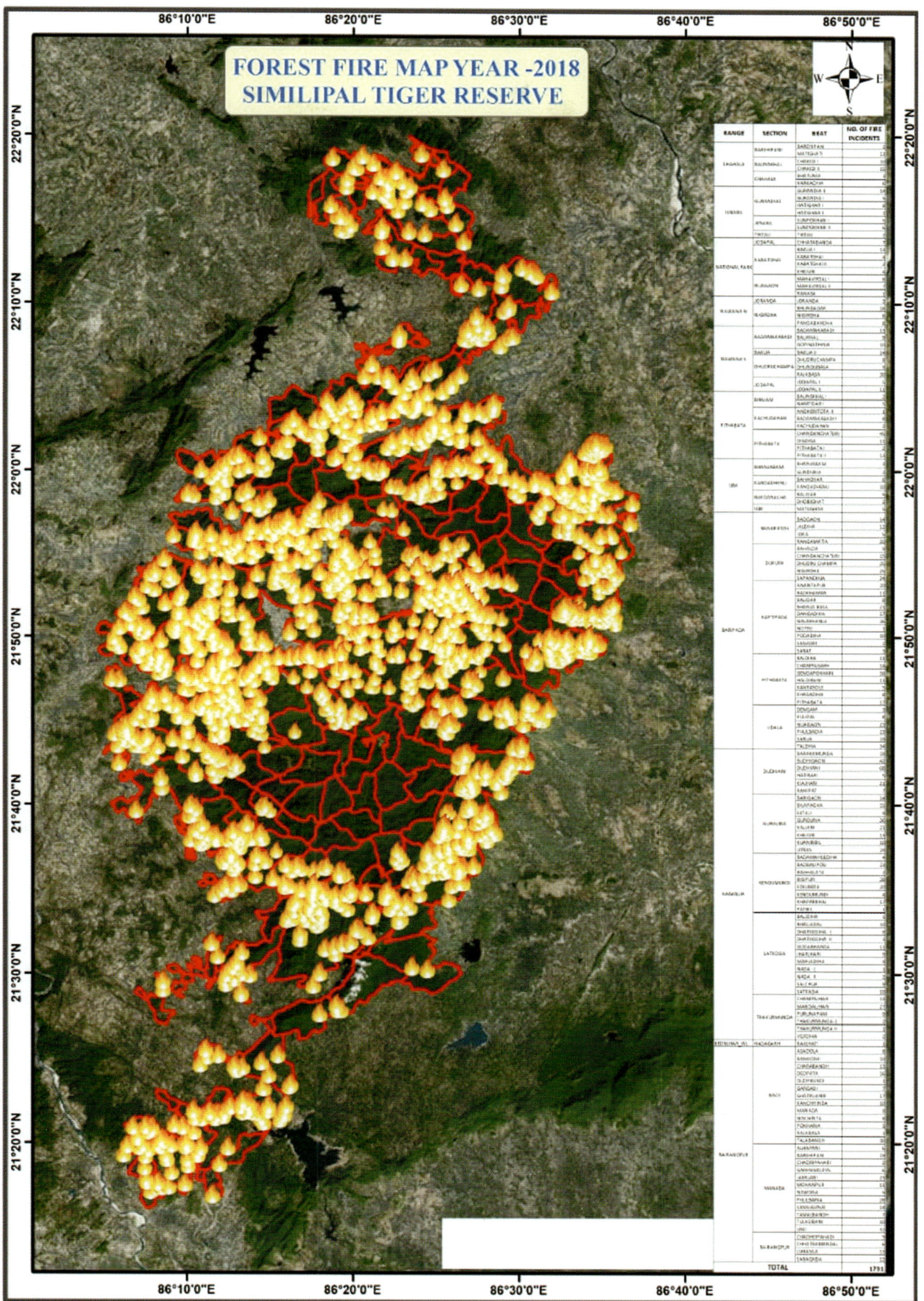

Figure 5.1. Fire incidences during 2018 in Similipal, Odisha, India.

this is not checked, the species will face the fate of Rose wood sooner or later. However, illegal timber removal of all species in certain pockets near Talabandh, Pithabata, Baniabasa, Dudhiani is quite alarming (**Plate 5.4**).

Plate 5.4. Illegal timber removal of all species in certain pockets near Talabandh, Pithabata, Baniabasa, Dudhiani is quite alarming in Similipal, Odisha, India. Photo: D Swain.

Collection of Non-Timber Forest Produce

As described in Chapter IV, inhabitants of core, buffer and transitional zones subsist on collection of non-timber forest produce. Honey, gum, arrowroot and wild mushrooms are collected by Kharias. Besides bark of paja tree (*Litsea monopetala*) is collected by the professional smugglers. Collection of mahua flowers and seeds, sal seeds also cause disturbance to the area. Forests are deliberately burnt to clean the floor to collect these items. However, in the core area the intensity of disturbance is comparatively less than the buffer areas.

Cattle Grazing

Grazing is prohibited in the core and buffer areas of the Similipal sanctuary. But incidence of grazing is noticed around the villages of core area as well as the other villages located just adjacent to the core area. The livestock population in the buffer villages as per 1995 cattle census was 7,866 and in 2006 it became 16, 956, an increase of 115 per cent. The livestock population in the core area is also showing a similar increasing trend.

The livestock population fluctuates seasonally during a year with the lowest number of goats and sheep during the rainy season, as villagers sell these animals to earn their livelihood. The cows and buffaloes are less productive in respect of milk. Their utility is more for manuring (dung and urine) and as draught animals. The maximum number of livestock shows the social and economic status of a household in the tribal society. There exists a competition for food between domestic livestock and wild ungulates in the grasslands in the vicinity of the villages.

Akhand shikar and Poaching

Tribal people in and around Similipal indulge in mass hunting as a ritual and this is known locally as *Akhand shikar*. It is celebrated by groups of villagers from a number of villages, the number in each group varying from 100 to 800, on 'Pana Sankranti' day (14th April) every year. These groups of people enter the forest, camp there for two days and kill any number of herbivores they come across. This traditional hunting was based on the superstition of prosperity of the people. Now-a-days, it is no longer attached with traditions, but is directly linked with surplus time available with the people. It is now not restricted to mid-April; rather it begins from December and continues upto June every year. About 15 to 20 attempts are made in a year. It is mostly for economic gain for bush meat for own consumption and also for selling in the market.

Professional Poaching is not well known in the tract. Incidences of elephant poaching have increased in recent times. As many as seven elephants were killed in April/May, 2010 alone (Wright and Mohanty, 2010). Crop-raiding by the wild animals prompts poaching by poisoning and killing with arms. For hares and jungle fowl snares are used (**Plate 5.5**).

Plate 5.5. A poacher with a number of killed and live animals and snares arrested in the fringe of Similipal, Odisha, India. Photo: G Mahanta

The pressure of poaching is more prominent on the south-east fringe of the sanctuary specifically from the foothill villages of Balma, Dengam, and Podadiha *etc.* (**Plate 5.6**).

Tourism

Before promulgation of Wildlife (Protection) Act, 1972, Similipal has been a shooting range of erstwhile

Plate 5.6. A Shikari (poacher) camera trapped in south Similipal, Odisha, India. Photo: Similipal TR archive

Plate 5.6A. A camping site of erstwhile Maharaja of Mayurbhanj and now abandoned since long in South Similipal, Odisha, India. Photo: D Swain.

Plate 5.7. Wildanimals at the arificial saltlick in the meadow at Debasthali inside Similipal, Odisha, India. Photo: J D Pati.

Maharaja of Mayurbhanj and other dignitaries (**Plate 5.6A**). Every year thousands of tourists visit Similipal to enjoy wildlife (**Plate 5.7**), to have a glimpse of natural bounties (**Plate 5.7A**) and to visit the deities (**Plate 5.8**) located inside the forests. Therefore, there exists a lot of pressure of tourism on Similipal.

Plate 5.7A. Sun rays through the forest of Similipal, Odisha, India. Photo: D Swain.

Plate 5.8. A place of worship of Bathudi tribe at Athardeuli inside Similipal, Odisha, India. Photo: D Swain.

The meadows at Chahala and other places **(Plates 5.9 and 5.10)**, magnificent waterfall at Barehipani (**Plate 1.4**), Nawana valley (**Plate 5.11)**, the bank of river Budhabalanga near Jamboo, the waterfalls and majestic hills at Joranda (**Plate 1.5**), Uski and Sitakund, and the shrine at Deokund attract the tourists. The scenic bounties of Similipal are beyond description (**Plates 5.12A and B**).

Every year Similipal is visited by numerous tourists from India and abroad. Proximity to big cities like Kolkata, Jamshedpur and Bhubaneswar makes it an ideal hunt for holidaying, recreation and nature study. There are excellent forest rest houses (**Plates 5.13 and 5.14**) managed by Similipal authorities for accommodation of the tourists. The people visiting Similipal for sightseeing as per official records are produced in Table 5.1. The list does not include the large number of people visiting the shrines at Deokund and Athar Deuli located inside Similipal.

Plate 5.9. Sunrise at Chahala inside Similipal, Odisha, India. Photo: D Swain.

Plate 5.10. A meadow inside Similipal, Odisha, India. Photo: J D Pati.

Plate 5.11. Mustard field with bright yellow flowers at Nawana valley inside Similipal, Odisha, India. Photo: D Swain.

Plate 5.12A. Forests changing colour at Barhakamuda, inside Similipal, Odisha, India. Photo: Similipal TR archive.

Plate 5.12B. Sunset at Meghasani hill-top inside Similipal, Odisha, India. Photo: D Swain.

Plate 5.13. Forest Rest House at Chahala inside Similipal, Odisha, India. Photo: D Swain.

Plate 5.14. A bamboo hut at Gudgudia inside Similipal, Odisha, India. Photo: D Swain.

Invasive Species

Lantana and *Eupatorium* sp., are two invasive species present in Similipal. While *Lantana* is seen to be confined to some of the villages inside Similipal, *Eupatorium* is invading into the core area of the forest. Though immediately they are under control, seen sporadically here and there, it poses a grave threat to the reserve.

Table 5.1. The people visiting Similipal Sanctuary Mayurbhanj, Odisha for sightseeing as per official records. The list does not include the large number of people visiting the shrines at Deokund and Athar Deuli located inside Similipal

Year	*Period*	*Indian*	*Foreigner*	*Total*	*Tourism Closed Season*
1980-81	Apr./March	5,979	39	6,018	None
1981-82	Apr./March	4,632	36	4,668	None
1982-83	Apr./March	5,601	46	5,647	None
1983-84	Apr./March	7,270	34	7,304	None
1984-85	Apr./March	5,078	23	5,101	None
1985-86	Apr./March	8,414	35	8,449	None
1986-87	Apr./March	8,458	44	8,500	None
1987-88	Apr./March	11,248	54	11,302	None
1988-89	Oct./June	14,994	51	15,045	16 June - 15 Oct
1989-90	Oct./June	15,176	81	15,257	16 June - 15 Oct
1990-91	Oct./June	14,002	88	14,090	16 June - 15 Oct
1991-92	Oct./June	12,579	87	12,656	16 June - 15 Oct
1992-93	Oct./June	19,260	72	19,332	16 June - 15 Oct
1993-94	Nov./June	17,493	132	17,625	16 June - 31 Oct
1994-95	Nov./June	16,908	148	17,056	16 June - 31 Oct
1995-96	Nov./June	20,236	134	20,370	16 June - 31 Oct
1996-97	Nov./June	21,133	140	21,273	16 June - 31 Oct
1997-98	Nov./June	24,413	161	24,574	16 June - 31 Oct
1998-99	Nov./June	19,377	163	19,530	16 June - 31 Oct
1999-00	Nov./June	13,403	84	13,487	16 June - 31 Oct
2000-01	Nov./June	22,142	129	22,271	16 June - 31 Oct
2001-02	Nov./June	22,508	146	22,654	16 June - 31 Oct
2002-03	Nov./June	21651	172	21,823	16 June - 31 Oct
2003-04	Nov./June	17125	192	17,317	16 June - 31 Oct
2004-05	Nov./June	19,677	171	19,848	16 June - 31 Oct
2005-06	Nov./June	14,064	189	14,253	16 June - 31 Oct
2006-07	Nov/June	13,256	240	13,496	16 June - 31 Oct
2007-08	Nov/June	15,105	164	15,269	16 June- 31 Oct
2008-09	Nov/June	20,561	182	20,743	16 June- 31 Oct
2009-10	Nov/June	Nil	Nil	Nil	Closed due to Maoist attack
2010-11	Nov/June	5,693	44	5,737	16 June - 31 Oct
2011-12	Nov/June	14,367	85	14,452	16 June - 31 Oct
2012-13	Nov/June	24,902	43	24,945	1 July - 31 Oct
2013-14	Nov/June	23,450	29	23,479	1 July - 31 Oct
2014-15	Nov/June	25,437	22	25,459	16 June- 31 Oct
2015-16	Nov/June	30,422	09	30,431	16 June-31 Oct.

Insect Attacks and Pathological Issues

Insect attacks and pathological issues are not properly documented as far as the intensity of their damage is concerned. The trees like *Madhuca longifolia*, *Buchanania lanzan*, and *Terminalia tomentosa* are occasionally infested by *Dendrophthoe falcata*. During May-July a green caterpillar defoliates the sal tree. However, the damage caused is not so serious and alarming. Grasshoppers occur in huge swarms during the rainy season but their damage is not so acute. A healthy population of birds may be the reason for low insect damage. A detailed entomological and pathological investigation is quite essential to ascertain the real status of these two groups of organisms.

Increasing Human Population

Because of the human population both inside and outside of Similipal and their dependence on forest resources, the richness of biodiversity is depleting fast. The presence of 62 villages inside, honeycombing the entire Similipal will become the nemesis of its biodiversity depletion. Table 5.2 shows how the population of these villages increased over the year from 6,036 in 1961 to 15,259 in 2011.

Poor Economic Condition of the People

The state of economy of the people, who influence the forest most, is very poor. They have got very low level of economy, agriculture is not well developed, and employment opportunities are very less. Except Kharia and Mankadia tribes, who are landless, the rest of the other tribal people depend on agriculture as their main occupation to duly supplement their income by collection of NTFP, firewood and timber. They use the land for agriculture, which is mostly rain-fed and they get only one crop with poor yield except inside the sanctuary where they also go for a second crop by irrigating the land through water from rivers and canals. Outside the sanctuary, they hardly go in for the second crop except where there are dams or water harvesting structures provided on streams. Thus as yield from agriculture is hardly sufficient to meet their own requirement, they find it easy to engage themselves round the year in collections of NTFP, cutting of trees for firewood and timber and hunting of the animals.

The conservation of biodiversity due to land use and resource dependency is affected. For the agricultural practices, more and more land are required which result in encroachment and clearing of forest. This is mostly confined within the sanctuary in the village forests. Diversion of natural water course as well as from water harvesting structure *etc.* for irrigation purpose not only affects natural flow of water affecting natural growth in this region but also wild animals become prone to hunting when they come to the artificial water sources of the villages. Movement of tractors for ploughing and other modern means of agricultural practices affect the peace and tranquility of wildlife. Use of fertilizer, insecticides and pesticides are also causing damage to wildlife. Large-scale dependence on firewood and timber is totally changing the crop composition and depleting the biodiversity.

Maoists Attack

In 2009 there were a series of Maoist insurgency attack on the infrastructures of Similipal reserve which disrupted the connectivity and ruined the residential accommodations of staff jeopardizing the management system of the reserve (Pandav *et al.*, 2009). Tiger reserves are safe havens for Maoist insurgents in the states of Odisha, Jharkhand, Andhra Pradesh and Chhatisgarh (**Plate 5.15**). Maoist movement started in Similipal since 2006 and it peaked in 2009. After stationing of armed forces at Nawana and Gudgudia, insurgency activities reduced inside Similipal.

IMPACT OF PRESENT MANAGEMENT SYSTEM

Human Settlement in 1935, 1995 and 2016

Forest cover map for three-time period namely 1935-40, 1995 and 2016 was prepared to estimate the

Table 5.2. Population trend of enclaved 62 villages of Similipal biosphere reserve, Mayurbhanj, Odisha from 1961 to 2011

Census Year	*Population Inside Similipal both Core and Buffer together*	*Decadal Growth Rate in Per cent*
1961	6036	
1971	7700	+27.567(1961-1971)
1981	8171	+11.428(1971-1981)
1991	10151	+18.309(1981-1991)
2001	12449	+22.637(1991-2001)
2011	15259	+28.10 (2001-2011)

Plate 5.15. A woman Maoist camera trapped inside Udanti-Sitanadi tiger reserve of Chhatisgarh state, India. Photo: Udanti-Sitanadi TR archive.

human seetlement area inside Similipal. To assess the modifications in forest cover, topo maps and remotely sensed satellite data were analysed using remote sensing and GIS technique. For the estimation of historical forest extent, a toposheet obtained from Survey of India, of 1:250,000 was used. Landsat raster images of 30 m resolution were used for extracting the forest information for the years 1995 and 2016, downloaded from the USGS (www.earthexplorer.com). Data from Landsat thematic mapper (TM-5) were downloaded for the year 1995, and for the year 2016, data were procured from the Landsat-8 satellite. It is noticed from Table 5.3 that area under human settlement has been reduced over the years inside Similipal (Swati Saini *pers.comm.*).

Table 5.3. Area under human settlement in 1935, 1995 and 2016 inside Similipal

Data Set Used	*Year of Assessment*	*Area under Human Settlement Area (km²)*
Based on Toposheet	1935	85.31
Landsat-TM(30m)	1995	67.42
Landsat-TM(30m)	2016	51.71

People Participation

Various authors have advocated the involvement of local people in the protection of natural resources such as forests and wildlife (*e.g.*, Gadgil, 1991; Kothari, 1996; Murty, 1996; Poti, 1995; Raju, 1996; Rathore, 1996; Suri, 1996; Wade, 1987; Wani and Kothari, 2007). Even India's National Forest Policy of 1988 emphasises the involvement of local communities in sustainable management of forests in order to minimise the conflict and maximise the harmonious relations. The state of Odisha and Mayurbhanj district are pioneers in joint forest management (Swain, 1997). I have discussed methods of implementation, funding, and a legal framework to involve the local community in the protection and management of the sanctuary and national parks in Swain (2001).

I reported the changes in forest cover of Similipal sanctuary for the year 1997, 2003 and 2006 in Swain (2014). The satellite data utilised were of IRS-1C/1D, sensor LISS-III, spatial resolution 23.5m*23.5m and spectral resolution 4 bands. Dense forest is having canopy density more than 40 per cent ; open forest is having canopy density 10 – 40 per cent and degraded forest is having canopy density less than 10 per cent.

The position of dense forest, open forest and degraded forest for the year 1997, 2003 and 2006 is presented in Table 5.4. It is evident from the Table 5.4 that Similipal witnessed decrease of 252.71 km² dense forest in six years from 1997 to 2003 and increase of 347.87 km² in three years from 2003 to 2006. It indicates that management between 2003 and 2006 was conducive to the protection of forest. During this time, people of fringe villages were involved in the management of Similipal.

Table 5.4. Forest cover changes for the year 1997, 2003 and 2006 in Similipal sanctuary, Mayurbhanj, Odisha (Swain 2014)

Category	*Assessment Year (Area in km²)*		
	1997	*2003*	*2006*
Dense forest	1846.18	1593.47	1941.34
Open forest	53.36	397.18	66.62
Degraded forest	187.03	152.47	125.00
Others	124.18	67.63	77.79
Total	**2210.75**	**2210.75**	**2210.75**

As would be seen from Chapter IV, people in and around Similipal lead a subsistence living. The people's livelihoods depend upon the collection of NTFPs, agriculture and rearing cattle, goats, and poultry. Almost all of the households in core, buffer, and transitional zones depend upon the Similipal forests for fuel wood and small timber requirements. Some of the landless, marginal and small farmers resort to illegal felling of trees for fuel wood and timber to eke out their living. They consider tiger conservation detrimental to their economic development. Even collection of NTFPs is also legally banned. Prior to 2006, the tiger reserve was a management concept without any legal backing. Since 2006, the amendment of the 1972 Wildlife (Protection) Act provides stringent punishment for any violation of the provisions inside a tiger reserve. This alienates the people depending on the reserve further.

I was field direct of Similipal tiger reserve from April 2002 to November 2007. In 2003, as part of involving local youths in protection of Similipal I introduced the formation of Sabuja Vahinee (Green Brigade) in which enthusiastic youths of fringe villages of Similipal participated to guard against the felling of trees and killing of animals. For the formation of Green Brigade, a series of meetings were conducted in fringe villages, explaining to the community the role of forest cover in soil and water conservation in the watershed of the river passing through their village. The importance of maintaining the biodiversity of Similipal was also discussed. The volunteers were provided with identity cards issued by the divisional forest officer. By 2006, 37 Green Brigade groups, consisting of 323 volunteers, were formed in and around Similipal. Even women participated as volunteers. The volunteers were successful in convincing people against the pernicious practice of *Akhand shikar* by indigenous people. As the volunteers were members of the same community, people listened to them, and felling of trees and hunting of animals reduced substantially **(Plate 5.16A and B)**. In recognition of such volunteer service, eco-development activities were adopted in their villages with funds available from Project Tiger. In 2006, the dense forest cover of ten out of eleven watersheds in Similipal sanctuary was improved, as compared to 2003 (Table 5.5) (Swain, 2009). The dense forest cover in the watershed of Sanjo was decreased, however, as the fringe villages of the watershed responded poorly. No *Akhand shikar* occurred in 2006 and 2007, as it was successfully prevented with the help of the volunteers. In 2005, 237 people were arrested at an incidence inside the reserve because of timely intervention by the volunteers. Prior to this, arresting people engaged in community hunting was impossible even with a contingent of armed police. The key factor in the formation of such groups as the Green Brigade is regular activities in the form of eco-development. But such activities are limited due to paucity of funds.

Policy Outlook

Alienating the people depending on the biosphere/ tiger reserves, conservation of biodiversity including the flagship species, the tiger and its habitats is not only futile but also detrimental to the socio-cultural-economic fabric of the ecosystem depended people. To elicit people participation in the conservation of the tiger and its habitat, needs of the local people must be addressed earnestly. A feeling of ownership over the reserve should come naturally to them. A paradigm shift in policy outlook is necessary. Fuentes-Quezada (1996) rightly states that unless the gap between global

Plate 5.16A. Green Brigade Corps in uniform ready to be used against poaching and tree felling in Similipal, Odisha, India. Photo: D Swain.

Plate 5.16B. People surrendered their saws and took oath not to resort to tree felling inside Similipal. Photo: D Swain.

Table 5.5. Dense forest cover change from 2003 – 2006 in 11 watersheds of the Similipal reserve, Mayurbhanj, Odisha (area in km^2) (Swain 2009)

Name of Watershed	*Area*	*Dense Forest Cover 2003*	*Dense Forest Cover 2006*
1.Bhandan	164.51	106.56	141.51
2. Khairi	322.81	275.12	297.86
3. Balanga	419.54	269.49	349.15
4. Khadkei	105.15	63.52	95.92
5. West Deo	314.15	254.69	287.64
6. Palpala	200.28	162.40	181.91
7. Salandi	176.19	94.55	157.46
8. Thakthaki	98.81	61.92	91.21
9. Sanjo	188.35	156.07	150.73
10.East Deo	126.72	91.06	103.93
11.Telsim	94.29	58.07	84.03
Total	**2210.75**	**1593.47**	**1941.34**

biodiversity benefits and the needs of local people is narrowed, the required economic growth will occur at the expense of much of the planet's biodiversity. Furthermore, he suggests that people maintaining biodiversity in the productive fields should be perceived and compensated for being the "world's in situ curators of biological diversity". To ensure the participation of local communities and the long term conservation of wilderness areas, the following policy measures are essential.

(i) Empower the people in the decision making process and management of the reserve. Integrate the reserve, and for that matter, all the protected areas into the social, aesthetic, and economic life of the community, so that conservation and development become parts of a single process (Swain, 2001).

(ii) Institute a programme of poverty alleviation in the fringe villages on a priority basis. Tiger reserves are located in underprivileged regions of the country. As the United Nations' Sustainable Development Goals aim to eradicate poverty and hunger by 2030, all antipoverty measures must start from the ecosystem living people who are the poorest of the poor.

(iii) Educate people, especially women, about the importance of forests and biodiversity. Shoulder responsibility for health care and hygiene in the villages.

(iv) Fulfill the energy needs of the villages by introducing modern cooking fuels and stoves, like biogas and LPG (liquefied petroleum gas) to reduce health risks and time-loss suffered by women and children, and halt the forest loss due to fuel wood demand. Launch a massive tree plantation programme in the transitional zone.

(v) Allow sustainable collection of NTFPs (non-timber forest products) in the buffer area.

ACTIONS CALLED FOR SIMILIPAL

A tiger reserve has a core or critical tiger habitat which serve as a 'source population' for tigers. The core area is surrounded by buffer which protects the source population of tiger against biotic interference. The surplus population of the core moves out to buffer and subsequently to the corridors to find its way to other protected area. The outer buffer and corridors act as sink for tiger population if adequate protection is not provided. Based on empirical data, Gopal (2015) suggests that an area of 800-1200 km^2 of inviolate space as core area is needed for maintaining a viable tiger population (20 breeding tigresses) under the Indian conditions. Considering sex ratio of tiger (1male:2 female), social dynamics, 50 per cent neonatal mortality, floaters, transients, old males expelled from territories and sub adult males striving to establish a territory, in a habitat with 20 breeding females, a population of 60-65 tigers would be present in the said area (Gopal, *supra*). On the basis of the above tiger land tenure dynamics following actions are suggested for Similipal.

Organisational Framework

There are 139 beats covering an area of 2210.75 km^2 of Similipal sanctuary (**Figure 5.2**). The average size of a beat is 15.90 km^2 with the smallest and largest beats at 6.80 km^2 and 49.61 km^2 respectively. Of the 139 beats in the sanctuary, 75 are in the core area of the tiger reserve and 64 in the buffer area. Area of beats varies, with about half the beats in the buffer area being larger than the largest beat in the core area. Core area beats have a median area of 11.3 km^2 with first and third quartiles at 10.0 and 13.8 km^2 in contrast to the buffer beats, which have a median area of 19.8 km^2 and first and third quartiles of 13.0 and 24.8 km^2. It is necessary to rationalise the area of beats in Similipal particularly that of buffer area beats where problems encountered are more than that in core area.

The core area of Similipal which used to be under the administrative control of a Deputy Director has been divided into seven Ranges, each one under the supervision of a Range Officer. The buffer area of the reserve divided into three parts, each one was looked after by a separate Divisional Forest Officer (DFO), namely Baripada, Rairangpur and Karanjia. Besides the buffer area of the reserve, the DFOs had to manage vast forest areas outside the reserve. Because of the concentration of activities in the management of forests outside the reserve, they could not find time for the management of Similipal buffer. Therefore, now, the entire reserve has been divided into two halves, namely, Similipal North and Similipal South, each one under the charge of a Deputy Director. This administrative change has been introduced recently in 2019 and it is hoped that protection and management will improve significantly.

Hadgarh and Kuldiha sanctuaries are part of the Similipal landscape. These two sanctuaries are linked to Similipal. Unless these are properly managed as Similipal, these sanctuaries will serve as sinks for the tiger population of Similipal. Therefore, these two sanctuaries should be included within the Similipal biosphere reserve and tiger reserve.

People Issue

To address the issues confronted in the management

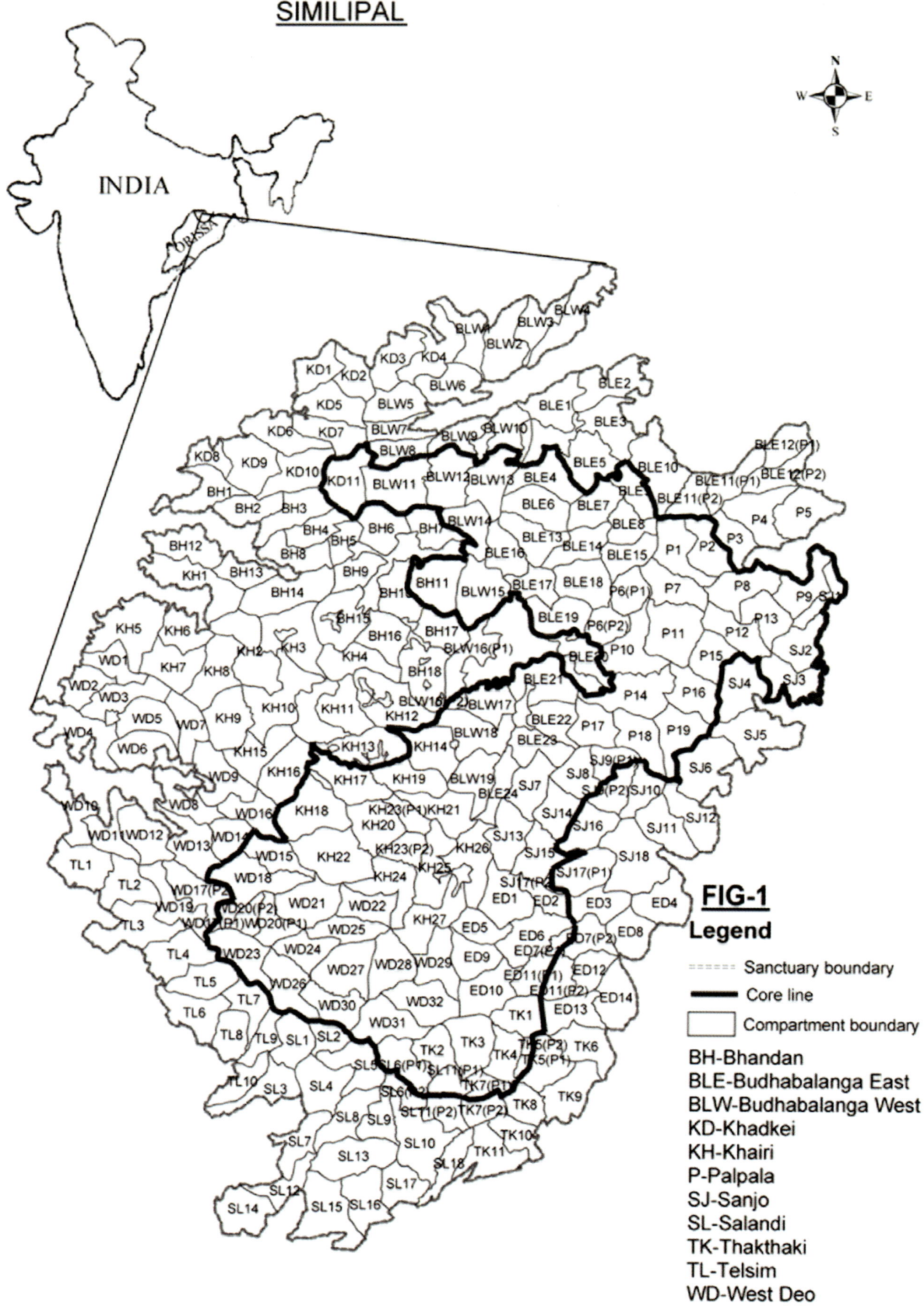

Figure 5.2. Location of 139 beats covering an area of 2210.75 km^2 of Similipal sanctuary, Odisha, India.

of Similipal, people residing in the forest fringes, particularly the villagers inside Similipal, should be made part of the management process and they should never feel alienated. To involve the people in the management of Similipal reserve, the following steps may be taken:

(i) A group comprising of about 18 to 20 youths selected from the villages inside Similipal may be formed to manage ecotourism, health services for human and cattle and all developmental activities inside the reserve. These youths need to be trained to take up all these activities professionally. If necessary, the group may be registered under Societies Registration Act.

(ii) People are alienated because they are not allowed to collect NTFPs even from the buffer area of Similipal as it comes within the sanctuary. Collection of NTFPs in the buffer areas adjoining the villages should be allowed by demarcating the area and incorporating the same in the management plan of Similipal tiger reserve. Memorandum of Understanding may be signed with the villagers so that NTFPs will be allowed to be collected provided they observe basic tenets of conservation.

(iii) Natural bamboos are very few in Similipal. People have planted bamboos (*Bambusa nutans* and *Bambusa vulgaris*) in their homestead land. But such bamboos are not allowed to be transported out of Similipal considering it as an item of NTFP. Project authorities should rather facilitate sale of bamboos by the people to outside agencies so that they get proper price and get encouraged to plant more and more bamboos in their homestead land.

(iv) Forest Department should take up activities to provide alternate remunerative avocations so that there is no or less dependence on forest. One of such activities is providing irrigation to the villages by constructing small water harvesting structures on the streams coming out of the forest. Such water harvesting structures on the fringe of forests will help both wild animals and villagers.

(v) There is unbridled falciferum malaria, gastroenteritis and other communicable diseases which need immediate shifting of the patients to a hospital at Jashipur/Baripda. There should be ambulances in charge of the Forest Department stationed at Nawana/Gududia to shift people to hospitals outside the reserve.

(vi) The system of deploying able bodied local youth (Green Brigade) in the vulnerable pockets of Similipal round the year should be followed without exception. This will prevent illicit felling of trees, *Akhand shikar* and forest fire.

(vii) There should be regular drive to create awareness and surveillance in all the surrounding weekly markets against *Akhand shikar*. A dossier of *Dehuries* (tribal priests) residing around Similipal may be made and they should be sensitized against the evils of *Akhand shikars*.

(viii) A chain of informers should be planted to transmit in advance the preparation for *Akhand shikar*, poaching, and illicit felling, so as to prevent such occurrences before the damage is done.

Habitat Improvement

Natural bamboo is rare in Similipal. Abundance of bamboo has been mentioned in Hatt's Working Plan. But subsequently because of forest fire, bamboo was wiped out except at a few places like Barehipani and along the fringes of Similipal. Bamboo is the principal food plant for prey species like gaur, sambar and cheetal. Bamboo is also staple food plant for elephants. Bamboo clumps help in hiding of tiger cubs when their mothers go for hunting. Tigers are prolific breeder. But cub mortality is very high. Cubs are killed by male tigers, and other co-predators like leopards, wild dogs, hyenas and even jackals. As tigress has to travel long distance for hunting, bamboo clumps and even lantana bushes serve as hiding places for cubs. Therefore, bamboo forest being a right kind of habitat for tigers and elephants, it is required to plant/broadcast seeds of *Dendocalamus strictus* (Salia baunsa) and *Bambusa bambus* (Kanta baunsa) in Similipal.

Monitoring

There should be annual sign survey of wild animals following protocol developed by the Wildlife Institute of India, Dehradun preferably during second/third week of January. This will not cost much and this will show which compartment/beat is harbouring tiger and co-predators. Non-availability of signs in a particular beat/compartment should be properly enquired into and corrective measures should be immediately taken. The monitoring of tiger signs provides useful information about the beats which are of the highest concern, where disturbance in the form of illicit felling required to be reduced, where prey animals poaching are occurring, and where the present level of protection to be continued. Such prescriptions based on sign survey and occupation analysis has been presented in Table 5.6.

Table 5.6. List of beats grouped by management strategies identified in Similipal tiger reserve, Odisha, India

Highest Concern	*Reduce Disturbance*	*Increase Prey Protection*	*Maintain Present Protection*
Haldibani BLE-12(P2)	Ghatakuanri BLW-1, 2	Nigirdha BLE-19	Barehipani BH-11, BLW14
Kontasole BLE-11(P1)	Chadheipahadi BH-12	Fulbaria BH-16, 18	Chhatadanda BLW-19, BLE-24
Bahalda SJ-6	Nuagaon SJ-11, 12, 10, 9(P2)	Charabandha BLW-8, 9, 10	Gurandia-I KH-21
Phulabadia ED-3, 8	Dengam SJ-16, 18, 17(P1), ED-4	Gurguria KH-3, 10	Palasibeda P-12
Kanchhinda BLW-3, 4	Rangamatia BLE-1, BLE-2	Kuanribil KH-4, 11, 12	Devasthali-II WD-22, 25
Balidiha P-5	Uski BH-9, 10, 15	Dudhiani WD-7, 9	Ranasa KH-22
Sapanchua SJ-5	Bisipur TL-1, 2	Baghalata WD-10	Mahavirsal-II WD-18
Gendapokhari BLE-12(P1)	Badgaon BLE-10, BLE-11(P2)	Utras KH-6, 7, 8	Baunskhal BLE-8, 9
	Tulasibani BH-1, 2, 3	Khaparkhai TL-7	Kachudahan P-11
	Kaliani KH-1, 2	Ranipat WD-4	Gurandia-II KH-20
	Dhudruchampa BLE-20	Badkhamana TK-8, 9	Jamunagarh KH-27
	Champagarh P-4	Purunapani SL-4, 5, 7	Jodapal-I SJ-7
	Garsimilipal BLW-17, BLW-16(P2)	Barhakamuda WD-5, 6	Sunpokhari-I SJ-13
	Taldiha ED-7(P2), 12, 13, 14, 11(P2)	Sanasialinai KD-8, 9	Gunduria TK-1
	Jamuani BH-4, 8	Badbaliposi TL-8, 9, 10	Bhajam P-1
	Pithabata-II P-9	Barigaon KH-9, 15, 16	Jodapal-II SJ-14
	Ektali KH-5	Chakidi-I BLE-4	Khadkei BLE-14, BLE-15
	Kiajhari WD-1, 2, 3	Hatibadi WD-11, 12	Dhundubasa SJ-8, 9(P1)
	Namti-I P-2	Pithabata-I P-8	Dhobighat TK-2, SL-6(P1)
	Jaldiha BLE-5, BLE-3	Mohanpur BH-13, 14	Bhanjabasa TK-4, TK-5(P2), TK-7(P1)
	Dangadiha-II SL-8, 9, 13, 15	Tamalabandha KD-6, 10	Kusumbani BLE-13
	Kendumundi TL-4, 3	Chakidi-II BLE-6	Kandadhanu WD-17(P1), 20(P1), 20(P2)
	Dangadiha-I SL-16, 17, 10, 18, SL-11(P2), SL-6(P2)	Chandanchaturi SJ-4	Bengapani WD-30
	Pithabata P-3	Bhatunia BLW-13	Patbil WD-24
		Dhudruchampa BLE-23	Baunskhal BLE-7
		Anantapur TK-6, TK-5(P1)	Sarua ED-6, ED-7(P1)
		Khejuri KH-19	
	Barehipani BH-5, 6, 7	Joranda BLE-16, BLE-17	Hatisal-II ED-5, ED-9
	Talabandha BLW-5, 6, 7	Bankidihi KD-1, 2, 5	Jenabil KH-26
	Mituani SL-3	Gopinathpur P-14	Sunpokhari-II SJ-15, 17(P2)
	Badmahuldiha TL-5, 6	Badamakabadi-I P-19	Tarinibilla WD-29
	Digdiga P-13, SJ-1	Kabatghai-II KH-18	Balikhal P-10
	Khejuri KH-13	Bakua-I KH-14	Pokharibadi WD-23
	Asadola KD-3, 4	Nawana BH-17	Meghasini WD-32
		GarSimilipal BLW-16(P1)	Chahala KD-11
		Kabatghai-I KH-17	Devasthali-I WD-21
		Karkachia BLW-12	UBK WD-28
		Hatighar-II KH-25	Bahaghar WD-26
		Nuagaon KH-24	Matughar WD-31
		Chandanchaturi SJ-2, SJ-3	Tiktali ED-10, ED-11(P1)
		Champajhar SL-14, 12	Tinadiha WD-27
		Matighati BLW-15	Balidar TK-3, SL-11(P1)
		Andharituta-II P-16	
		Rajabasa BLE-21, 22	

Highest Concern	Reduce Disturbance	Increase Prey Protection	Maintain Present Protection
		Bhundadar P-6(P1)(P2)	
		Hatighar-I KH-23(P1), (P2)	
		Budhigaon WD-8, 13, 14, 16	
		Mahavirsal-I WD-15	
		Allapani KD-7	
		Bakua-II BLW-18	
		Edelbeda WD-19, 17(P2)	
		Hatisal-I ED-1, 2	
		Kairakacha BLW-11	
		Pandabandha BLE-18	
		Podadiha TK-10, 11, TK-7(P2)	
		Mandaljhari SL-1, 2	
		Andharituta-I P-15	
		Namti-II P-7	
		Badamakabadi-II P-17, 18	

Once in every two years vegetation cover of each beat/compartment should be studied by using satellite imagery. Any decrease in forest cover in a particular beat/compartment has to be properly analysed and if necessary responsibility has to be fixed. Range officers and all subordinate staff should be made sensitized of this particular tool which shall be quite efficient to locate the illicit felling areas inside Similipal.

Phase IV monitoring using camera trap methodology introduced by National Tiger Conservation Authority should be taken up every year. The entire reserve should be covered and photo database of all the tigers of the reserve should be maintained. In the first such survey, pug mark census should be simultaneously carried out to establish a correlation between camera trap and pug mark census methods.

Ecotourism

Every year thousands of tourists visit Similipal and there exists a lot of pressure of tourism on this reserve. A four year study (Prusty and Singh, 1997; Srivastava and Singh, 1998) conducted in the reserve on profile of tourists and the vehicles they used offered several managerial tools aimed at making tourism in the tiger reserve eco-friendly and orient the drivers of hired vehicles to make the trips safe, interesting and memorable. It also highlighted the benefits due to tourism transferred to the people and the consumer market existing at the entry points of Similipal.

World Ecotourism Summit (2002) states that ecotourism is a complex activity, often seeking to meet a range of objectives, involving a variety of stakeholders and taking place in environmentally and economically fragile locations. It therefore needs careful planning. Ecotourism embraces the principles of sustainable tourism concerning the economic, social and environmental impacts of tourism. Singh(2015) states that wildlife tourism may fall into one of the types of ecotourism, but it constitutes the most fragile and sensitive type of nature tourism. Hence extreme care and restraint is necessary while promoting ecotourism in area otherwise created and managed for wildlife conservation.

In Similipal tourism activity is extended over 130 km road length, out of which 38 km falls within core area in north Similipal and rest 92 km comes under buffer area. A small area of 0.05 km^2 in core zone at Chahala, where sighting of wild animals like spotted deer, elephants, peafowls and other birds is possible, has been opened to day-visitors. Besides, the tourism route passes through stretches of forests in core zone between Bhajam-Nigirdha over 18 km, Haldia chhak to Chahala via Brundaban over 15 km and a small transit route near Kalikaprasad gate over 5 km have been used for tourism. There are many picturesque and wildlife abundant areas in the core zone, like Debasthali, Upper Barhakamuda, Jenabil, and Bhanjabasa, are out of bound to the tourists. The reserve is open to tourism from November to mid-June of succeeding year. Maximum 60 tourist vehicles are allowed each day during tourism season to ply inside Similipal. There are ecotourism complexes with modest accommodation available at Kumari, Jamuani and Barehipani. Besides, accommodatins are available at Ramatirth near Jashipur, Gurguria and Nawana inside Similipal. Revenue collected is not much as mentioned yearwise in Table 5.7.

Table 5.7. Revenue collected due to tourism in Similipal tiger resrve

Year	*Revenue Collected in INR*
1992-93	95,926
1993-94	92,994
1994-95	2,28,200
1995-96	3,57,494
1996-97	6,25,778
1997-98	9,00,873
1998-99	8,42,743
1999-00	6,12,629
2000-01	9,40,893
2001-02	8,46,049
2002-03	9,44,109
2003-04	9,55,900
2004-05	12,15,739
2005-06	11,21,271
2006-07	22,61,646
2007-08	25,50,259
2008-09	26,98,952
2009-10	1,64,469
2010-11	6,84,020
2011-12	13,36,450
2012-13	20,38,052
2013-14	28,37,250

The fact remains that only seven and half months are available to tourists to visit Similipal. It is natural that there is pressure of visitors during tourism season on Similipal. Because of sparsity of tiger population, visitors rarely see any tiger and sighting of other wild animals is also occasional. Therefore there is exasperation on the part of visitors. Ecodevelopment committees (EDCs) are maintaining ecotourism complexes and the members of EDCs earn their livelihood from accommodation and running of resturants. During lean season when the reserve is closed for visitors the establishment of ecotourism complex remains idle and local people are deprived of earning livelihood from such activities. Similipal is known for existence of melanistic tigers in wild. There are a few melanisic cubs born in Nanadan Kanan zoo recently and on pedigree analysis of these cubs, it appears that they are progeny of a Similipal tigress 'Rani'. Therefore a 'melanistic tiger safari' at Jashipur is imperative to be established in accordance with the National Tiger Conservation Authority guidelines to showcase the presence of such tigers in Similipal. It will generate revenue to supplement financially for conservation efforts in Similipal in line with that of a 'lion safari' near Gir National Park, Gujarat. It generates revenue of about 120 million INR (equivalent to 1.7 million US$) annually **(Plate 5.17A and B)**. Such a safari will keep the local people engaged in ecotourism complexes round the year which is now available for about seven months during the tourism season of Similipal.

Plate 5.17A. Lion Safari at Gir, Gujarat. Photo: D Swain.

Plate 5.17B. Two lions inside lion safari at Gir, Gujarat. Photo: D Swain.

CONCLUSION

Similipal reserve forest is a sanctuary, and the core area of the sanctuary is a proposed National Park under Wildlife (Protection) Act, 1972. It has been constituted as a tiger reserve since 1973. It formed part of Mayurbhanj elephant reserve in 1992 and declared as a biosphere reserve in 1994. In the mean time considering the importance of biodiversity of this reserve, UNESCO recognised it as an international biosphere reserve in 2009. While biosphere and elephant reserves are of management concept, sanctuary, national park and tiger reserve have legal sanctity. Any offence in a sanctuary/

national park/tiger reserve is punishable under Wildlife (Protection) Act, 1972.

The conservation issues for Similipal are encroachment, fire, illegal felling for timber and firewood, illegal collection of NTFPs including medicinal plants, grazing of cattle, poaching and tradition of *Akhand shikar* (mass hunting of animals), unsustainable tourism, increasing human population inside Similipal, and invasive species like *Eupatorium* and *Lantana*. There was settlement of 112.38 ha inside Similipal during 1997, and it increased to 155.18 ha in 2006. Similarly, agricultural area increased from 1778.54 ha to 1808.21 ha during that period. The survey of livestock per household in the core area of Similipal also indicates increase of more than 100 per cent during the 10 years period of 1997 to 2006. Increase of human population in the 62 enclaved villages of Similipal poses a serious threat. It increased from 6,063 in 1961 to 15,259 in 2011. In the last 50 years human population has more than doubled which is higher than the state and national average. Recent increased incidences of elephant poaching have raised alarm to the management. Besides, the traditional *Akhand shikar* by local people has depleted the herbivore population in the reserve. The Maoist insurgency attack on the infrastructures of the reserve in 2009 has disrupted the connectivity and ruined the residential accommodations of forest staff.

Study of Similipal forests using the data collected from field and land use data for 1997 and 2006, it was found that removal of fuelwood from compartments P (Palpala) 3, P4 and P5 of Similipal biosphere reserve is not sustainable (Swain and Behura, 2013). Such unsustainable collection of fuelwood has a high adverse impact on the biodiversity of the compartments. In all these compartments there is loss of dense forest varying from 10.59 to 22.19 percent and the highest loss is in P5 which is the outermost compartment bordering villages.

The experiment of engaging local community in the protection and management of the reserve in the form of deploying able bodied youth as *Sabuja Vahinee* or Green Brigade has yielded result. The position of dense forest, open forest and degraded forest for the year 1997, 2003 and 2006 was studied and the result shows that there was decrease of 252.71 km^2 dense forest in six years from 1997 to 2003 and increase of 347.87 km^2 in three years from 2003 to 2006 (Swain, 2014). It indicates that management between 2003 and 2006 was conducive to the protection of forest, as during this time locals were involved in the protection and management of Similipal.

There are 870 protected areas (551 Sanctuaries, 104 National Parks, 127 community reserves and 88 conservation reserves) in India as on May, 2019 (WII, 2020). The case of other protected areas across the country may be similar to Similipal. An estimated 36 percent of total forest area of the world is classified as primary forests, *i.e.*, forests of native species in which there are no clearly visible indications of human activity and ecological processes are not significantly disturbed. But about six million ha of these forests were lost or modified each year since 1990, and there is no indication that the rate of change is slowing down (FAO 2006). Most of the primary forests in India and world over are the source of fuelwood, small timber and NTFPs to the poor natives for household use and for sale to eke out a living. If the issues of conservation are not addressed earnestly, sooner than later, we stand to lose the prime forests and associated biodiversity forever.

Dr J P DAS
Dr Siris Behera
Dr D. Das
JD
Medical Camp
7.3.84

Chapter 6

The Corridor

"At the national level we are faced with essentially two problems: Poaching of the animals by traders for economic exploitation, and (ii) the problem of inbreeding leadinmg to their possible extinction in future. Both of these problems are of equal concern."

– A. P. J. Abdul Kalam (15 October 1931 – 27 July 2015), former President of India (25 July 2002 – 25 July 2007) during his address at the dedication of the laboratory for the conservation of endangered species, Hyderabad, India on 01.02.2007.

INTRODUCTION

In 2012, I founded Save Elephant Foundation Trust for conservation of elephants and its habitat, and for reducing human-elephant conflict. The trust received prestigious 'State level Biju Patnaik award for conservation of wildlife' during 2020 from Government of Odisha. It organised a corridor walk between two tiger reserves of the state, namely Similipal and Satkosia tiger reserves. The walk spanned over 11 days from 11 January to 21 January 2020 with twin objectives of assessing the present state of affairs of the corridor and to make aware the stakeholders about the importance of such a corridor (**Plate 6.1**).

Plate 6.1. The Save Elephant Foundation Trust organised a corridor walk between Similipal and Satkosia tiger reserves of Odisha, spanning over 11 days from 11 January to 21 January 2020. Photo: DFO, Keonjhar WL

Satkosia tiger reserve of 963.87 km^2 spread on both sides of Satkosia gorge of river Mahanadi **(Plate 6.2A)** is comprised of Satkosia gorge sanctuary and Baisipalli sanctuary in the districts of Angul, Cuttack, Nayagarh and Boudh. The tiger reserve is a part of Mahanadi elephant reserve (8036.32 km^2). Satkosia is the meeting point of two bio-geographic regions, namely the Deccan Peninsula and the Eastern Ghats, contributing immensely to the biodiversity of this reserve. The forest is mostly moist deciduous and riverine forest. Dominant species are sal (*Shorea robusta*) and its assosiates. Bamboos, namely *Dendrocalamus strictus* and *Bambusa bambos*, are abundantly available. Wild animals in Satkosia include tiger (*Panthera tigris*), elephant (*Elephas maximus*), leopard (*Panthera pardus*), sambar (*Rusa unicolor*), bison (*Bos gaurus*), Chausingha (*Tetracerus quadricornis*), chittal (*Axis axis*), Malabar giant squirrel

(*Ratufa indica*), dhole (*Cuon alpinus*), porcupine (*Hystrix indica*) and many others. Mugger (*Crocodylus palustris*) and Gharial (*Gavialis gangeticus*) are present in this stretch of Mahanadi. It is the southen most home of Gharial **(Plate 6.2B).**

Plate 6.2A. Satkosia tiger reserve of 963.87 km^2 spread on both sides of Satkosia gorge of river Mahanadi in Odisha, India. Photo: A P Das.

Plate 6.2B. Gharial (*Gavialis gangeticus*) in river Mahanadi, Odisha, India. Photo: D Swain.

Cobden-Ramsay (1910) states that the extensive and almost unpopulated tracts of the Similipal forest was a sanctuary of elephants. Probably most of the elephants in Odisha visited this magnificent elephant-forest at sometime or other in the course of their existence. This statement implies that all elephant lands were contiguous with Similipal forest and elephants freely migrated from one habitat to another. In fact the predicted forest cover map from 1800 shows a continuous patch of forest from Similipal to central India (Ellis *et al.*, 2010). Now the forests have been fragmented and the migratory routes have been devastated and are hardly used by elephants.

PRESENT STATUS OF SIMILIPAL – SATKOSIA CORRIDOR

Presently, the corridor between Satkosia and Simlipal is tenuous, long and consists of fragmented forest patches interspersed with agriculture, mines and varied human uses like thermal power plant. Major restorative inputs are required in this landscape to keep this corridor functional (Qureshi *et al.*, 2014). There are two routes joining Similipal tiger reserve and Satkosia tiger reserve **(Figure 6.1)**. One route is meandering through the forest divisions of Baripada, Karanjia, Keonjhar (Wildlife), Keonjhar, Deogarh, Bonai, Bamra (Wildlife), Rairakhol, Athmallik and Angul. The other route passes through Baripada, Karanjia, Keonjhar (Wildlife), Cuttack, Dhenkanal, Angul and Athmallik forest divisions.

Location and Importance

The Similipal-Satkosia corridor passing through eight districts in the north-central part of Odisha connects the Similipal tiger reserve (TR) of Mayurbhanj district (21°35'35.22"N and 86°17'40.34"E) with Satkosia TR of Angul District (20°30'25.74"N and 84°51'58.39"E). The population of the tiger is very less in Similipal TR (17-19 Tiger) and Satkosia TR (2-4 tigers) (Jhala *et al.*, 2015). Hence their long term survival would depend on gene flow between these populations and the results of possible reintroduction of gene pool from different wild populations. The movement of wildlife between Similipal and Satkosia TRs is facilitated by wildlife sanctuaries and reserve forests (RFs) connected by a chain of corridors: Similipal-Hadgarh, Baula-Kuldiha, Kanheijena-Anantpur, Anantpur-Awakhola, Aswakhola-Sunajhari, Nuagaon-Barauni, and Tal-Kholgarh.

Similipal is connected to Hadgarh sanctuary in Keonjhar district and Hadgarh in turn is linked to Kuldiha sanctuary in Balasore district. There is a 6-7 km corridor that connects Similipal tiger reserve with the Hadgarh wildlife sanctuary through the Noto and Satkosia reserve forests. Baula-Kuldiha corridor, a 15-16 km long corridor connects the Kuldiha wildlife sanctuary with the Hadgarh wildlife sanctuary through small hillocks in the Garsahi, Gaguapahar, Balihudi, and Baula hills. This corridor is being regularly used by herds of 20- 25 elephants. Thus a connection between Similipal tiger reserve, Hadgarh wildlife sanctuary, and Kuldiha wildlife sanctuary forms a L-shaped path that provides a potential movement corridor to tigers and elephants **(Figure 6.2)**. Hence, the degradation of this connectivity (corridor) will hinder the movement of elephants and other wildlife from Similipal to the rest of the landscape.

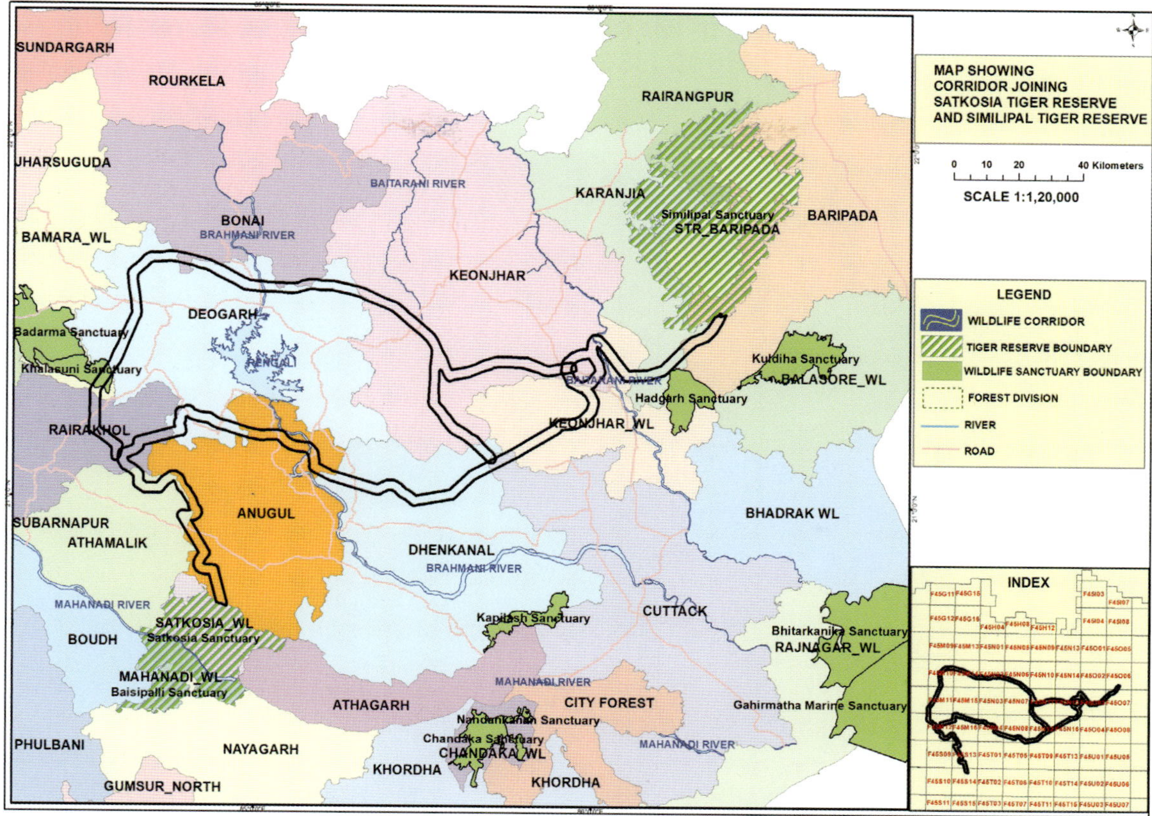

Figure 6.1. Corridor between Similipal-Satkosia tiger reserve, Odisha, India.

Methodology

Two teams consisting of scholars from Wildlife Institution of India, Indian Institute of Science, and North Odisha University, covered around 400 km (Route 1) and 350 km (Route 2) on foot and vehicle from Thakurmunda (Similipal TR) to Tikarpada (Satkosia TR) from 11 to 21 January 2020 **(Figure 6.1)**. The teams visited villages to determine potential route of elephant movement and traversed a distance of 08-10 km per day in various forest range to determine presence of wild animals. Various elephant movement areas were visited and noted the barriers, where existed. In various villages, meeting cum awareness programmes were held with people's participation. Communities were interacted about their dependency, conflict with wild animals especially elephants and other issues related to wildlife **(Plate 6.3)**.

Plate 6.3. Meeting with women group of Rebena village, Keonjhar district, Odisha, India, located along Similipal-Satkosia corridor. Photo: D Swain.

RESULTS

Vegetation Types and Flora

The Similipal-Satkosia corridor is a mosaic of forest, villages, mines (active and closed), and water bodies (streams and reservoirs). The forest is represented by sal (*Shorea robusta*) forest, sal-dominated mixed forest, and mixed forest with or without sal **(Plate 6.4)**. The corridor supports many fruit and timber yielding trees along with local herbs, creepers, and shrubs. *Shorea robusta, Terminalia tomentosa, Mangifera indica, Alstonia scholaris, Anogeissus latifolia, Diospyros melanoxylon, Buchanania lanzan, Cassia fistula, Lagerstroemia parviflora, Semecarpus anacardium, Careya arborea, Kydia calycina, Radermachera xylocarpa, Melia azedarach, Azadirachta indica* were frequently encountered tree species. Common

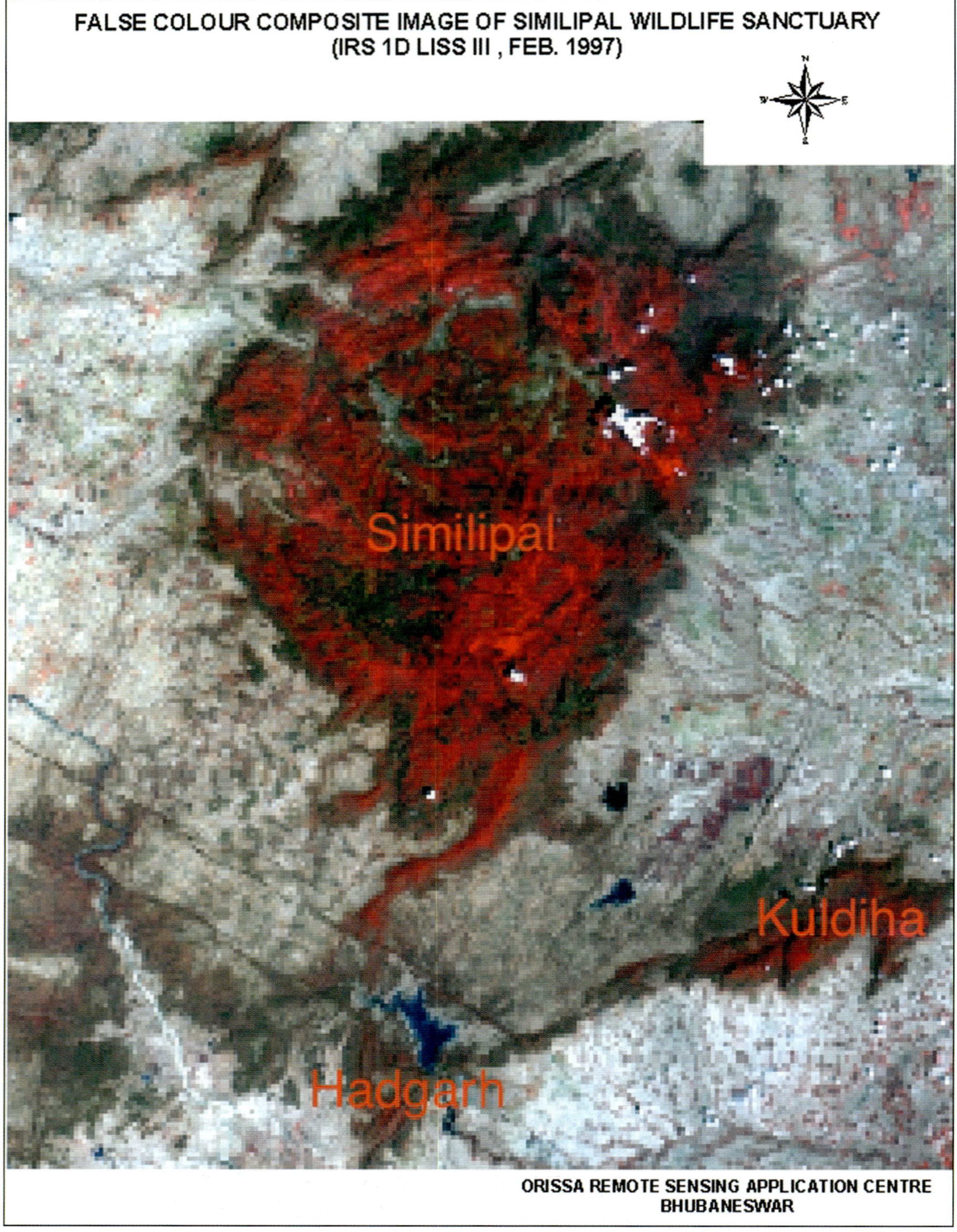

Figure 6.2. Satellite imagery of Similipal-Hadgarh-Kuldiha landscape. Connectivity between Similipal tiger reserve, Hadgarh wildlife sanctuary, and Kuldiha wildlife sanctuary that forms a L-shaped path that provides a potential movement corridor to tigers and elephants.

shrubs around the mining lease area were *Flemingia strobilifera, Woodfordia fruticosa, Helicteres isora, Ziziphus oenoplia, Clerodendrum infortunatum, Vitex negundo.*

Wild Fauna

Direct sighting and sign evidences of digging, pugmarks, hoof marks, scats and faeces of 11 mammals belonging to eight families were recorded in Similipal-

Plate 6.4. Sal (*Shorea robusta*) dominated mixed forest in Similipal-Satkosia corridor, Odisha, India. Photo: D Swain.

Plate 6.5. Buff striped keelback (*Amphiesma stolatum*) snake killed in road accident in Similipal-Satkosia corridor, Odisha, India. Photo: T Ahmed.

Satkosia corridor. They are wild boar, jungle cat, leopard, sloth bear, Asian palm civet, Rhesus macaque, langur, Asian elephant, Indian crested porcupine, Indian palm squirrel and Indian giant squirrel. Of the mammalian species recorded, one is endangered (Asian elephant), and two (leopard and sloth bear) are vulnerable, as per IUCN red list assessment (IUCN, 2020). A total of 20 species of herpetofauna representing seven amphibians and 13 reptiles (Lizard and snakes) were encountered during the survey. Amphibians represented by frogs (n=5) and toads (n=2) belong to six genera and two families; Bufonidae and Dicroglossidae. All species recorded are 'Least Concern' in the IUCN Red List of Threatened Species (IUCN, 2020). Lizards belonging to three families, Agamidae, Gekkonidae, and Scincidae were represented by five genera. Snakes of three families; Colubridae (5 species), Elapidae (1 species), and Viperidae (2 species) were recorded. Except Russell's viper (*Daboia russelii*), Checkered keelback (*Xenochrophis piscator*) and Green keelback (*Macropisthodon plumbicolor*), all the snakes were encountered as road kills **(Plate 6.5)**.

Avifauna recorded during the reconnaissance survey were comprised of 53 species representing 33 families. Numerically family Columbidae dominated the area with the maximum number of species (n=5). It was followed by Campephagidae, Ardeidae, Psittaculidae, with three species of each family. Most species of birds recorded in the area are enlisted as 'Least Concern' in the IUCN Red List of Threatened Species (IUCN, 2020). One Near Threatened species, *viz.*, Alexandrine parakeet (*Psittacula eupatria*), was also recorded around the mines at Daitari **(Plate 6.6)**.

Plate 6.6. Alexandrine parakeet (*Psittacula eupatria*) in Similipal-Satkosia corridor, Odisha, India. Photo: T Ahmed.

During the survey, 55 species of butterflies were recorded, representing 42 genera and five families. Most species of family Nymphalidae (45.4 per cent) were recorded, and it was followed by members of family Lycaenidae (18.1 per cent) and Papilionidae. The species of butterfly recorded are categorized either as 'Least concern' or 'Not Evaluated' category of IUCN Red List of Threatened Species (IUCN, 2020). Spiders including 25 species representing 19 genera and eight families were also found. Family Araneidae numerically dominated the area with ten species, followed by Salticidae.

Socio-economic Characteristics in Similipal-Satkosia corrdior

A number of villages and hamlets, that included Taramara, Gajapathara, Kaduabahali, Dargarisila,

Nuadihi, Rail, Bagawanpur, Arjuna Bania, Kimbitarangiri, Rebana, Nagada, Budhibil, Padapatia, Hinjirida, Khola regeda, Korab and Tikarpada, spread over eight revenue districts, were visited to assess the socio-economic condition and related issues. The teams found that the primary livelihood sources of the people are agriculture, livestock grazing, manual labour, and mining labour. Most people in tribal populations do not have any formal education, and only a few are educated up to the primary level. The majority of the population lives in informal settlements and 'Kuccha' houses.

Most of the people have less than an acre of agricultural or ancestral land. The land is mostly used for agriculture **(Plate 6.7)**, both as subsistence and kitchen gardening. The principal crops cultivated are rice and vegetables. The important sources of water in the area are streams, open wells, taps, and tube wells. Most village Panchayats in the area have access to schools in their vicinity; however, healthcare centers are sporadic. The main livestock in the region are cows, buffalos, goats, and poultry.

Plate 6.7. Agricultural land in Kholaregeda village inside Norhern RF of Handapa range, Athmallik forest division, Odisha, India. Photo: D Swain.

Shifting cultivation is prevalent in Keonjhar wildlife division and Cuttack forest division, particularly around Nagada village, recently infamous for malnutrion and child mortality. Forest department has taken steps to dissuade the tribal inhabitants from this pernicious practice of slash-and-burn agriculture. District administration initiated developmental activities of road construiction, electrification and piped water supply at the cost of forest and wilderness of this corridor **(Plate 6.8)**.

Plate 6.8.A. Nagada village, in Similipal-Satkosia corridor, Odisha, India. B. Village sign post. C. Forest used for shifting cultivation in Nagada village. D. Forest cleared for road and electrification. Photo: D Swain.

Cost and Benefits to Local Community

Similipal-Satkosia corridor is a source of livelihood and sustenance for the local people as they collect fuel wood, fodder, traditional medicinal plants, and NTFPs like Mushrooms from the nearby forest. Wood from the forest serves as the primary source of fuel to supplement kerosene supplied by the Government, as there is no formal gas supply in the area. The main types of wood extracted for fuel are Asan (*Terminalia tomentosa*) and Sal wood (*Shorea robusta*). Sal twigs are sold in the market as teeth cleaning brush or *datun*. Livestock are usually left in the forest for grazing. However, local community is prone to crop damage and livestock depredation. Elephants and wild pigs frequently damage crops in the area and livestock are majorly depredated by hyaena and leopard.

Barrier to Wildlife Movement in the Corridor

Various developmental programmes such as National highways, railway tracks, mining activities, irrigation canals and habitat fragmentation due to human settlements are serious challenges to wildlife movement in the corridor.

Elephant Movement

There are about 2,000 elephants in Odisha. Similipal accounts for about 500 of these; where as there are about 200 elephants in Satkosia TR. Between Similipal and Satkosia, about 600 elephants are seen in the districts of Angul, Dhenkanal, Sambalpur, Deogarh, Keonjhar, Jajpur and Cuttack. There are infrastructural developments like roads, railway lines and power lines. There are also iron ore and coal mines, thermal power projects, water reservoirs and hydro-electric projects. The movements of elephants in Route-1 and Route-2 are described as follows.

Route-1: Elephants from Satkosia RF of Similipal cross the road connecting Thakurmunda of Mayurbhanj district with Anandapur of Keonjhar district (State Highway-53) between Chinbeda and Satkosia villages to go to Santoshpur RF of Keonjhar (Wildlife) division. There is no elephant pass made on this important road. Elephants cross river Baitarani around Singnali village of Keonjhar district. Sometimes elephants cross the river Baitarani downstream and thereafter Musali stream near village Kaduabahali and enter into Atei RF (Keonjhar forest division), and reach National Highway (NH-20, earlier NH-215), two km from Ghatagaon towards Anandapur near Balijodi village. On this particular place two adult females and a female calf died being hit by a truck and a bus during the early hours of 22 August 2019 **(Plate 6.9)**. Two under-passes constructed on NH-20 at a distance of 100 to 150m apart from each other are inadequate **(Plate 6.10)** and not as per specification mentioned in WII (2016). After traversing NH-20 elephants enter Ghatgaon RF and then to Balipokhari RF of same Keonjhar forest division crossing over a major district road, Ghatgaon-Harichandanpur road, near the village Pauchia. Thereafter, negotiating Harichandanpur-Dhenkikote road and Vyasanagar-Keonjhar railwayline they enter into Kalimati RF and adjoining revenue forest. Vyasanagar-Keonjhar rail line has become death trap for elephants. Several elephants died at different places on this track. During 2013, an elephant died at this crossing point as explained by the local forest staff. From Kalimati RF the elephants go to Barabanka RF passing over Harichandrapur-Basantpur road and from there to Taluri Pahar and Kalapat RF of Telkoi forest range.

Plate 6.9. Two adult females and a female calf died being hit by a truck and a bus during the early hours of 22 August 2019 on NH-20, near Ghatgaon, Keonjhar district, Odisha, India. Photo: DFO, Keonjhar.

Plate 6.10. One of the two under-passes constructed on NH-20. Elephants hardly use these underpasses. If these underpasses were constructed as per specification mentioned in WII (2016), tragic deaths of three elephants at Plate 6.9 could be avoided. Photo: C K Sar.

From Telkoi forest the elephants move either to Malyagiri forest of Palalhada forest range or cross the National Highway-49 (earlier named, NH-6) at Rugudihi and Golabandh villages (Deogarh forest division). No underpasses exist at these points on this interstate National Highways. Elephants from Malyagiri forest reach Palalhada-Banarpal National Highway (NH-149, earlier NH-23) through Ranja proposed RF (PRF) and after the NH the pachyderms go to Pita RF. The crossing point on this NH-49 is partially blocked by side metal fencing of the NH. The proposed Talcher-Bimalagarh railway line will also create further obstacle for elephant movement at this point. Elephants traverse river Brahmani near Gandamula village and proceed to Reamal forests, and then to Landimal and Rail RFs of Naktideul forest range of Rairakhol forest division. Elephants in Rairakhol division use Tal-Kolgarh corridor to go to Athmallik forest division. National Highway-55 (earlier NH-42) and Angul-Sambalpur railway line pass through this corridor. Further there are several construction activities beyond the NH and rail track in the heart of this corridor. A metalled road runs in the middle of the corridor. One godown and a dumping yard are present in the corridor **(Plate 6.11)**. Rairakhol municipal area is extended into the corridor. However, elephants are still using the corridor braving the obstacles. Elephants pass through the East and West Barni RFs of Athmallik division and they move to Hatidhara forests of Satkosia TR. About five km upstream of Tikarpada, they also sometimes cross the river to go to the southern side of Mahanadi, near Binikei temple on the left bank, close to Hatidhara RF.

Plate 6.11. A godown close to Rairakhol municipality in the Tal-Kolgarh corridor in Sambalpur district, Odisha, India. Photo: C K Sar.

Route-2: Apparently, there is no issue of preventing movement of elephants from Noto RF of Baripada forest division to Satkosia RF of Karanjia forest division within Similipal TR. Thakurmunda-Udala, a major district road, passes through this corridor. There is a proposal for improvement of this road, converting it to be a National Haighway, which shall seriously jeopardise the movement of elephants and other wild animals. There are series of metalled roads and settlement which may further hamper the movement of wild animals from Similipal sanctuary to Satkosia RF of Similipal TR and from there to Hadgarh sanctuary. From Hadgarh sanctuary elephants enter into Santoshpur RF, then to Atei RF and Rebena RF. There are several locations, where elephants cross NH-20. Atei, Rebena and adjoining Daitari forests are having good population of elephants. An iron ore mine of Odisha Mining Corporation exists in the midst of the corridor **(Plate 6.12A)**. A long distance conveyor belt inside Daitari/ Rebena forests, constructed at the ground level, runs 3.8 km clearly fragmenting the corridor **(Plate 6.12B)**. Daitari forests extending over Keonjhar wildlife division and Cuttack forest division are connected to Ranjagada forests of Dhenkanal forest division.

Plate 6.12A. Daitari iron ore mine in Similipal-Satkosia corridor in Odisha, India. Photo: D Swain.

Plate 6.12B. 3.8 km long conveyor belt of Daitari iron ore mine in Similipal-Satkosia corridor in Odisha, India. Photo: D Swain.

Nagada villages conglomerate (villages of Nagada Panchayat) exists in the heart of Daitari forest of Sukinda range in Cuttack forest division. In the recent past there were child mortalities in these villages due to malnutrition; and government is taking all efforts for development in these villages. As a result, the roads are made deep inside forest on the hills and the corridor has been devastated. Ranjagada forests (Kamakshyanagar range) of Dhenkanal forest division are connected with Telkoi forests of Keonjhar forest division. Dandadhar reservoir is situated at the edge of both Ranjagada and Telkoi forests. Elephants migrate to Mahabirod Range from Telkoi and Ranjagada forests at the tail end of this resrvoir which is connected to Anantapur RF of Kamakshyanagar (West) and Mahabirod ranges of Dhenkanal division.

As per 2017 census, there are 169 elephants in Dhenkanal, most of them are congregated in Kamakshyanagar (West) and Mahabirod forest ranges. Mahabirod range is worst affected. Elephants are unable to migrate to adjoining Angul forest division due to the presence of irrigation canal (Rengali irrigation project), highways and railway tracks. There are presence of coal mines, thermal power stations and townships in Angul forest division, and there are 45 elephants (2017 census) locked in this division and creating havoc in Talcher, Kaniha and Chhendipada ranges.

Elephants are present in Northern and Handapa RFs of Handapa ranges under Athmallik forest division adjoining to Chhendipada range of Angul division. But migratory route from Chhendipada range to Handapa range has been disrupted due to roads, urbanisation, settlements and agricultural lands. Elephants from Handapa range have to cross Angul-Sambalpur railway track and National Highway-55 (earlier NH-42) to reach Satkosia TR atleast at two locations where no underpass exists for their movement.

Tiger Movement

Hadgarh sanctuary which falls in the Similipal-Satkosia corridor is close to Similipal TR. Tiger movement is usually camera trapped in this sanctuary. Last tiger was camera trapped during 2017; there after not a single tiger could be traced, not even during all India tiger estimation of 2018. A cattle was killed by a tiger during 2017 in the village Kaduabahali on the right bank of river Baitarani adjoining to Atei RF of Keonjhar forest division in Route-1 of Similipal-Satkosia corridor. But no trace of tiger was observed after the said incidence. Similarly casual movement of tiger was observed in the Daitari forest around iron ore mining area of Orissa Mining Corporation until 2017. Now tiger movement is not noticed in the entire corridor of Similipal-Satkosia, neither of the Route-1 and Route-2.

IMPROVING THE CORRIDOR

Hadgarh sanctuary and several reserve forests (Santoshpur, Atei, Telkoi, Malyagiri, Rebena, Daitari, Ranjagarh, Northern, Handapa, *etc.*) exist in the Similipal-Satkosia corridor. They are the habitats for many wild animals. Though no tiger movement is noticed, elephant herds are found to use these forests. But the pity is that forests are being depleted and degraded because of reasons listed above. Bamboos useful as a habitat constituent for a variety of animals including elephants and tigers, are scarcely present in the forests of the corridor. Conserevation and improvement of bamboo groves, sowing of salia bamboo (*Dendrocalamus strictus*) seeds, removal of encroachments, shifting of villages disturbing the movement of animals, are some of the actions needed for improvement of Similipal-Satkosia corridor. There are linear infrastructures like roads (National highways, State highways, major district roads and Pradhan Mantri Grama Sadaka Yojana roads), railway tracks and canals inhibiting movement of wild animals in this corridor. WII (2016) has set out general guidelines to maintain connectivity across a corridor as follows:

i. If the width of the corridor through forest habitat is one km or less, the construction of flyovers (underpass) should be undertaken in such a way that the entire stretch of forest remains connected.

ii. If the width of the corridor is one to two km, one underpass of 750 m should be provided across the landscape. The exact location of the underpass should be based on topographic features of the area and information about customary animal crossing zones.This 750 m stretch of elevated road could also be divided into two parts of minimum 300 m each,located within that corridor. Their location would depend upon the terrain, characteristic of the particular species and its movement patterns.

iii. If the width of the corridor is three km or more, or if the forest landscape is to be dissected by either a new road or upgrading of an existing road, 300 m underpasses are suggested within every km stretch of the road. The exact location of the underpass should be based on topographic features, crossing zones, and the particular ecological requirements of the affected species.

iv. Other than maintaining connectivity for larger mammalian species, for amphibians or reptiles across the landscape, small pipe culverts or bridges should be constructed in every 100 m stretch of road.

The guidelines (WII, 2016) further state that the height of viaduct for elephant passage should be thrice the height of an adult bull elephant (i.e, height of the pillar should be 8-10 m above the ground), so as to provide safe passage for elephants. In the event of constraint of fund for the projects like major district roads and PMGSY roads, a minimum span of 50 m with a height of 6-8 m and a width of 10-12 m is desirable for movement of elephants. National highways-20, 49, 149, 53 and 55, besides a number of State highways, major district roads and PMGSY roads are present in Similipal-Satkosia corridor. None of the roads has the passage for elephants and other wild animals as per the guidelines of WII (2016).

Railway tracks are death traps for elephants in Odisha. As on November, 2018, there were 24 elephant deaths in the state since 2010 (Senapati, 2018). In Similipal-Satkosia corridor, there are railway tracks, namely Vyasanar-Keonjhar and Naraj-Sambalpur, and several elephants have been killed in these tracks recently due to lack of adequate mitigation measures. Further two railway lines have been planned in this region. They are Talcher-Bimalagarh and Talcher Jakhapura. More such rail tracks may come up to carry minerals like coal, iron, manganese and chrome ores. There is presently doubling of existing Naraj-Sambalpur railway line in progress. Therefore mitigation measures for safe passage of wild animals on these railway tracks in this important corridor should be of prime importance of the authorities concerned. Underpasses prescribed for linear infrastructures like roads are applicable for railway tracks. Until required passages are made, mitigation measures need to be undertaken to stop killing of elephants and other wild animals on the railway tracks. A train can not stop abruptly by locating elephants on the track and it can not brake hard on curves as they would topple over. Again it can not navigate steep slopes by reducing their speed. Further railway lines enclosed on both sides by steep slopes cut through hill sides of undulating terrain become dangerous for elephants as they can not climb out or exit such areas quickly. Therefore necessary mitigation measures need to be in place to avoid accidents of killing elephants and wild animals. First, wildlife biologists, forest personnel and Save Elephant Foundation Trust volunteers who have local knowledge of movement of elephant herds, resource needs (food and water), and behaviour, should be consulted to locate elephant crossing point and type of mitigation measures to be adopted. Second, level crossing for elephants should be provided at regular intervals. These crossings should ideally have smooth slopes with the railway tracks embedded in the cement. Third, to prevent elephants being trapped in railway tracks between steep embankments, their entry into such areas should be prevented by installing cattle-proof barriers or fences made of iron railings of rail track. These fences should be reinforced immediately as and when damaged. Fourth, as it is unsafe for a train to halt suddenly on a curve, the presence of elephants in these places should be avoided by installing physical barriers on both sides of the curve. Fifth, barricading the major length of the track, and giving access to elephants at the required number of locations is the optimum long-term mitigation measure. Delay in implementing the measures, especially those related to construction of underpasses, can negate the benefits of securing connected habitats (WII, 2016).

Rengali dam located across Brahmani river in Angul district is of 1040 m in length and 70.5 m in height. It has installed capacity of 250 MW power production. A barrage has been constructed at Samal across the same river, 35 km downstream of Rengali dam. The water released from the Rengali power house is picked up in the barrage to provide industrial water use and extend irrigation facilities on either side of river Brahmani in an area of 2,35,500 ha lying in Angul, Dhenkanal, Cuttack, Jajpur and Keonjhar district (Government of Odisha, not dated). Both left and right canals (depth of canal being 25m and bed width being 19.70 m) **(Plate 6.13)** restrict the movement of elephants across the river Brahmani for which human-elephant conflicts in the Dhenkanal and Angul districts have accentuated and there are rising fatalities of humans and elephants. The overpasses on these canals are of little use. Except a few daring bull elephants, elephant herds rarely use these crossings. Sometimes they prefer to use the canal where canal is either damaged or some modifications have been made for elephant crossing endangering the lives of young ones. Singh and Chalisgaonkar (2006) state that (i) the width of the bridge be kept at least 15-25 m for movement of herd; (ii) the side wall or railing be made blind up to a height of 2.75 m so that the elephants are not visually disturbed by the turbulent water in the canal. They further suggest that the existing overpass can be modified to be used by elephant herds for crossing over to the other side of the corridor. The central pier and abutments be extended on both the banks so that the width of the bridge at the centre be made as 15 m and at the start/end as 25 m. The additional width shall only be available for facilitating the movement of the elephants on the bridge. On both sides of the bridge 30 cm earth cushion should be provided so that local vegetation and grass will be grown on the top surface of the bridge to provide camouflage and congenial environment to the elephants for adapting the path for crossing over. Side railings of 2.75 m height should be constructed of reinforced concrete with 50 per cent perforations and suitable pockets for growing local vegetation and creepers. Similar mitigation measures are inevitable for canals of Manjore irrigation project in Athmallik forest division of Angul district.

Plate 6.13. Rengali irrigation canal in Similipal-Satkosis corridor, Odisha, India. Photo: D Swain.

Odisha is known for its mineral resources. It has large deposits of coal, iron ore, chromite and manganese and therefore mines of these minerals exist in the Similipal-Satkosia corridor. The important mines located in this corridor impeding the movement of wild animals are Daitari iron ore mine of Odisha Mining Corporation Ltd. and the Talcher coal field of Mahanadi Coalfields Ltd. About eight km^2 forest area in Keonjhar wildlife and Cuttack forest divisions of this corridor has been leased out for Daitari iron ore mine. Extreme care needs to be taken during mining and sufficient forest land should be left for movement of elephants and other wild animals. The conveyor belt of the Daitari iron ore mine that is 3.8 km long should be raised at least eight meter above ground to allow large mammal like elephants to pass underneath the conveyor belt. Carriage of minerals in the conveyor belt should not be allowed during night to facilitate movement of animals.

Talcher coalfield is the largest repository of power grade coal in India. It occupies a basin in the south-eastern part of the Mahanadi Valley belt of Gondwana Basin and covers an area of 1800 km^2 in Angul district. The strike length of the coalfield in east-west direction is about 80 kms and the width in north-south direction is about 26 kms (Anonymous, not dated). It has a number of underground and open-cast mines managed by Mahanadi Coalfields Ltd. Because of these mines human-elephant conflicts have increased in Angul forest division. Over 80 elephants have been land-locked because of canals and railway lines in the east and State highway 63 (Angul-Chhendipada-Deogarh) and coal mines in between. It is required to provide safe passage for these animals to Ahmallik division (Northern RF and Handapa RF) in the west and to Dhenkanal division (Anantapur RF) in the east. A corridor of at least one km width from Anantapur RF to Northern/Handapa RF, passing through Bulajhar RF, Kanheijena RF, Badkathia RF and Chhendipada RF has to be carved out with the help of local forest officials and wildlife biologist to make Similipal-Satkosia functional. This way only about 100 km^2 of Talcher coalfield, out of 1800 km^2, will be required, which is not a big deal.

CONCLUSION

Similipal-Satkosia corridor is the lifeline for connecting wildlife populations of two tiger reserves (Similipal and Satkosia) and two elephant reserves (Mayurbhanj and Mahanadi). They hold about 35 per cent of elephant population and 90 per cent of tiger population in the state. The elephant population is an important constituent of the Eastern Range habitat for Indian elephant population. It is a fact that the corridor is about 400 km long, and is tenuous due to the presence of mines, settlements, roads, railway lines, *etc.* Still then, it is possible to improve the corridor for movement of wild animals, if adequate mitigation measures are undertaken. The guidelines for ecofriendly measures to mitigate impacts of linear infrastructure on wildlife laid down in WII (2016) should be followed, with appropriate micro-level improvements. The overpasses on canals at present are not wide enough for use by the elephant herds. They are required to be increased to at least 15 m at the centre and 25 m at both ends. A safe passage of one km width has to be planned for elephants in Talcher coalfield area using Anantapur RF, Bulajhar RF, Kanheijena RF, Chhendipda RF and Northern/Handapa RF. All the forest habitats along the corridor have to be augmented with food and water for wild animals.

Satkosia has probably lost all of its breeding tigers and therefore reintroduction of tigers is underway. There are as usual initial setbacks and we should not be deterred for such teething problems. The programme for augmenting tiger population in Satkosia should continue without fail. Similipal-Satkosia corridor will help in gene flow of tigers between the two tiger reserves and the issue of melanism in Similipal may be addressed in one hand and in the other hand movement of tiger in the corridor from Similipal may reinforce tiger population in Satkosia.

Corridors are to be protected and regular monitoring system of biotic and abiotic factors should be in place. Participation of all stake-holders is also equally important. Securing the corridors for free movement of wild animals will ensure genetic flow from one habitat to the other, thereby will offset possible long term negative impact of inbreeding and genetic isolation. Corridors will reduce human-wildlife conflict and will restore ecological processes in the lanscape.

Plate 6.14. Elevated stretches of the NH-44 (earlier NH-7) in Pench tiger landscape, Maharashtra which have become win-win situation for tiger conservation and development. Photo: Pench (Maharashtra) TR archive.

When I was Regional IG, NTCA at Nagpur, in collaboration with scientists from WII, we planned five underpasses and four minor bridges on 37 km strech of NH 44 (earlier NH 7) passing through Pench tiger landscape in Maharashtra part of the NH during 2014-15. Two of the animal underpasses are of 750m long, and others are 300m, 100m and 80m. I personally monitored the location and construction of these underpasses between 2015-18. Now these elevated streches of the NH have become win-win situation for tiger conservation and development (**Plate 6.14**). Earlier large number of mortalities of wild animals including tigers were observed on this NH. Habib *et al.* (2020) report that 18 species of wild animals were found to use the crossing structures, with varying frequencies. Seven species of small mammals, which include Indian hare, jungle cat, and the rare rusty spotted cat were the most frequent users of the underpasses. Among wild ungulates, the

Plate 6.15. A tiger found crossing in one of the underpass structure of NH-44 (earlier NH-7) in Pench tiger landscape, Maharashtra, India. Photo: Akanksha, WII.

Plate 6.16. Barriers against sound and light along elevated streches of the NH-44 (earlier NH-7) in Pench tiger landscape, Maharashtra, India. Photo: Pench (Maharashtra) TR archive.

five major species *viz.*, spotted deer, gaur, nilgai, sambar and wildboar were found to use the structures. Tiger, leopard, wild dog, sloth bear and jackal, the major carnivore species in the landscape, are found using the structures with varying frequencies. Wild dogs are found to use the structures the most, followed by tigers (**Plate 6.15**). A total of 89 tiger crossings were recorded from six of the nine structures, by 11 individual tigers. The height of these underpasses are five meters as the largest animals in Pench landscape are gaurs. We need such elevated structures for National Highways passing through Odisha with height of underpasses should be such that elephants can pass easily. Further these elevated stretches should have barriers against sound and light as has been done in NH-44 (**Plate 6.16**).

Chapter 7

The Epilogue

"Forests are inseparable from wild animals. Both are mutually complementary. Destruction of one leads to destruction of the other."

– Narendra Damodardas Modi (born 17 September 1950-), Honourable Prime Minister of India, in his address to the delegates in the third Asia Ministerial Conference on Tiger Conservation held at New Delhi from 12-14 April 2016.

INTRODUCTION

Before the Aryans set their feet in India, it was full of forests. They entered in this land and cleared forests to make settlements. They started cultivation and Aryan civilization and culture took root. But there was no commercial exploitation of forest until the British established their colony here. The British conquerors first enacted Indian Forest Act and designated reserved forests for sustained exploitation of forest resources. Working Plans were prepared for the reserved forest blocks and annual coup working was carried out engaging contractors to harvest timber for ship making and laying railway tracks. In response to famine in the early 1940s, the Grow More Food' campaign further strained the forest resources. As an incentive for the expansion of cultivation, the Government often conceded proprietary rights to cultivators who encroached on forest land. The regularization of encroachments continued after independence with the practice of granting community lands to poor individuals as part of anti-poverty programmes. Just as food production remained an urgent concern after the Second World War, strategic and commercial exploitation of forests increased after independence in 1947. The building of roads and bridges, which may be claimed to link remote areas with modern India, have served to facilitate extraction of timber from forests once too remote to be reached. It would be clear that the "scientific system of forest management" introduced by the British was intensified by the National Government after independence in 1947, has failed to save the forests from devastation and to provide protection to its denizens, the wild life. India's population in 1901 was about 238.4 million, which has increased by more than four times in 110 years to reach a figure of 1,210 million in 2011 (MHA, 2012). With this ever increasing population the destruction of forests and killing of wild animals are imminent.

BIOSPHERE RESERVES

From the inception of the human civilization, nature and people are inseparable part of the life support system. In the course of time, human pressure on the forest wealth is worsening the quality of natural resources and hence their balance is quite disturbed. The biosphere has been adversely affected and many species have become extinct. Therefore, conservation and development of biodiversity and natural resources are important. UNESCO initiated the idea of the biosphere reserve in 1974 under the aegis of Man and Biosphere (MAB) programme. The objective of the programme was to obtain international cooperation for the conservation of the biospheres. In the first council meeting, the idea of a biosphere reserve was mooted to conserve biodiversity.

Biosphere reserves are areas designated to serve as places to test different approaches to integrated management of terrestrial, freshwater, coastal and marine resources and biodiversity. As a result, they are sites for experimenting with and learning about sustainable development. Therefore, biosphere reserves should fulfill the following three objectives:

(i) In-situ conservation of biodiversity of natural and semi-natural ecosystems and landscapes.

(ii) Contribution to sustainable economic development of the human population living within and around the biosphere reserve.

(iii) Provision of facilities for long-term ecological studies, environmental education and training and research and monitoring.

In order to fulfill the above objectives, biosphere reserves are classified into zones like the core area, buffer and transitional zone. The system of functions is prescribed for each zone.

Zonation of the Biosphere Reserves

One or more core zones: Securely protected sites for conserving biological diversity, monitoring minimally disturbed ecosystems and undertaking non-destructive research and other low-impact uses (such as eco-tourism and education).

A well defined buffer zone(s): Usually surrounds or adjoins the core zones and is used for cooperative activities compatible with sound ecological practices, including environmental education, recreation and applied and basic research.

A flexible transition area or area of cooperation: May contain a variety of agricultural activities, settlements and other uses and in which local communities, management agencies, scientists, non-governmental organizations, cultural groups, economic interests and other stakeholders work together to manage and sustainably develop the area's resources.

To fulfill the main objectives of a biosphere reserve, the support of local people is essential. In 1994, UNESCO recommended 10 important points for this purpose. They are:

(i) Recognize that local support is fundamental to the long-term conservation and sustainable use of biodiversity.

(ii) Ensure that local populations participate as true partners in designing and managing conservation programmes.

(iii) Allow local populations to identify their own socio-economic needs.

(iv) Ensure that people who bear the costs of conservation projects (*e.g.* restrictions on fishing) also receive a huge proportion of benefits (*e.g.* tourist revenues).

(v) Initiate research activities that identify options for sustainable use of biodiversity.

(vi) Use indigenous knowledge to manage protected areas to the possible extent.

(vii) Ensure that local populations have maximum stewardship over local resources (rather than government agencies at the regional and national levels).

(viii) Offer income-earning activities and/or services (*e.g.* improved access to markets, low interest credit, controlled access to resources) to local populations and others with a stake in conservation-development projects.

(ix) Provide local population with the skills and resources needed to make life-style changes necessitated by conservation measures.

(x) Educate local populations about the rationale for conservation and the relationship between conservation actions and benefits.

Biosphere Reserves of India

The first biosphere reserve of India, Nilgiri was established in 1986 and so far 18 reserves have been established in the country. The largest biosphere reserve in the country is the Gulf of Mannar (10,500 km^2) and the smallest one is the Nokrek (820 km^2). The list of biosphere reserves, their area, date of designation and location are furnished in Table 6.1.

World Network of Biosphere Reserves (WNBR)

The WNBR of the MAB Programme promotes North-South and South-South collaboration and represents a unique tool for international co-operation through sharing of knowledge, exchanging experiences, building capacity and promoting best practices (UNESCO, 2012). As on 29 October 2020, there are 714 biosphere reserves in 129 countries in the UNESCO's WNBR (https://en.unesco.org/news/panna-india-fuvahmulah-and-addu-atoll-maldives-join-unescos-world-network-biosphere-reserves).

The WNBR consists of a dynamic and interactive network of sites of excellence. It fosters the harmonious integration of people and nature for sustainable development through participatory dialogue; knowledge sharing; poverty reduction and human well-being improvements; respect for cultural values and society's

Table 6.1. List of biosphere reserves, their area, date of designation and its location in India

Sl. No.	Name of the Biosphere Reserve and Area in km² given in Parentheses	Date of Designation	Location of the Biosphere Reserve Site in the State(s)/Union Territory
1	Nilgiri (5520)	1.8.1986	Part of Wynad, Nagarhole, Bandipur and Madumalai, Nilambur, Silent Valley and Siruvani hills in Tamil Nadu, Kerala and Karnataka
2	Nanda Devi (5860.69)	18.1.1988	Part of Chamoli, Pithoragarh and Almora districts in Uttarakhand
3	Nokrek (820)	1.9.1988	Part of East, West and South Garo Hill districts in Meghalaya.
4	Manas (2837)	14.3.1989	Part of Kokrajhar, Bongaigaon, Barpeta, Nalbari, Kamprup and Darang districts in Assam.
5	Sunderbans (9630)	29.3.1989	Part of delta of Ganges and Brahamaputra river system in West Bengal.
6	Gulf of Manner (10500)	18.2.1989	India part of Gulf of Manner extending from Rameswaram island in the North to Kanyakumari in the South of Tamil Nadu.
7	Great Nicobar (885)	6.1.1989	Southern most island of Andaman and Nicobar Islands.
8	Similipal (5569)	22.6.1994	Part of Mayurbhanj district in Odisha
9	Dibru-Saikhova (765)	28.7.1997	Part of Dibrugarh and Tinsukia districts in Assam.
10	Dehang-Dibang (5111.5)	2.9.1998	Part of Upper Siang, West Siang and Dibang Valley districts in Arunachal Pradesh.
11	Pachmarhi (4981.72) (Area extended on 30.7.2009)	3.3.1999	Part of Betul, Hoshangabad and Chhindwara districts in Madhya Pradesh.
12	Khangchendzonga (2619.92)	7.2.2000	Part of North and West districts in Sikkim.
13	Agasthyamalai (3500.36)	12.11.2001	Part of Thirunelveli and Kanyakumari districts in Tamil Nadu and Thiruvanthapuram, Kollam and Pathanmthitta districts in Kerala.
14	Achanakmar-Amarkantak (3835.51)	30.3.2005	Part of Anuppur and Dindori districts of Madhya Pradesh and Bilaspur district of Chattisgarh.
15	Kachchh (12454)	29.1.2008	Part of Kachchh, Rajkot, Surendranagar and Patan districts in Gujarat.
16	Cold Desert (7770)	28.8.2009	Pin Valley National Park and surroundings; Chandratal and Sarchu; and Kibber Wildlife sanctuary in Himachal Pradesh.
17	Seshachalam (4755.997)	20.9.2010	Seshachalam hill ranges in Eastern Ghats encompassing part of Chittor and Kadapa districts in Andhra Pradesh.
18	Panna (2998.98)	25.8.2011	Part of Panna and Chhattarpur districts in Madhya Pradesh

ability to cope with change - thus contributing to the Millennium Development Goals. Accordingly, the WNBR is one of the main international tools to develop and implement sustainable development approaches in a wide array of contexts.

The mission of the WNBR is to ensure environmental, economic and social (including cultural and spiritual) sustainability through:

(i) the development and coordination of a worldwide network of places acting as demonstration areas and learning sites with the aim of maintaining and developing ecological and cultural diversity, and securing ecosystem services for human well-being;

(ii) the development and integration of knowledge, including science, to advance our understanding of interactions between people and the rest of nature;

(iii) building global capacity for the management of complex socio-ecological systems, particularly through encouraging greater dialogue at the science-policy interface; environmental education; and multi-media outreach to the wider community.

Government of India notified Similipal biosphere reserve on 22 June 1994. Since then it was not included in WNBR. As director of Similipal biosphere reserve, I took initiative to enter Similipal in WNBR and prepared necessary documents in collaboration with Regional Plant Resources Centre, Bhubaneswar and submitted to UNESCO through Ministry of Environment and Forests. UNESCO included the reserve in WNBR during 2009 (22 May 2009). Other biosphere reserves of India in

WNBR are Nilgiri, Nanda Devi, Nokrek, Sunderban, Gulf of Mannar, Great Nicobar, Pachmarhi, Achanakmar-Amarkantak, Agasthyamalai, Khangchendzonga and recently (2020) Panna.

Similipal so far is the only biosphere reserve in Odisha. Mahendragiri hill complex of over 4000 km^2 is now under active consideration of Government of Odisha to propose as the second biosphere reserve of the state. With its unique flora, fauna and heritage, Orissa Environmental Society is campaigning since a decade to bring Mahendragiri to the network of biosphere reserves.

TIGER RESERVES

With the decline of tiger population in the country, the Government of India constituted the Tiger Task Force headed by Karan Singh which submitted its report in 1972. Based on this report tiger reserves were established across India. Initially in 1973-74, there were only nine tiger reserves. Two reserves were added in 1978-79, followed by seven between 1982 to 1990, and 10 between 1992 to 2000. Therefore by the end of 2000, there were 28 tiger reserves in the country. Further 11 tiger reserves were established between 2008 to 2010 and 11 more between 2011 to 2016, making the total tiger reserves so far (Febuary, 2021) established to 50 covering core area or critical tiger habitat of 39,778.82 km^2. The list of tiger reserves, the state to which it belongs and its area (core, buffer and total) are furnished in Table 6.2. As evident from the said Table, tiger reserves account for 72,086.30 km^2 area of the country, amounting to 2.2 per cent of its geographical area. In Odisha there are two tiger reserves, namely Similipal and Satkosia; and Sunabeda is a proposed tiger reserve.

Table 6.2. The name of tiger reserve, the state to which it belongs and its area in India (as on February 2021)

Sl. No.	Name of Tiger Reserve	State	Area of the Core/ Critical Tiger Habitat (in km^2)	Area of the Buffer/ Peripheral (in km^2)	Total Area (in km^2)
1	Bandipur	Karnataka	626.19	287.82	914.02
2	Corbett	Uttarakhand	821.99	466.32	1288.31
	Amangarh (buffer of Corbett TR)	Uttar Pradesh		80.60	80.60
3	Kanha	Madhya Pradesh	917.43	1134.361	2051.791
4	Manas	Assam	526.22	2310.88	2837.10
5	Melghat	Maharashtra	1500.49	1268.03	2768.52
6	Palamau	Jharkhand	414.08	715.85	1129.93
7	Ranthambhore	Rajasthan	1113.364	297.9265	1411.291
8	Similipal	Odisha	1194.75	1555.25	2750.00
9	Sunderbans	West Bengal	1699.62	885.27	2584.89
10	Periyar	Kerala	881.00	44.00	925.00
11	Sariska	Rajasthan	881.1124	332.23	1213.342
12	Buxa	West Bengal	390.5813	367.3225	757.9038
13	Indravati	Chhattisgarh	1258.37	1540.70	2799.07
14	Namdapha	Arunachal Pradesh	1807.82	245.00	2052.82
15	Dudhwa	Uttar Pradesh	1093.79	1107.9848	2201.7748
16	Kalakad-Mundanthurai	Tamil Nadu	895.00	706.542	1601.542
17	Valmiki	Bihar	598.45	300.93	899.38
18	Pench	Madhya Pradesh	411.33	768.30225	1179.63225
19	Tadoba-Andhari	Maharashtra	625.82	1101.7711	1727.5911
20	Bandhavgarh	Madhya Pradesh	716.903	820.035	1536.938
21	Panna	Madhya Pradesh	576.13	1021.97	1598.10
22	Dampa	Mizoram	500.00	488.00	988.00
23	Bhadra	Karnataka	492.46	571.83	1064.29
24	Pench	Maharashtra	257.26	483.96	741.22
25	Pakke	Arunachal Pradesh	683.45	515.00	1198.45
26	Nameri	Assam	200.00	144.00	344.00
27	Satpura	Madhya Pradesh	1339.264	794.04397	2133.30797
28	Anamalai	Tamil Nadu	958.59	521.28	1479.87

Sl. No.	Name of Tiger Reserve	State	Area of the Core/ Critical Tiger Habitat (in km²)	Area of the Buffer/ Peripheral (in km²)	Total Area (in km²)
29	Udanti-Sitanadi	Chattisgarh	851.09	991.45	1842.54
30	Satkosia	Odisha	523.61	440.26	963.87
31	Kaziranga	Assam	625.58	548.00	1173.58
32	Achanakmar	Chattisgarh	626.195	287.822	914.017
33	Dandeli-Anshi	Karnataka	814.884	282.63	1097.514
34	Sanjay-Dubri	Madhya Pradesh	812.571	861.931	1674.502
35	Mudumalai	Tamil Nadu	321.00	367.59	688.59
36	Nagarahole	Karnataka	643.35	562.41	1205.76
37	Parambikulam	Kerala	390.89	252.772	643.662
38	Sahyadri	Maharashtra	600.12	565.45	1165.57
39	Biligiri Ranganatha Temple	Karnataka	359.10	215.72	574.82
40	Kawal	Telangana	892.23	1123.21	2015.44
41	Sathyamangalam	Tamil Nadu	793.49	614.91	1408.40
42	Mukandra Hills	Rajasthan	417.17	342.82	759.99
43	Nawegaon-Nagzira	Maharashtra	653.674	1241.27	1894.94
44	Nagarjun Sagar	Andhra Pradesh	2595.72	700.59	3296.31
45	Amrabad	Telangana	2166.37	445.02	2611.39
46	Pilibhit	Uttar Pradesh	602.7980	127.4518	730.2498
47	Bor	Maharashtra	138.12	678.15	816.27
43	Rajaji	Uttarakhand	819.54	255.63	1075.17
49	Orang	Assam	79.28	413.18	492.46
50	Kamlang	Arunachal Pradesh	671.00	112.00	783.00
	TOTAL		**39.778.82**	**32,307.48**	**72,086.30**

Project Tiger and National Tiger Conservation Authority

Project Tiger is an ongoing Centrally Sponsored Scheme of the Ministry of Environment, Forests and Climate Change (MoEF&CC) providing central assistance to the tiger States for tiger conservation in designated tiger reserves. As per the Tiger Task Force Report, 2005, the National Tiger Conservation Authority (NTCA) came into existence. It is a statutory body of MoEF&CC, with an overarching supervisory/coordination role as provided in the Wildlife (Protection) Act, 1972. Minister, MoEF&CC is the Chairperson and ADG (Project Tiger) serves as the Member Secretary of NTCA. Regional offices of the NTCA have been established at Bengaluru, Guwahati and Nagpur, each headed by an Inspector General of Forests (IGF) and assisted by an Assistant Inspector General of Forests (AIGF). I was IGF at Regional office of NTCA at Nagpur from October, 2014 to February, 2017 and later on became ADG (Project Tiger) and Member Secretary, NTCA during February, 2017 and continued in that post upto April, 2018.

Functions of NTCA

The Wildlife (Protection) Act, 1972 was amended in 2006 to constitute NTCA as per Chapter IVB of the Act. The important functions of NTCA are as follows:

- Approve *Tiger Conservation Plan* as mandated under Section 38 V of the Act.
- Laying down *normative standards* for tiger conservation and tourism activities under Section 38 (O) (1)(c).
- *Assess and disallow* any ecologically unsustainable land use *e.g.* mining, industry *etc.* under Section 38 (O)(1) (b) and 38 (O)(1) (g).
- Ensure measures to address *man-tiger conflicts* and *fostering co-existence* with local people.
- Providing technical inputs on several aspects which include *protection, tiger estimation, patrolling, etc.*
- Any directions issued by NTCA to any person, officer or Authority are legally binding as per Section 38 (O) 2.

Supports under Project Tiger

NTCA under the centrally sponsored scheme of Project Tiger provides technical and financial support to the states for conservation of tiger, its habitat and corridors linking one tiger reserve with another. It has issued normative guidelines for tourism and day to day management of the tiger reserves. National level tiger estimation is taken up once in four years with technical and financial support from NTCA. Eight standard operating procedures (SOPs) have been formulated as on December, 2020 to deal with emergency arising due to straying of tigers in human dominated landscapes, to deal with tiger deaths, for disposing tiger/leopard carcass/body parts, to deal with tiger depredation on livestock, for active management towards reintroduction of tigers from source areas at the landscape level, to deal with orphaned/abandoned tiger cubs and old/injured tigers, interstate coordination for interstate tiger reserves, and to deal with stray/feral dogs in tiger reserves.

Tiger Population Estimation

Tiger numbers were estimated in 2006, 2010, 2014 and 2018 (Table 6.3) based on double sampling technique using camera traps as mentioned in Chapter III (Box-3.1). For 2018 tiger estimation, the process

Table 6.3. Estimated Tiger numbers (>1 year of age) in 2018 for landscapes and States compared with estimates for 2006, 2010 and 2014. Numbers in parentheses are standard error limits (Jhala *et al.*, 2019)

Sl. No.	*State*	*Tiger Population*			
		2006	*2010*	*2014*	*2018*
A.	**Shivalik Hills and Gangetic Landscape:**				
1.	Bihar	10 (7-13)	8	28 (25-31)	31 (26-37)
2.	Uttarakhand	178 (161-195)	227 (199-256)	340 (299-381)	442 (393-491)
3.	Uttar Pradesh	109 (91-127)	118 (113-124)	117 (103-131)	173 (148-198)
	Shivalik-Gangetic	**297 (259-335)**	**353 (320-388)**	**485 (427-543)**	**646 (567-726)**
B.	**Central Indian Landscape and Eastern Ghats:**				
4.	Andhra Pradesh	95 (84-107)	72 (64-79)	68 (58-78)	48 (40-56)#
5.	Telengana	-	-	-	26 (23-30)#
6.	Chhatisgarh	26 (23-28)	26 (24-27)	46 (39-53)*	19 (18-21)
7.	Jharkhand	-	10 (6-14)	3*	5
8.	Madhya Pradesh	300 (236-364)	257 (213-301)	308 (264-352)*	526 (441-621)
9.	Mahararashtra	103 (76-131)	168 (155-183)	190 (163-217)*	312 (270-354)
10.	Odisha	45 (37-53)	32 (20-44)	28 (24-32)*	28 (26-30)
11.	Rajasthan	32 (30-35)	36 (35-37)	45 (39-51)	69 (62-76)
	Central India and Eastern Ghats	**601 (486-718)**	**601 (518-685)**	**688 (596-780)**	**1033 (885-1193)**
C.	**Western Ghats Landscape:**				
12.	Goa	-	-	5*	3
13.	Karnataka	290 (241-339)	300 (280-320)	406 (360-452)	524 (475-573)
14.	Kerala	46 (39-53)	71 (67-75)	136 (119-150)	190 (166-215)
15.	Tamil Nadu	76 (56-95)	163 (153-173)	229 (201-253)	264 (227-302)
	Western Ghats	**402 (336-487)**	**534 (500-568)**	**776 (685-861)**	**981 (871-1093)**
D.	**North East Hills and Brahmaputra Plains Landscape:**				
16.	Arunachal Pradesh	14 (12-18)	-	28*	29*
17.	Assam	70 (60-80)	143 (113-173)	167 (150-184)	190 (165-215)
18.	Mizoram	6 (4-8)	5	3*	0
19.	Nagaland	-	-	-	0
20.	Northern West Bengal	10 (8-12)	-	3*	0
	North East Hills and Brahmaputra	**100 (84-118)**	**148 (118-178)**	**201 (174-212)**	**219 (194-244)**
20.	**Sundarbans (West Bengal)**	**-**	**70 (62-96)**	**76 (62-96)**	**88 (86-90)**
	TOTAL	**1411 (1165-1657)**	**1706 (1507-1896)**	**2226 (1945-2491)**	**2967 (2603-3346)**

** Estimated through scat DNA.*

 # For comparison with previous estimates of Andhra Pradesh, combine Andhra Pradesh and Telengana population estimate of current year.

started since October, 2017. To improve the precesion in tiger estimation it was planned to install camera traps/ two sq km area. In earlier cycles one camera trap was put in four sq km area. It was a huge task and as the Member Secretary, NTCA, I had to plan and execute training exercises and workshops across the tiger states of the country to sensitise the field staff. With the financial and technical support of NTCA and WII (Wildlife Institute of India), sign survey of tigers and prey estimation started over 3,81,400 sq km forest area of the country in November 2017; and camera trapping exercises at 26,838 locations in 141 sites over 121,337 sq km started from December/January, 2017-18. This is a record for using such a large number of camera traps for wildlife survey anywhere in the world and ultimately it entered into 'Guiness World Records'.

Success Stories of NTCA

- In 2018 there are 2967 (2603-3346) tigers compared to 1411 (1165-1657) tigers in 2006, thereby doubling the tiger numbers four years before 2022, commitment made by world leaders at the Tiger Summit at St. Petersburg in 2010. The tiger population of India is the largest for any country, accounting for >80 per cent of the Global population of 3159 adult free-ranging tigers (Jhala *et al.*, 2019).
- Tigers in two of the Indian tiger reserves, namely Sariska and Panna wiped out due to poaching and anthropogenic pressure. Reintroduction of tigers were taken up in these two reserves and now Sariska has 14 tigers and Panna has 42 tigers. Reintroduction of tigers under process in Satkosia tiger reserve of Odisha, Buxa tiger reserve of Bengal, western part of Rajaji tiger reserve of Uttarakhand, and Mukundara tiger reserve of Rajasthan.
- People are engaged in the management of tiger reserves and tiger tourism generating 2.5 million man days for benefit of local people.
- Tiger reserves conserve forest stock, some of India's (about 1/3rd) highly dense forests.
- Verma *et al.* (2015, 2017) identify 25 ecosystem services in selected six tiger reserves in the country and establish that monetary value of flow benefits is Rs.8.3 to17.6 billion/annum (equivalent to US$ 127.9 to 271.1 million/annum) from these six tiger reserves, which accounts for flow benefits of Rs. 50,000 to 1,90,000/ ha/annum (equivalent to US$ 769 to 2923/ha/ annum). The report further states that these six tiger reserves protect and conserve stock value of Rs. 22 to 656 billion (equivalent to US$ 344.5 to 10088.8 million).
- More than 350 rivers originate from tiger reserves.
- Setting a world standard in tiger conservation and developing capacity of management of tiger range countries.

Tiger Reserves and Water Security

The primary objective of tiger conservation is elucidated in the report of the first tiger task force, headed by Karan Singh, as to ensure maintenance of a viable population of tiger in India and to preserve, for all time, areas of biological importance as a national heritage for the benefit, education, and enjoyment of the people (Tiger Task Force Report, 1972). Little was imagined at that time that these tiger reserves would form the catchments/watersheds of many major and minor streams and rivers in the country. They are now the backbone of the irrigation networks and drinking water for many cities and villages. This fact is echoed in the report of the tiger task force headed by Sunita Narain, which states that without the reserves, water security would certainly be further compromised, and this ecological service rendered by the reserves is not rewarded (Tiger Task Force Report, 2005). The report states that the dam in Pench tiger reserve provides water to the city of Nagpur in Maharashtra. The dam in Periyar tiger reserve, in Kerala, is used for towns and agriculture in the neighbouring state of Tamil Nadu. The town of Sawai Madhopur in Rajasthan obtains water through deep tube-wells at the edge of the Ranthambhore tiger reserve. The city of Tirunelveli in Tamil Nadu accesses its water from the dam in the Kalakad-Mundanthurai tiger reserve. Table 6.4 summarises the catchments/ watersheds of river systems coming under major tiger reserves.

Table 6.4. Tiger reserves in India and name of the river system forming the catchments

Sl. No.	Name of the Tiger Reserve	Rivers/Streams
1	Anamalai	1. Puvar nadhi 2. Thorakadavu Ar
2	Bandhavgarh	1. Charanganga 2. Damnar 3.Umrar 4.Chamkubi 5. Banbei *etc.*
3	Bandipur - Nagarhole	1. Nugu 2. Kabini 3. Moyar

Sl. No.	Name of the Tiger Reserve	Rivers/Streams
4	Bhadra	1. Bhadra 2. Somavahini
5	Buxa	1. Sankosh 2. Rydak 3. Jainti 4. Bala 5. Dima 6. Pana
6	Corbett	Catchment of river Ganges through 1. Ramganga 2. Sona 3. Mandal 4. Palain 5. Kosi
7	Dampa	1. Saza 2. Kelsalam 3. seling 4. Aivapui
8	Dudhwa - Katarniaghat	1. Mohana 2. Suheli 3.Ghaghra
9	Indravati	1. Indravati
10	Kali	1. Kali 2 Kaneri 3. Pandhari 4. Vaki
11	Kalkad - Mundathurai	1. Tambraparani 2. Ramanadi 3. Karayar 4. Servalar 5. Manimuthar 6. Pachayar 7. Kodaiyar 8. Kadnar 9. Kallar *etc.* 14 rivers/streams and 12 tributaries there of
12	Kamlang	1. Kamlang 2. Lati Lang 3. Tawa river 4. Tawai river
13	Kanha	1. Halon 2. Surpan 3. Banjar 4. Phen
14	Kaziranga	1. Brahmaputra 2. Difalu 3. Moradifalu
15	Manas	1. Sankosh 2. Dhansiri 3. Manas 4. Pagladia
16	Melghat	Tributaries of river Tapti, *e.g.*, 1. Khandu 2. Khapra 3. Sipna 4. Gadga 5. Wan 6. Dolar
17	Mukundara	1. Chambal 2. Kalisindh 3. Aru
18	Nagarjunasagar - Srisailam	1. Krishna 2. Baireni 3. Gundlakamma 4. Sagileru 5. Paleru 6. Thunga
19	Namdapha	Catchment of Brahmaputra through the Noa-Dehing river system
20	Nameri	1. Jia Bhoreli 2. Bordikarai
21	Orang	1. Brahmaputra
22	Pakke	1.Kameng 2. Pakke 3. Papu 4. Nameri 5. Denai 6. Diji 7. Upper Dekorai 8. Khari 9. Lalung 10. Doigurung 11. Romoni

Sl. No.	Name of the Tiger Reserve	Rivers/Streams
23	Palamu	1. North Koel
		2. Burha
		3. Auranga
24	Panna	1. Ken
25	Pench (M.P.)	1. Pench
26	Pench (Mharashtra)	1. Pench
		2. Dhobi Gotta
		3. Kukada
27	Periyar	1. Periyar
28	Pilibhit	1. Sharda
		2. Mala
		3. Katna
29	Ranthambhor	1. Kalisil
		2. Chakal
		3.Balra
		4. Kandali
		5. Banas
30	Sahyadri	1. Koyana
		2. Solashi
		3. Kandati
		4. Varna
		5. Ram
31	Satkosia	1. Mahanad
		2. Bhimdhara
32	Satpura	1. Narmada
		2. Tawa
		3. Denwa
		4. Nagdwari
		5. Sonbhadra
		6. palak
		7. Koti
33	Similipal	1. Budhabalanga
		2. Palpala
		3. East Deo
		4. West Deo
		5. Khairi
		6. Bhandan
		7. Khadkei
		8. Salandi
		9. Sanjo
		10.Thakthaki
		11.Tel and Sim
34	Sundarbans	Innumerable tidal Rivers
35	Tadoba - Andhari	1. Andhari
		2. Human
		3. Kolsa
36	Valmiki	1. Great Gandak
		2. Masan
		3. Panchnad
		4. Manor
		5. Bhapsa
		6. Kapan

Forests conserved due to establishment of tiger resrves play an important role in soil and water conservation in the origin/catchments of rivers and streams. FAO (2005) emphasises the role of forests to help maintain healthy aquatic ecosystems and provide reliable supplies of clean freshwater. The report further says that not only do they filter and clean water — forests also help prevent soil erosion, reduce sedimentation in reservoirs, and mitigate the risks of mudslides and floods, all problems that can threaten downstream water supplies. Furthermore, while forests themselves consume water, they also improve infiltration rates, thereby helping recharge underground aquifers. Loss of tiger reserve can adversely affect freshwater supplies, threatening the food security of millions of people and hindering their ability to earn a living for themselves and their families.

Economic Valuation of Similipal Tiger Reserve

The Similipal tiger reserve has eleven major watersheds as enumerated in Table 6.4. These rivers provide water to millions of people in Eastern India, comprising parts of Odisha, West Bengal and Jharkhand. They are sources of water to such townships as Baripada (population as per 2011 census: 1,16,849), Balasore (1,77,751), Bhadrak (1,29,228), Rairangpur (47,294), Karanjia (22,865), and Udala (13,152). There are as many as seven irrigation dams on these rivers- namely, Kalo, Gangahar, Sunei, Haldia, Nedam, Hadgarh and Deo. However, none of the townships and people, located at a significant distance from the reserve utilising the fruits of conservation, pay for the sacrifices made by the people of the ecosystem. Besides conserving the soil and moisture, the Similipal forests are home to a variety of flora and fauna, both rare and endemic as enumerated in Chapter II and III. They are also the source of numerous wood and non-wood forest products, the valuation of which is difficult to determine. In fact, no mechanism currently exists to account for all of the services which standing forests and protected areas provide to the nation.

Verma *et al.* (2019) identify the following ecosystem services for Similipal and calculate the economic valuation as per the price prevailing during 2017-18 while 1 US $ = Rs.65/- as follows.

i. Employment Generation

The total number of man-days generation from labour activities in Similipal is about 11,000 man-days per year which account for the economic value of employment generation to be Rs.602.25 million per annum assuming a wage rate of Rs.150/- per day.

ii. Fodder/Grazing

The total annual quantity of fodder harvested is equal to 89,831 tonnes per year. Assuming an average price of Re 1 per kilogram of fodder, the grazing benefits come to Rs.89.83 million per annum.

iii. Standing Timber (Stock)

From the forest inventory data base of the Forest Survey of India, it is estimated that approximately 14.18 million cubic meters of standing stock of timber are there in Similipal tiger reserve. Using an average price of Rs.25,000 per cubic meter of timber after discounting for transportation and maintenance cost, the standing stock has value of Rs.354.52 billion.

iv. Genepool Protection

Using estimates of economic value of genepool protection for tropical forests (Rs.100,122/ha/annum) and cropland (Rs.68,772/ha/annum) from global meta-analysis study, the annual economic value of this service from 256,579.37 ha of forests and 7,874.81 ha of cropland is estimated to be Rs.26.23 billion/annum.

v. Carbon Storage

Similipal tiger reserve stores approximately 25.72 million tonnes of carbon, the economic value of which is calculated as Rs.143.80 billion.

vi. Carbon Sequestration

The total carbon sequestered in the forests of Similipal tiger reserve to be 556.42 kilo tonnes annually which is in monetary terms calculated to be Rs.22.83 billion per annum.

vii. Water Provisioning

The total water yield volume from Similipal amounts to 3816.08 million cubic meters. Using the monetary value of Rs.18.43 per cubic meter, the economic value of water provisioning service is estimated to be Rs.70.33 billion per annum.

viii. Water Purification

Estimating around two million people depending on water from Similipal, it is proposed that the annual drinking water requirement to be about 5,159.64 kilo litres. With an average cost of treating water for domestic supply as Rs.10 per cubic meter, the avoided cost of water purification for drinking water is about Rs.29.2 million per annum.

ix. Soil Conservation/Sediment Retention

To estimate the economic value of soil loss avoided by the forests of Similipal, the cost of dredging/de-siltation has been considered. Presuming the cost estimate of Rs.60 per cubic meter for dredging/de-siltation with an assumed weight of soil as 1.2 tonnes per cubic meter, the economic value thus derived is Rs.1,160.28 million per annum.

x. Nutrient Retention

Using the total soil loss avoided and soil nutrient Nitrogen (N), Phosphorous (P) and Potassium (K), it is found out the total quantity of nutrients retained is about 53,837.36 tonnes of N, 1,021.05 tonnes of P and 19,1447.50 tonnes of K annually. Taking the substitutes of these nutrients as their respective fertilizers, *i.e.* Urea for N, DAP for P and Muriate of Potash for K, the monetary value has been derived and the total value of nutrient loss prevented by the forests of Similipal reserve comes out to be Rs.2,614.12 million per annum.

xi. Biological Control

Using estimates of economic value of biological control for tropical forests (Rs.726/ha/annum) and cropland (Rs.2,178/ha/annum) from a global meta-analysis study, it is estimated that the economic value of biological control service to be Rs.203.48 million per annum.

xii. Pollination

Using estimates of economic value of pollination for tropical forests (Rs.1,980/ha/annum) and cropland (Rs.1,452/ha/annum) from a global meta-analysis study, the economic value of pollination service becomes Rs.519.46 million per annum.

xiii. Habitat for Species

Using estimates of economic value of habitat service for tropical forests (Rs.2,574/ha/annum) from a global meta-analysis data, the economic value of habitat for species service comes to Rs.660.44 million per annum.

xiv. Gas Regulation

Using estimates of economic value of gas regulation service for tropical forests (Rs.792/ha/annum) from a global meta-analysis data, the economic value of this gas regulation services is Rs.203.21 million per annum.

xv. Climate Regulation

Using estimates of economic value of climate regulation service for tropical forests (Rs.134,904/ha/annum) and cropland (Rs.27,126/ha/annum) from a global meta-analysis study data, it is estimated that the economic value of climate regulation is Rs.34.82 billion per annum.

Therefore, Similipal reserve provides flow benefits (employment generation, fodder, carbon sequestration, water provisioning, genepool conservation, water purification, soil conservation, nutrient retention, habitat for flora and fauna, biological control, pollination, gas regulation *etc.*) worth Rs.160.30 billion per annum, *i.e.* 0.59 million per hector per annum. Similarly stock benefits (standing timber and carbon storage) amount to Rs.498.33 billion. The report further states that approximately 5.15 per cent of flow benefits accrue at the local level, 20.71 per cent at the national level and 74.13 per cent at the global level.

However, Verma *et al.* (2019) have not considered important ecosystem services of Similipal reserves, namely fishing, fuel wood, timber flow, non-timber forest produce, micro-nutrient retention, moderation of extreme events, nursery function, cultural heritage, recreation, spiritual tourism, research, education and nature interpretation due to data deficiency and in some cases non availability of proper mechanism to value these services.

PERSONS WHO BROUGHT SIMILIPAL IN THE LIMELIGHT

Many frontline staff, forest officers, researchers, *Vana Suraksha Samitee* and Eco-development Committee members, *Sabuja Vahinee* volunteers, and well wishers worked hard for the development, protection and management of Similipal. It is pertinent to mention the contribution briefly of following persons who brought laurels to Similipal for which it remained in the limelight.

Saroj Raj Choudhury is undoubtedly the forerunner among them. He was the founder field director of the reserve from 1973 until his sad demise in 1982. While he was in the Forest Research Institute, Dehradun, heading wildlife training, he pioneered the all India census of tiger in 1972 using pugmark technique (Srivastava, 1979). Result of this census showed that the tiger population was dwindling; and therefore he along with Kailash Sankhala drafted proposal for Project Tiger with the blessings from the then Prime Minister, Indira Gandhi. That is how Project Tiger came into existence and Kailash Sankhala became the first director of the Project at Delhi. Initially, nine tiger bearing forests including Similipal were notified as tiger reserves in 1973. Sri Choudhury wrote the first management plan of Similipal tiger reserve (Choudhury, 1974) which is the cornerstone for preparation of its subsequent management plans. In 1974 he rescued a tiger cub from Kharias (nomadic tribe inside Similipal) and hand-reared it to a full-grown tigress. The then Chief Minister Smt. Nandini Satpathy named the cub as 'Khairi', after river Khairi in Similipal. Khairi lived freely in the residence of Sri Choudhury at Jashipur, on the outskirt of Similipal and it was also visiting wilderness areas of Similipal often (Choudhury, 1999). Such an experiment of living with a laissez-faire tigress was first of its kind and Similipal drew the attention of people world over and tourist inflow to Similipal was the highest in comparison to other tiger reserves during that time. Foot-falls to idyllic Jashipur increased and it became eminent due to visit of celebrities, governors, and people at high places. Sri Choudhury was invited to national and international seminars and workshops to talk about Similipal and its princess Khairi. Unfortunately, Khairi had to be euthanised in 1981 when it showed the symptoms of rabies. Sri Choudhry could not bear the loss of his beloved tigress and he expired due to heart attack in 1982. Sri Choudhury has the distinction of becoming the first forest officer in the country receiving *Padma Shri* posthumously, the fourth highest civilian honour by Government of India. A documentary film on Khairi was televised by Door Darshan, the official TV channel of Government of India. E. P. Gee once stated on Similipal that he saw very little wildlife during his sojourn to this forest as all the grassy valleys were occupied by villagers with their cattle and cultivations (Gee, 1964). But the sincere efforts of Saroj Raj Choudhury, and successive field directors (who followed the foot steps of Sri Choudhury), have made Similipal teeming with wild animals, more so in south Similipal.

Lala Aswini Kumar Singh is continuously working in the field of wildlife conservation, research and education for over 44 years, from 1975, starting from the Gharial project at Tikarpada in Satkoshia landscape. In 1979 he conducted an on-foot survey of rivers in Similipal for possible rehabilitation of mugger crocodiles as it was thought to be extinct here since 1975. On return from Government of India deputation in 1987, he permanently shifted to Similipal, and started staying at the campus of crocodile project at Ramatirtha near Jashipur till 2003. He developed techniques for improved captive breeding, better survival, planned territorial resource partitioning among adult crocodiles, and devised methods for capture and translocation of crocodiles moving near human habitations. Now, mugger crocodiles are a part of Similipal. Simultaneously he worked for monitoring of tigers in Similipal. In 1999, the WWF-India recognized Dr Singh with WWF medal and award for significant and innovative contributions in tiger conservation. The WWF-India published his works on improved pugmark tracking and colour aberrations in tiger (Singh 1999, 1999a). In August 1987, Dr Singh proposed the first plan for elephant conservation in Odisha in Project mode to address issues relating to human interface, design of elephant reserves and corridors. The project is viable till date. Some of the other contributions of Dr Singh from Similipal are the checklist on birds and large herbivores, studies on tourism pattern, training of students, public and tourist guides through nature education, planning of schemes on Similipal tiger reserve, Similipal biosphere reserve, Mayurbhanj elephant reserve, ecodevelopment, eco-tourism, the distribution of fauna at higher canopy, and the innovated method for estimation of large herbivores. He plannned method for assessing elephant population in Similipal from 1989, which later with changes became the method for estimating elephants in eastern region and the country. In most of his work Dr Singh used simple and refined traditional knowledge. He produced more than 300 publications, and made applied research intelligible at grass root level, for which he was recognized by the Odisha Government with Biju Patnaik Award for Wildlife Conservation, 2013.

Sarat Chandra Misra, a civil engineer by profession (superannuated as Chief Engineer, Civil of Odisha in 1998) is an orchidologist par excellence. He has established himself through his books, Orchids of Orissa (2004), Orchids of India – a glimpse (2007), Orchids of Odisha – a handbook (2014), and Orchids of India – a handbook (2019). Prof. Dariusz L. Szlachetko of University of Gdansk, Poland, a world authority on orchids, writes in the Foreword of Misra (2019) that he is indubitably a leading expert on Indian orchids. Dr Misra studied orchids of Similipal from 1974 to 2016 and recorded 41 terrestrial and 53 epiphytic species of orchids (Misra 2014). This included *Eria meghasaniensis*, a new species, and *Goodyera thailandica* a new record for India. His posting in the erstwhile Similipal Forest Development Corporation between 1986-1987 as Project Manager (Engineering), has helped much in the pursuit of orchid study. He constructed roads and bridges to improve connectivity inside Similipal. Dr Misra has published 60 papers in various journals in India and abroad. He has recorded 1484 taxa of orchids in India, 143 taxa in Odisha and 94 in Similipal. His passion for orchids earned him Doctorate of Science (*Honoris causa*) in Orchidology from Utkal University in 1996. He was also awarded the Gopinath Panigrahi Memorial Award in Taxonomy (2007) instituted by the Odisha Botanical Society, Usha Vij Memorial Award (2014) and Dr T. N. Khoshoo Memorial Lecture Award (2017), both instituted by the Orchid Society of India. The Odisha Bigyan Academy awarded him the Senior Scientist Award for the year 2016. He has continued studies on orchid of Similipal and discovered a new record for Similipal, namely *Eulophia mackinnonii* (Misra and Kar, 2017).

Amarendra Lal Bose, a veteran journalist of Mayurbhanj made sustained campaigning in the print media over 40 years for conservation of wildlife, emphasizing flora and fauna of Similipal. His writings in the Telegraph, the Statesman and the Times of India published from Kolkata (then Calcutta) about Khairi, tigers of Similipal, and its enchanting landscape have enticed tourists and academicians to this reserve. He was made Honorary Wildlife Warden thrice for Mayurbhanj district by Government of Odisha and he was recognized by the Odisha Government with Biju Patnaik award in 2010, the highest award for wildlife conservation in Odisha for his lifetime contribution making Similipal well-known within and outside Odisha. He was quite active with his pen writing well-researched pieces about Similipal and biodiversity until he breathed his last in 2014.

Prasanna Kumar Dash, a member of the Odisha Legislative Assembly from Mayurbhanj for record nine times spanning over 40 years between 1952 to 2000, out of which he was Speaker of the Assembly for six years (1984 to 1990) and a Cabinet Minister looking after science, technology and environment for four years (1995 to 1999). He raised the conservation issues of Similipal in the Odisha Assembly, and a House Committee was formed to suggest measures to address the issues. On his insistence, felling of trees by Similipal Forest Development Corporation was stopped in Similipal and a proposal was mooted to declare Similipal as a biosphere reserve. While he was the Speaker of the Odisha Assembly, he organised a national level workshop on Similipal at Baripada, the district Headquarters of Mayurbhanj on May 22 and 23, 1985. The Union Minister on Environment and Forest, Chief Minister of Odisha and several other ministers, high level officials of the centre and the State along with scientists and academicians participated in this two days workshop, emphasizing Similipal to be a biosphere reserve, and the dream of Shri Dash became true when Similipal was notified as the 8th biosphere reserve of India on June 22, 1994. Prasanna Kumar Dash was the President of Odisha

Environmental Society since its inception in 1982 and remained so up to 2000. Similipal lost its fervent political adherent on his demise in 2005.

Sundara Narayana Patro, teaching Botany and Environmental Science to degree students in the Maharaja Purna Chandra College at Baripada (1975 to 1980) was known for his quality of teaching and love for Similipal and its biodiversity. He joined Prasanna Kumar Dash to campaign for the protection of Similipal. The camaraderie continued even after the transfer of Dr Patro to Phulbani, where he formed Orissa Environmental Society in 1982 with Prasanna Kumar Dash as President of the Society, and he himself became the Secretary. Odisha Environmental Society is the oldest non-governmental organization in the State of Odisha in the field of environment, forest and wildlife. Dr Patro compiled and edited two books on Similipal, namely Patro and Misra (1985), and Tripathy and Patro (1997). He was the NGO-force for recognition of Similipal biosphere reserve. For his sustained life-time contribution and campaign for the cause of conservation of wildlife and biodiversity of Odisha he received Bijupatnaik Award for Wildlife Conservation, 2009. He also received Ashoka Fellow for his innovative ideas and his relentless efforts to bring Similipal to the biosphere reserve network of the country. He is still active and at present he is the President, Orissa Environmental Society.

CONCLUSION

There are 18 biosphere reserves and 50 tiger reserves in India. Biosphere reserves are special entities (sites) for both the people and nature and are living examples of how human beings and nature can co-exist while respecting each other's needs. It may be noted that biosphere reserves are not a substitute or alternative, but a re-enforcement to the existing protected areas. Tiger reserves are protected areas having viable tiger population where the species can persist for the future with necessary managerial inputs. Initially, tiger reserve was a management concept like biosphere reserve, but in 2006, the Wildlife (Protection) Act, 1972 was amended to provide legal support to tiger reserves. The nine tiger reserves including Similipal were first established in India in 1973. Nilgiri biosphere reserve established in 1986 is the first biosphere reserve in the country and it includes Bandipur tiger reserve. Now as many as seven tiger reserves are part of biosphere reserves in India.

It is now established that conservation efforts of the tiger in India have helped maintain the country's water security as well. At the same time, the protection of tiger reserves results in asymmetry in the distribution of costs and benefits. While the benefits accrue most significantly to people living outside of the reserves, the costs are borne primarily by local communities. Therefore, such conservation efforts are not sustainable. Instead, some arrangement must be made to redistribute these benefits to the population equitably and only then will the world's biodiversity have a chance to survive the future.

Similipal is the eighth largest biosphere reserve and fourth largest tiger reserve in the country. The geologic formation consisting of three concentric cups of metamorphic rocks inter-bedded with sub-metamorphic layers provides immense water holding capacity to Similipal and for that matter moist deciduous and semi-evergreen categories of vegetation together constitute most part of its forest which give rise to multitude of flora and fauna.

Similipal is located in the most underdeveloped and underprivileged backward area of India and is a bright example of coexistence of poverty and biodiversity where people are basically tribal and lead traditional ways of lifestyle and biodiversity-based livelihood. Now, because of over exploitation, there is a great threat to biodiversity of Similipal.

The experiment of engaging local community in the protection and management of the reserve in the form of deploying able bodied youth as *Sabuja Vahinee* or Green Brigade has yielded result. As stated earlier, there was decrease of 252.71 km^2 dense forests in six years from 1997 to 2003 and increase of 347.87 km^2 dense forests in three years from 2003 to 2006. It indicates that management between 2003 and 2006 was conducive to the protection of forest as during this time locals were involved in the protection and management of Similipal. It is suggested that instead of considering the large multitude of human population in and around Similipal to be detrimental to conservation of flora, fauna and landscape of Similipal, they may be used as resources for the management of the reserve. The local people have to be engaged in monitoring of keystone species, and protection and management of forest and its wild denizens, failing which we stand to lose this unique ecosystem forever.

References

Achard F, Eva H D, Stibig H-J, Mayaux P, Gallego J, Richards T and Malingreau J-P (2002). Determination of deforestation rates of the world's human tropical forests. *Science* **297**: 999–1002.

Achard F, Eva H D, Mayaux P, Stibig H-J and Belward A (2004). Improved estimates of net carbon emissions from land cover change in the tropics for the 1990s. *Glob. Biogeochem. Cycles* **18** GB2008 doi: 10.1029/2003GB002142.

Agrawal A and Redford K (2009). Conservation and displacement: An overview, *Conserv Soc* **71**: 1-10.

Alam A, Behera K K, Vats S and Iqbal M (2013). A preliminary study on bryodiversity of Similipal biosphere reserve (Odisha), India. *Archive for Bryology* **157**: 1 - 9.

Ali S M, Chattopadhyaya S, Kankane P L and Gayen N C (1988). A new species of palm civet from Orissa. *India. Rec. Zool. Surv. India*, **85**: 45-48.

Anderson T M, vonHoldt B M, Candille S I, Musiani M, Greco C, Stahler D R, Smith D W, Padhukasahasram B, Randi E, Leonard J A, Bustamante C D, Ostrander E A, Tang H, Wayne R K and Barsh G S (2009). Molecular and evolutionary history of melanism in North American Gray Wolves. *Science* **323**: 1339-1343.

Andheria A P, Karanth K U and Kumar N S. (2007). Diet and prey profiles of three sympatric large carnivores in Bandipur tiger reserve, India. *Journal of Zoology* **273**: 169–175. doi: 10.1111/j.1469-7998.2007.00310.x

Andrews H V. (1999). Status and distribution of mugger crocodile in Tamilnadu. *ENVIS (Wildlife and Protected Areas)* **2** (1): 44-57.

Anonymous (1990). Improved Technique used to estimate population of major carnivores during 1990. *Technical Report-4*. ORI/N/SIM. May, 1990 Similipal Tiger Reserve, Orissa, pp. 24

Anonymous (1999): *Crocodile-Research: Conservation.* Based on the compilation by S.S.Srivastava and L.A.K.Singh, Similipal Tiger Reserve, pp. 21.

Anonymous (2003). *Wildlife Conservation in Odisha.* Wildlife Organisation, Forest Department, Odisha.

Anonymous (not dated). *Annual environmental monitoring report of Talcher coalfields for 2014-15.* A report of Mahanadi Coalfields Limited, Sambalpur and Central Mine Planning and Design Institutes Ltd, Bhubaneswar

Ashton P (2014). *On the forests of tropical Asia: Lest the memory fade.* Kew publishing, Royal Botanic Gardens, Kew, UK.

Baliarsingh B K, Kosygin L, Swain S K and Nayak A K (2013). Species diversity and habitat characteristics of freshwater fishes in the Similipal biosphere reserve, Odisha with some new records. Biological Forum **5** (2): 64-70.

Barik B (2003). *Tribal population and settlement inside Similipal biosphere reserve: Their impact and management.* Dissertation submitted in partial fulfillment for the award of Master of Philosophy in Geography, Post Graduate Department of Geography, Utkal University, Bhubaneswar, Odisha, pp. 26.

Barnes R F W (2001). How reliable are dung counts for estimating elephant numbers? *African Journal of Ecology* **39**: 1-9

Barnes R F W, Blom A, Alers M P T and Barnes K L (1995). An estimate of the numbers of forest elephants in Gabon. *Journal of Tropical Ecology* **11**: 27-37.

Barsh G S (1996). The genetics of pigmentation: From fancy genes to complex traits. *Trends Genet* **12**: 299-305.

Behera D, Rath C C and Mohapatra U (2013). Medicinal orchids in India: A review. *Floriculture and Ornamental Biotechnology* **7**(*1*): 53-59.

Behura B K (1987). The Langurs of India. *Science Reporter*. November, CSIR, New Delhi, Pp. 574-579.

Behura B K (2004). Preface. In *Asian elephants: Past, present and future* (Author: Swain D), International Book Distributors, Dehradun, India.

Beermann F, Orlow S J and Lamoreux M L (2004). The *Tyr* (*albino*) locus of the laboratory mouse. *Mamm. Genome* **15**: 749-758.

Benson S B (1933).Concealing coloration among some desert rodents of the Southwestern United States. *Uni. Calif Publication Zool.* **40**: 1-69.

Bhakta S, Pattanaik L, Dutta P, Sahu E and Bastia A K (2014). Diversity of corticolous algae from Similipal biosphere reserve, Mayurbhanj, Odisha. *Phykos* **44** (*1*): 9-16.

Borges R M (1989). *Resource heterogeneity and the foraging ecology of the Malabar Giant Squirrel* (*Ratufa indica*). Ph.D. dissertation. University of Miami, Florida.

Borges R M, Mali S, Ranganathan S (1992). *The status, ecology and conservation of the Indian giant squirrel (Ratufa indica).* Draft Technical Report No,1 (submitted to Wildlife Institute of India). Indo-US Project.

Bose S (1980). *Revised working plan for the reserved forests of Baripada forest division for the period 1973-74 to 1992-93*. Government of Odisha, Forest Department, Pp. 355+xiii.

Boswell T and Takeuchi S (2005). Recent developments in our understanding of the avian melanocortin system: Its involvement in the regulation of pigmentation and energy homeostasis. *Peptides* **26**: 1733-1743.

Brahmachry R L, Sarkar M P and Dutta J (1990). The aroma of rice ... and tiger. *Nature* **344**: 26.

Brandon-Jones D (2004). A taxonomic revision of the langurs and leaf monkeys (Primates: Colobinae) of south Asia. *Zoos' Print Journal* **19**(8): 1552-1594.

Brockington D and Igoe J (2006). Eviction for conservation. A global overview. *Conserv Soc* **43**: 424-470.

Brown S (1997). Estimating biomass and biomass change of tropical forests: A primer. *FAO Forestry Paper no. 134* Rome.

Buckland S T, Thomas L, Marques F F C, Strindberg S, Hedley S L, Pollard J H, Borchers D L and Burt M L (2002). Distance sampling: Recent advances and future directions. In *Quantitative Methods for Current Environmental Issues* (eds. V. Barnett, A. El.Shaarawi, C. Anderson and P. Chatwin), Springer-Verlag, Pp.79-97.

Burnham K P and Anderson D R (2002). Model *selection and multimodel inference* [2nd ed.], Springer-Verlag, New York, Pp. 488.

Carbone C, Christie S, Conforti K, Coulson T, Franklin N, Ginsberg J R, Griffiths M, Holden J, Kawanish K, Kinnaird M, Laidlaw R, Lynam A, Macdonald D W, Martyr D, McDougal C, Nath L, O'Brien T, Seidensticker J, Smith D J L, Sunquist M, Tilson R and Wan Shahruddin W N (2001). The use of photographic rates to estimate densities of tigers and other cryptic mammals. *Animal Conservation* **4**: 75-79.

Castle W E and Allen G (1903). The heredity of albinism. *Proc Am Acad Arts Sci* **38**: 603-621.

Champion H G and Seth S K (1968). *A revised survey of the forest types of India*, Government of India, New Delhi, Pp.404+xxvii.

Chandrashekharan K and Cheeran J V (1996). Use of antiandrogen in controlling musth in captive elephants.*Zoos' Print* 11(4): 25-26.

Chintala S, Li W, Lamoreux M, Ito S, Wakamatsu K and Sviderskaya E (2005). Slc7a11 gene controls production of pheomelanin pigment and proliferation of cultured cells. *Proc Natl Acad Sci* **102**: 10964-10969.

Choudhury S R (1970). Let us count tiger. *Cheetal* **14**(*2*): 41-51.

Choudhury S R (1974). *Management plan of Similipal tiger reserve,* Unpublished, Similipal Tiger Reserve, Baripada, Odisha, India.

Choudhury S R (1999). *Khairi: The beloved tigress.* Natraj Publishers, Dehra Dun, India, Pp. 192.

Choudhari A B and Chakrabarti K (1972). Wildlife biology of the Sundarbans forests: Observations on tigers. *The Cheetal* **15** (1): 11-30.

Cobden-Ramsay L E B (1910). *Feudatory States of Orissa*. The university of Michigan, Reprint Firma KLM, 1982.

Colchester M (2004). Conservation policy and indigenous peoples. *Cult Surviv Q* **28**: 17-22.

Cuenot L (1904). L'heredite de la pigmentation chez les souris. *Arch Zoo Exp Get* **2**: 45-56.

Dasgupta B N (1946). *Working plan for the Similipahar and Noto Reserves, Mayurbhanj State (1947-48 to 1966-67)*. Unpublished, Pp.40.

Dash J (1998). *Human ecology of foragers (A study of the Kharia (Savara), Ujia (Savara) and Berhor in Similipal hills)*. Commonwealth, New Delhi, pp.483+xx.

Das I and Chanda S K (1997). *Philautus sanctisilvaticus* (Anura: Rhacophoridae), a new frog from the sacred groves of Amar- kantak, central India. *Hamadryad* **22**: 21–27.

Das R N and Das N (1968). Living conditions of tribals of Similipal hills. *Adivasi* (special issue) **10**(2): 1-69.

Das S K, Ahmed R A, Sajan S K, Dash N, Sahoo P, Mohanta P, Sahu H K, Rout S D and Dutta S K (2012). Diversity, distribution and species composition of odonates in buffer areas of Similipal tiger reserve, Eastern Ghat, India. *Academic Journal of Entomology* **5**(*1*): 54-61.

Datta A (1993). *Space-use patterns of the Indian giant squirrel (Ratufa indica) in relation to food availability in Bori Wildlife Sanctuary, Madhya Pradesh, India*. M.Sc. thesis submitted to Saurashtra University.

Davis G E (2005). National Park stewardship and 'vital signs' monitoring: a case study from Channel Islands National Park, California. *Aquatic Conservation : Marine and Freshwater Ecosystems* **15**: 71-89.

Dawson S (1990). *A model to estimate density of Asian elephants (Elephas maximus) in forest habitats*. MSc thesis, University of Oxford.

Dawson S and Dekker A J F M (1992). *Counting Asian elephants in forests*. Regional Office for Asia and the Pacific. F.A.O., Bangkok.

DeFries R S, Houghton R A, Hansen M C, Field C B, Skole D and Townshend J (2002). Carbon emissions from tropical deforestation and regrowth based on satellite observations for the 1980s and 1990s. *Proc. Natl Acad. Sci. USA* **99**: 14256–61.

Deepak V, Narayanan S, Sarkar V, Dutta S K and Mohapatra P P (2019). A new species of *Ahaetulla* Link, 1807 (Serpentes: Colubridae: Ahaetullinae) from India. *Journal of Natural History*, **53**(*9-10*): 497-516.

Dice L R and Blossom P M (1937). Studies of mammalian ecology in Southwestern North America, with special attention to the colors of desert mammals. *Publ Carnegie Inst Washington* **485**: 1-25.

Dinerstein E, Loucks C, Wikramanayake E, Ginsberg J, Sanderson E, Seidensticker J, Forrest J, Bryja G, Heydlauff A, Klenzendorf S, Leimgruber P, Mills J, O'Brien G T, Shrestha M, Simons R and Songer M (2007). The fate of wild tigers. *BioScience*, **57**(6)**:** 508-14.

Dodson K M (1982). *Genetic linkage relationship among several coat color mutations in the deer mouse (Peromyscus maniculatus)*. Masters thesis, University of South Carolina.

Donovan T M and Hines J (2007). Exercises in occupancy modeling and estimation. University of Vermont, Burlington, VT. Available from http: //www.uvm.edu/envnr/vtcfwru/spreadsheets/occupancy/occupancy.htm (Accessed 2009 March 12).

Dutta S K (2003): A new species of Rhacophorid frog from Similipal biosphere reserve, Odisha, India. *Russian Journal of Herpetology* **10** (1): 25-32.

Dutta S K, Nair M V, Mohapatra P P and Mahapatra A K (2009). Amphibians and reptiles of Similipal biosphere reserve. Regional Plant Resource Centre, Bhubaneswar, India, Pp.174.

Eizirik E, Yuhki N, Johnson W E, Menotti-Raymond M, Hannah S S and O'Brien S J (2003). Molecular genetics and evolution of melanism in the cat family. *Curr Biol* **13**: 448-453.

Ellerman J P (1940). The *families and genera of living rodents*. Vol-1. British Museum Natural History). London, Pp. 383.

Ellis E C, Goldewijk K K, Siebert S and Lightman D (2010). Anthropogenic transformation of the biomes, 1700 to 2000. *Glob. Ecol. Biogeogr.* **19**, 589–606

Evans K and Guariguata M R (2008). *Participatory monitoring in tropical forest management: A review of tools, concepts and lessons learned*. Center for International Forestry Research (CIFOR), Bogor, Indonesia.

Favre D S (1989). *International trade in endangered species: guide to CITES*. Martines Nijhoff Publishers.

Fearnside P M (2000). Global warming and tropical land-use change: greenhouse gas emissions from biomass burning, decomposition and soils in forest conversion, shifting cultivation and secondary vegetation. *Clim. Change* **46:** 115–58.

Ferraro P J (2008).Protected areas and human well-being economics and conservation in the tropics: A strategic dialogue. *RFP Conference Paper*, Jan 31.

Field C R and Ross I C (1976). The Savanna ecology of Kidepo Valley National Park, II. Feeding ecology of elephant and giraffe. *E. Afr. Wildl. J.* **14**: 1-15.

Food and Agriculture Organization (FAO) (2005). *Forests and freshwater: Vital connections.* FAO, Rome. Available at http: //www.fao.org/english/newsroom/focus/2003/wfc2.htm

Food and Agriculture Organization (FAO) (2006). *Global forest resources assessment 2005: Progress towards sustainable management.* FAO Forestry Paper 147, Rome.

Forest Survey of India (FSI) (2006). *Forest cover in tiger reserves of India – status and changes.* Forest Survey of India, Dehradun, India.

Forest Survey of India (FSI) (2009). *State of forest report 2009.* FSI, Ministry of Environment and Forest, Government of India, Dehradun, Pp.199+xxiv.

Frankel O.H. and Soule M.E. (1981). *Conservation and Evoluation.* Cambridge University Press, Cambridge.

Fuentes-Quezada E (1996). Economic growth and long-term carrying capacity: how will the bill be split? *Ecological Applications* **6**(1): 29-30.

Gadgil M (1991). Conserving India's biodiversity: The social context. *Evolutionary Trends in Plants* **5**(1): 3-8.

Gee E P (1964). *Wild life of India, with a foreword by Jawaharlal Nehru.* E P Dutton and Company, Inc, New York.

Gibbs H K and Brown S (2007a). Geographical distribution of woody biomass carbon stocks in tropical Africa: an updated database for 2000. Available at http: //cdiac.ornl.gov/epubs/ndp/ndp0555/ndp05b.html from the Carbon Dioxide Information Center, Oak Ridge National Laboratory, Oak Ridge, TN.

Gibbs H K and Brown S (2007b). Geographical distribution of biomass carbon in tropical Southeast Asian forests: an updated database for 2000. Available at http: //cdiac.ornl.gov/epubs/ndp/ndp068/ndp068b.html from the Carbon Dioxide Information Center, Oak Ridge National Laboratory, Oak Ridge, TN.

Gibbs H K, Brown S, Niles J O and Foley J A (2007). Monitoring and estimating tropical forest carbon stocks: making REDD a reality. *Environ. Res. Lett.* **2** (2007) 045023 (13pp) doi: 10.1088/1748-9326/2/4/045023.

Gill R M A, Thomas M L and Stocker D (1997). The use of portable thermal imaging for estimating deer population density in forest habitats. *Journal of Applied Ecology* **34**: 1273-1286.

Gittleman J L (1985). Carnivore body size: Ecological and taxonomic correlates. *Oecologia* **67**: 540-554.

Global Land Cover (GLC) Database (2000). European Commission, Joint Research Centre 2003. http: //www-gem.jrc.it/glc2000.

Gopal R (2015). *Dynamics of tiger management in priority landscapes.* Natraj, Dehradun.

Government of Odisha (not dated). *Annual report 2016-17 of Head Works Division, Samal.* Department of Water Resources, Odisha, Government of Odisha

Graham K, Beckermaman A P and Thirgood S (2005). Human-predator-prey conflicts: Ecological corelates, prey losses and pattern of management. *Biological Conservation* **122:** 159-171.

Griffiths M G and van Schaik C P (1993). Camera-trapping: A new tool for the study of elusive rain forest animals. *Trop. Biodiv.* **1**: 131-135.

Habib B, Saxena A, Jhala Y V and Rajvanshi A (2020). *Monitoring of animal underpasses on National Highway 44 (Earlier 7) passing through Pench Tiger Reserve, Maharashtra, India.* Wildlife Institute of India, Dehradun, TR. No. 2020/09, Pp 30.

Haemig P D (2012). Sympatric Tiger, Leopard and Dhole. *Ecology.info* 22. http: //www.ecology.info/tiger-leopard.htm, dt. 03.05.2015

Haines H H (1921-25). *The Botany of Bihar and Odisha.* 6 parts. Adlard and Son and West Newman Ltd., London.

Haldane J B S (1948). The theory of a cline. *J Genet* **48**: 277-284.

Haldane J B S, Sprunt A D and Haldane N M (1915). Redudplication in mice. *J Genet* 5: 133-135.

Hall J G (1981). A field study of the Kaibab squirrel in the Grand Canyon National Park. *Wildlife Monogr* **75**: 1-54.

Hayne D N (1949). An examination of strip census method for estimating animal populations. *J Wildlife Manage* **13**: 145-157.

Hendrichs H (1972). *A report on Sunderban tigers.* IUCN, Morges.

Hendrichs H (1975). The status of the tiger *Panthera tigris* (Linné, 1758) in the Sundarbans Mangrove Forest. *Saeugetierkundliche Mitteilungen* **23**: 161-199.

Heptner V G and Sludskij A A (1980). *Die Siiugetiere der Sowjetunion.* VolIIk Raubtiere (Feloidea). VEB Gustav Fischer Verlag, Jena, Germany.

Hight M F, Goodman M and Prychodko W (1974). Immunological studies of the Sciuridae. *Syst Zool* **23**: 12-25.

Hines J E (2006). *PRESENCE2- Software to estimate patch occupancy and related parameters.* USGS-PWRC. [Software] http: //www.mbr-pwrc.usgs.gov/software/presence.html.

Hoekstra H E (2006). Genetics, development and evolution of adaptive pigmentation in vertebrates. *Heredity* **97**: 222-234.

Houghton R A (1999). The annual net flux of carbon to the atmosphere from changes in land use 1850–1990. *Tellus* B **51**: 298–13.

Houghton R A, Lawrence K T, Hackler J L and Brown S (2001). The spatial distribution of forest biomass in the Brazilian Amazon: A comparison of estimates. *Glob. Change Biol.* **7**: 731–46.

IPCC (2006). *IPCC Guidelines for National Greenhouse Gas Inventories. Prepared by the National Greenhouse Gas Inventories Programme.* (Eds. H S Eggleston, L Buendia, K Miwa, T Ngara and K Tanabe), Institute for Global Environmental Strategies, Japan.

IUCN (2020). *The IUCN Red List of Threatened Species.* Version 2020-1. https: //www.iucnredlist.org. Downloaded on 19 March 2020.

Iyenger, S V P and Alwar, M A (1965). The Dhanori eugeosyncline and its bearing on the stratigraphy of the Singhbhum, Keonjhar and Mayurbhanja districts. *D.N. Wadia comm. Vol.* Min. Geol. Met Inst. Of India, Calcutta, Pp. 138-162.

Iyenger, S V P and Banerjee, S (1964). Magmatic phases associated with the Precambran tectonics of Mayurbhanj district, Odisha, India. *Proc. 22nd Intl. Geol. Cong.*, New Delhi, Section 10, Pp. 525-538.

Jairajpuri M S (1991): *Animal resources of India: Protozoa to mammalia: State of the art.* Zoological Survey of India. http: //indpaedia.com/ind/index.php/Mammals: _India

Jee S (1968). *Working plan for the reserved forests in the Mayurbhanj forest Divisions, district Mayurbhanj, Odisha, 1953-54 to 1972-73.* Odisha Government Press, Cuttack, Pp. 227+xii.

Jhala Y V, Gopal R and Qureshi Q (eds) (2008). *Status of the tigers, co-predators, and prey in India.* National Tiger Conservation Authority, Govt. of India, New Delhi, and Wildlife Institute of India, Dehradun, TR 08/001, Pp. 151.

Jhala Y V, Qureshi Q and Gopal R (2005). *Field guide: Monitoring tigers, co-predators, prey and their habitats.* Project Tiger, Union Ministry of Environment and Forests, New Delhi, India.

Jhala Y V, Qureshi Q and Gopal R (eds) 2015. *The status of tigers, copredators and prey in India 2014.* National Tiger Conservation Authority, New Delhi and Wildlife Institute of India, Dehradun. TR2015/021

Jhala Y V, Qureshi Q and Nayak A K (eds) (2019). *Status of the tigers, co-predators, and prey in India 2018. Summary report.* National Tiger Conservation Authority, Govt. of India, New Delhi, and Wildlife Institute of India, Dehradun, TR No/2019/05.

Johnsingh, A J T (1979) Evidence for a tiger eating a panther cub. *Journal of the Bombay Natural History Society* **76**: 152-153

Johnsingh, A J T and Joshua J (1991). *Ecology of endangered grizzled giant squirrel Ratufa macroura in TamilNadu.* Wildlife Institute of India, Dehra Dun.

Johnsingh A J T and Negi A S (2003). Status of tiger and leopard in Rajaji-Corbett Conservation Unit, northern India. *Biological Conservation* **111:** 385-393.

Joshua J (1992). *Ecology of the endangered grizzled giant squirrel (Ratufa macroura) in Tamil Nadu, South India.* Ph. D. dissertation, Bharatidasan University, Tiruchirapalli, TamilNadu.

Kaelin C B, Xu X, Hong L Z, David V A, McGowan K A, Schmidt-Küntzel A, Roelke M E, Pino J, Pontius J, Cooper G M, Manuel H, Swanson W F, Marker L, Harper C K, Dyk A v, Yue B, Mullikin J C, Warren W C, Eizirik E, Kos L, O'Brien S J, Barsh G S and Menotti-Raymond M (2012). Specifying and Sustaining Pigmentation Patterns in Domestic and Wild Cats. *Science* **337**(*6101*): 1536–1541

Karanth K U (1987). Tigers in India: a critical review of field censuses. Pages In *Tigers of the world: the biology, biopolitics, management and conservation of an endangered species*, Tilson, R L and Seal, U S (Eds). New Jersey: Park Ridge, Noyes, Pp. 118– 133.

Karanth K U (1991). Line transect method of censusing elephants in Nagarhole National Park. In *Censusing elephants in forest: Proceedings of an International Workshop, Southern India, 2-10 January, 1991* (eds. U Ramakrishnan, J A Santosh and R Sukumar), Asian Elephant Conservation Centre, Bangalore, India, Pp.51-53.

Karanth, K U (1993). How many tigers? Field censuses in India. *Hornbill* **1**: 2–9.

Karanth K U (1999). Counting tigers with confidence. *In Riding the tiger: Tiger conservation in human-dominated landscapes*, J. S. Seidensticker, S. Christie and P. Jackson (Eds), Cambridge University Press, London, Pp.350-353.

Karanth K U and Nichols J D (1998). Estimation of tiger densities in India using photographic captures and recaptures. *Ecology* **79**: 2852-2862.

Karanth K U, Nichols J D, Seidensticker J, Dinerstein E, Smith J L D, McDougal C, Johnsingh A J T, Chundawat R S and Thapar V (2003). Science deficiency in conservation practice: The monitoring of tiger populations in India. *Animal Conservation* **6**: 141-146.

Karanth K U and Stith B M (1999). Prey Depletion as a critcal determinant of tiger population viability. In *Riding the tiger: Tiger conservation in human-dominated landscapes*, J S Seidensticker, S Christie and P Jackson (Eds), Cambridge University Press, London, Pp. 100-113.

Karanth K U and Sunquist M E (1995). Prey selection by tiger, leopard and dhole in tropical forests. *Journal of Animal Ecology* **64:** 439-50.

Karanth K U and Sunquist M E (2000). Behavioural correlates of predation by tiger (*Panthera tigris*), leopard (*Panthera pardus*) and dhole (*Cuon alpinus*) in Nagarahole, India. *Journal of Zoology* **250:** 255-265.

Kar T, Gosavi K V C and Mandal K K (2018). *Coelorachis khasiana* and *Eulalia fastigiata* (Poaceae)- two new generic additions to the grass flora of Odisha, India. *Nelumbo* **60** (2): 125-129.

Kar T, Mohan M and Mandal K K (2017). *Disperis* and *Epipogium* (Orchidaceae): Two new generic record for the flora of Odisha. *Nelumbo* **59** (2): 159-163

Kar T, Nayak A K, Dash B R and Mandal K K (2014a). *Duchesnea indica* (Rosaceae): An addition to the flora of Odisha, India. *Bioscience Discovery*, **5**(2): 202-203.

Kar T, Nayak A K and Mandal K K (2014b). *Neanotis wightiana* (Rubiaceae): A new distributional record for Odisha, India and taxonomic notes on its variations. *Annals of Plant Sciences* **3** (09): 811-812.

Kar T, Nayak A K and Mandal K K (2014c). Occurrence of *Triadica cochinchinensis* Lour.(Euphorbiaceae) in Odisha, India. *Indian Journal Of Applied Research* **4** (10): 3-4.

Kar T, Nayak A K and Mandal K K (2014d). *Sloanea sterculiacea* (Elaeocarpaceae): a New Generic Record for Odisha, India. *International Journal of Scientific Research* **3** (7): 42.

Kawanishi K and Sunquist M E (2004). Conservation status of tigers in a primary rainforest of Peninsular Malaysia. *Biological Conservation* **120:** 329–344.

Khan M N I, Suwa R and Hagihara A (2009). Biomass and aboveground net primary production in a subtropical mangrove stand of *Kandelia obovata* (S., L.) Yong at Manko Wetland, Okinawa, Japan, *Wetlands Ecology and Management* **17**(6): 585–599.

Kitchener, A C (1999). Tiger distribution, phenotypic variation and conservation issues. In *Riding the tiger: Tiger conservation in human-dominated landscapes*, J S Seidensticker, S Christie and P Jackson (Eds), Cambridge University Press, London, Pp. 19-39.

Kitchener A C, Breitenmoser-Würsten Ch, Eizirik E, Gentry A, Werdelin L, Wilting A, Yamaguchi N, Abramov A V, Christiansen P, Driscoll C, Duckworth J W, Johnson W, Luo S-J, Meijaard E, O'Donoghue P, Sanderson J, Seymour K, Bruford M, Groves C, Hoffmann M, Nowell K, Timmons Z and Tobe S (2017). A revised taxonomy of the Felidae. The final report of the Cat Classification Task Force of the IUCN/SSC Cat Specialist Group. *Cat News Special Issue* 11, 80 pp.

Kothari A (1996). Is joint management of protected areas desirable and possible? In *People and protected areas: Towards participatory conservation in India,* (eds.) Kothari A, Singh N and Suri S, Sage Publications, New Delhi, Pp. 17-47.

Krishnamurthy V (1991). Elephant census: Total count method in Tamil Nadu. In *Censusing elephants in forest: Proceedings of an International Workshop, Southern India, 2-10 January, 1991* (eds. U.Ramakrishnan, J.A. Santosh and R. Sukumar), Asian Elephant Conservation Centre, Bangalore, India, Pp.30-32.

Lahiri-Choudhury D K (1991). Direct count of elephants in north-east India. In *Censusing elephants in forest: Proceedings of an International Workshop, Southern India, 2-10 January, 1991* (eds. U Ramakrishnan, J A Santosh and R Sukumar), Asian Elephant Conservation Centre, Bangalore, India, pp. 33-45.

Lasgorceix A and Kothari A (2009). Displacement and relocation of protected areas: A synthesis and analysis of case studies. *EPW* **49**: 37-49.

Luo S-J, Kim J-H, Johnson W E, van der Walt J, Martenson J, Yuhki N, Miquelle D G, Uphyrkina O, Goodrich J M, Quigley H B, Tilson R, Brady G, Martelli P, Subramaniam V, McDougal C, Hean S, Huang S-Q, Pan W, Karanth U K, Sunquist M, Smith J L D and O'Brien S J (2004). Phylogeography and genetic ancestry of tigers (*Panthera tigris*). *PLOS Biology* **2**: 2275-2293.

Mackenzie D I, Nichols J D, Lachman G B, Droege S, Royale J A and Langtimm CA (2002). Estimating site occupancy rates when detection probabilities are less than one. *Ecology* **83**: 2248-2255.

Mackenzie D I, Nichols J D, Sutton N, Kawanishi K and Bailey LL (2005). Improving inferences in population studies of rare species that are detected imperfectly. *Ecology* **86**: 1101–1113.

MacKenzie D I, Nichols JD, Royle J A, Pollock KH, Bailey LL and Hines JE (2006). *Occupancy estimation and modeling: Inferring patterns and dynamics of species occurrence*. Academic Press, Burlington, MA, Pp. 324.

Mahalik N K (1997). Geology and geomorphology of Similipal biosphere reserve. In *Similipal: A natural habitat of unique biodiversity*. Eds. P.C. Tripathy and S.N. Patro, Odisha Environmental Society, Bhubaneswar, India, Pp. 21-25.

Mahapatra A K, Tewari D D and Baboo B (2015). Displacement, deprivation and development: The impact of relocation on income and livelihood of tribes in Similipal tiger and biosphere reserve, India. *Environ Manage* **56**(2): 420-32.

Majerus M E N and Mundy N I (2003). Mammalian melanism: Natural selection in black and white. *Trends Genet* **19**: 585-588.

Maza'k V J (1996). *Der tiger*. Magdeburg, Westarp Wissenschaften. (Reprint of 1983 edn.).

McDougal C (1977). *The face of the tiger*. London, Rivington Books.

McDougal C (1988). Leopard and tiger interactions at Royal Chitwan National Park, Nepal. *Journal of the Bombay Natural History Society* **85**: 609–610.

McKay G M (1990). Behavior and ecology of the Asiatic elephant in South-eastern Ceylon. In *Asian elephants* (eds. J F Eisenberg, G M McKay and J Seidensticker), Smithsonian Institution, Washington, D.C., Pp.1-113.

Merz G (1986). Counting elephants (*Loxodonta africana cyclotis*) in tropical rain forests with particular reference to theTai National Park, Ivory Coast. *African Journal of Ecology* **24**: 61-68.

MHA (2012). *Census posts*. Ministry of Home Affairs, Government of India. Available at: http: // censusindia.gov.in/(seen on 01.02.2012).

Mills G and Hofer H (1998). *Hyaenas: status survey and conservation action plan*. IUCN/SSC Hyena Specialist Group. ISBN 2-8317-0442-1

Miquelle D G, Smirnov E N, Merrill T W, Myslenkof A E, Quigley H B, Hornocker M C and Schleyer B (1999). Heirarchical Spatial Analysis of Amur tiger relationships to habitat and prey. In *Riding the tiger: Tiger conservation in human-dominated landscapes*, J S Seidensticker, S Christie and P Jackson (Eds), Cambridge University Press, London, Pp. 71-99.

Mirza Z A, Warekar P, Mohapatra P P, Raju D, Patil P, Dutta S K and Pal S (2019). Endangered or an artifact of unsound taxonomy? Case of the critically endangered bush frog *Philautus sanctisilvaticus* Das and Chanda, 1997. *Zootaxa* **4683** (*4*): 563-576.

Mishra C G, Acharjyo L N and Choudhury L N (1982). Birth of white tiger (*Panthera tigris*) cubs to normal coloured tigers in captivity. *J. Bombay Nat. Hist. Soc.* **79** (*2*): 404-407.

Mishra C G, Patnaik S K, Sinha S K, Kar S K, Kar C S and Singh L A K (1996). *Wildlife Wealth of Odisha*. Wildlife Wing, Forest Department, Government of Odisha, Bhubaneswar.

Mishra S R, Nayak A K and Nandi D (2013). Population status of mugger crocodile (*Crocodylus palustris*) in Similipal tiger reserve, Odisha. India. *Int. Res. J. Environment Sci.* **2**(*5*): 92-94.

Mishra R (1978). *Revised working plan for the reserved forests of Karanjia forest division for the period 1973-74 to 1992-93*. Conservator of Forests, Working Plan Circle, Cuttack, Pp. 278+xiv.

Mishra R C (1997): The status of rare, endangered and endemic flora of Similipal biosphere reserve. In:

Tripathy, P C and Patro, S N (Ed.): *Similipal: A Natural Habitat of Unique Biodiversity,* Odisha Environmental Society, Bhubaneswar, Pp. 60-72.

Mishra R K, Upadhyay V P and Mohanty R C (2008): Vegetation ecology of the Similipal biosphere reserve, Odisha, India. *Applied Ecology and Environmental Research* **6**(2): 89-99.

Misra S (1997): Orchids of Similipal Forest. In: Tripathy, P C and Patro, S N (Ed.): *Similipal: A Natural Habitat of Unique Biodiversity,* Odisha Environmental Society, Bhubaneswar, Pp. 73-91.

Misra S (2004). Orchid flora of Similipal, Orissa, India. *J. Orchid Soc. India* **18** (*1-2*): 117-121.

Misra S (2014). *Orchids of Odisha: A handbook*. Bishen Singh Mahendra Pal Singh, Dehra Dun.

Misra S (2019). Orchids of India: A handbook. Bishen Singh Mahendra Pal Singh, Dehra Dun.

Misra S and Kar T (2017). Extended distribution of *Eulophia mackinnonii* (Orchidaceae), in eastern India. *Nelumbo* **59**(*1*): 61-65.

Misra R C and Sahoo H K (2014). *Thladiantha Bunge* (Cucurbitaceae): A new genus from Odisha and Central India. *Vegetos* **27** (*3*): 274-278.

Misra R C, Sahoo H K, Mahapatra A K and Reddy R N (2011): Additions to the flora of Similipal biosphere reserve, Odisha, India. *J. Bombay Nat. Hist. Soc.* **108** (1): 69-76.

Mohanty R C (2001). *Long-term biodiversity assessment of Similipal biosphere reserve.* Final Technical Report (unpublished), sponsored by Environment and Forests, Government of India, Pp. 158+iii.

Mohapatra P P, Behera S R, Dutta S K, Mahapatra C and Upadhyay H S (2019). Mammals of Odisha, India: Updated checklist, distribution and conservation status. *Pranikee - Journal of Zoological Society of Orissa,* **31**: 1-112.

Mohapatra P P, Dutta S K and Mahapatra C (2017). Herpetofauna of Similipal and their conservation. In *Facets of Similipal* (Editors, Thatoi H N, Sahu H K, Bastia A K, Biswal A K, Patra A K and Mohapatra C), Noth Orissa University, Baripada, India Pp. 164-179.

Montgomery S and Briggs E (2001). *The man-eating tigers of Sundarbans*. Houghton Mifflin Company, New York.

Mooney H F (1950). *Supplement to the Botany of Bihar and Odisha*. International Book Distributors, Dehradun (Reprinted in 1986), Pp. 294.

Moore J C (1959). Relationships among living squirrels of the Sciurinae. *Bull. American Museum Natural History* **118**: 153-206.

Morse H (1978). Introduction. In *Origins of inbred mice* (ed. Morse H), Academic, New York, Pp.1-31.

Mundy N I, Badcock N S, Hart T, Scribner K, Janssen K and Nadeau N J (2004). Conserved genetic basis of a quantitative plumage trait involved in mate choice. *Science* **303**: 1870-1873.

Murty M N (1996). Contractual arrangements for sharing benefits from preservations: Joint management of wildlife. In *People and protected areas: towards participatory conservation in India,* (eds.) Kothari A, Singh N and Suri S, Sage Publications, New Delhi, Pp. 127-39.

Nachman M W, Hoekstra H E, D'Agostino S L (2003). The genetic basis of adaptive melanism in pocket mice. *Proc Natl Acad Sci* **100**: 5268-5273.

Nascimento A A, Roland J T and Gelfand V I (2003). Pigment cells: A model for the study of organelle transport. *Annu Rev Cell Dev Biol* **19**: 469-491.

Nayak A K (not dated). *Tiger conservation plan of Similipal tiger reserve* (*2013 to 2022-23*). Forest Department, Government of Odisha, Bhubaneswar, India.

Nayak A K, Dash B R, Kar T and Mandal K K (2013). *Orthosiphon aristatus* (Lamiaceae): A new record for the flora of Odisha, India. *Plant Science Research* **35** (*1 and 2*) : 72-73.

Nayak A K, Kar T and Mandal K K (2014). *Trees of Similipal biosphere reserve.* Similipal Tiger Reserve, Baripada, Odisha, pp. 304.

Nayak A K and Naik S (2014). *Birds of Similipal biosphere reserve.* Similipal Tiger Reserve, Baripada, Odisha, Pp. 280.

Neumann-Denzau G and Denzau H (2010). Examining certain aspects of human-tiger conflict in the Sundarbans forest, Bangladesh**.** *Tigerpaper*, 37(3): 1-11.

Neville M K (1968). Ecology and activity of the Himalayan foot-hill rhesus monkeys (*Macaca mulatta*). *Ecology* **49**: 110-123.

Nowell K and Jackson P (1996). *Wild cats: Status survey and conservation action plan*, Gland, Switzerland, IUCN.

O'Brien T G, Kinnaird M F and Wibisono H T (2003). Crouching tigers, hidden prey: Sumatran tiger and prey populations in a tropical forest landscape. *Animal Conservation* **6**: 131-139.

Odisha (Odisha) Climate Change Action Plan (OCAP) (2011). Available at: http: //www.odisha.gov.in/ forest and environment/ActionPlan/CAP_Report_ Draft.pdf

Olivier R (1978). *On the ecology of the Asian elephant* (*Elephas maximus* Linn.) *with particular reference to Malaya and Sri Lanka*. Sidney Sussex College, Cambridge.

Overton W S (1971). Estimating the numbers of animals in wildlife population. In *Wildlife management technique* (ed. R H Giles), 3rd edn., revised, The Wildlife Society, Washington D.C., Pp.403-452.

Palei H S, Mohapatra P P and Sahu H K (2014). Dry season diet of the sloth bear (*Melursus ursinus*) in Hadagarh Wildlife Sanctuary, eastern India. *Proceedings of Zoological Society,* **67**: 67–71.

Palei H S, Sahu H K and Nayak A K (2016). Ungulate densities and biomass in the tropical moist deciduous forest of Similipal tiger reserve. *India. Natl. Acad. Sc. Lett.* **39**(*4*): 255-258.

Panda M K (2019). *Diversity and bioactivity studies on wild edible mushrooms of northern Odisha*, India. Ph.D. thesis submitted to North Orissa University, Baripada, India.

Pandav B, Mishra S K, Yadav S P and MD Madhusudan (2009). Similipal tiger reserve: A rapid field assessment of conservation status. Available at http: //projecttiger.nic.in/whtsnew/Similipal_ Appraisal_Report_FINAL[1]

Pandey A K, Rout S D and Pandit N (2002). Medicinal Plants of Similipal biosphere reserve-I. In *Perspectives of Plant Biodiversity* (ed. Das, A. P.), Bishen Singh Mahendra Pal Singh, Dehra Dun, Pp. 681-696.

Pandey A K, Varma S K, Pandit N and Singh L A K (2000). Medicinal Plants of Similipal biosphere reserve. *J. Indian Bot. Soc.* **79** (Suppl.): 52-53.

Panigrahi G, Chaudhury S, Raju D C S and Deka G K (1964). *A contribution to the botany of Odisha. Bull. Bot. Surv. India* **6**(2-4): 237-266.

Panwar H S (2003). Historical overview of relocation issues in conservation. In *Proceeding of National Seminar-cum-Consultation on Relocation of People from Protected Areas in India: Policy and Process*. Council for Social Development, New Delhi, September 18, 2003.

Patnaik S K (1997): Conservation and management of wildlife in Similipal. In: Tripathy, P C and Patro, S N (Eds): *Similipal: A Natural Habitat of Unique Biodiversity,* Odisha Environmental Society, Bhubaneswar, Pp. 122-128.

Patro S N and Misra M K (Eds) (1985). Conservation of Similipal in its wilderness. Proceedings of national conference on biosphere, Similipal, Baripada, May 19-21, 1985, Orissa Environmental Society, Bhubaneswar, India.

Patton D R (1975). Abert squirrel cover requirements in southwestern ponderosa pine. U.S. Dept. Agric. For. Serv. Res. Pap. RM-145. R. Mtn. For. and Range Exp. Sta. Fort Collins, Colo. Pp.12.

Payne J B (1979). Abundance of diurnal squirrels at the kuala Lompat post of the Krau Game Reserve, Peninsular Malaysia. In *The Abundance of animals in Malaysian rain forests*, A G Marshall (ed.). Department of Geography, University of Hull, Pp. 37-51.

Payne J B (1980). Competitors. In *Malayan Forest Primates*, David Chivers (ed.), Plenum Press, New York and London, Pp. 261-277.

Payra A, Das G N, Boruah B, Dash S K, Das U P and Sethy J (2016). Butterfly diversity in two selected fringe area of Similipal biosphere reserve, Odisha, India, with notes on some important sightings. *Journal of Wildlife Research* **4**(2): 17-25.

Pocock R I (1941). *The fauna of British India. Mammalia VoZ II*. Taylor and Francis, London.

Poti P S (1995). *Preliminary report on Rajaji national park*. Indian People's Tribunal on Environment and Human Right, Bombay.

Prater S H (1965). *The Book of Indian Animals*. Second (Revised) Edition, Bombay Natural History Society and Oxford University Press, Mumbai, Pp. 324.

Prinzivalli R (1992). Investigating the use of infrared detectors to count elephants in a forest environment. *African Journal of Ecology* **30**: 346-348.

Protas M E, Hersey C, Kochanek D, ZhouY, Wilkens H and Jeffery W R (2006). Genetic analysis of cavefish reveals molecular convergence in the evolution of albinism. *Nat Genet* **38**: 107-111.

Prusty B C and Singh L A K (1996). Nature camp in Similipal tiger reserve. *Zoos' Print*, XI(11): 36-37.

Qureshi Q, Saini S, Basu P, Gopal R, Raza R and Jhala Y (2014). *Connecting Tiger populations for long-term conservation*. National Tiger Conservation Authority and Wildlife Institute of India, Dehradun.

R Development Core Team (2009). *R: A language and environment for statistical computing*. R Foundation for Statistical Computing, Vienna, Austria. ISBN 3-900051-07-0, URL http: //www.R-project.org.

Rabinowitz A (1989). The density and behaviour of large cats in a dry tropical forest mosaic in Huai Kha Khaeng Wildlife Sanctuary, Thailand. *Natural History Bulletin of the Siam Society* **37**(2): 235-251.

Raj P, Dinesh K P, Das A, Dutta S K, Kar N B and Mohapatra P P (2018). Two new species of cricket frogs of the genus *Fejervarya* Bolkay, 1915 (Anura: Dicroglossidae) from the Peninsular India. Records of the Zoological Survey of India **118**: 1–21.

Raju G (1996). From joint forest management to joint protected area management: Can the concept be extended? In *People and protected areas: Towards participatory conservation in India,* (eds.) Kothari A, Singh N and Suri, S., Sage Publications, New Delhi, Pp. 93-113.

Ramachandran K K (1988). *Ecology and Behaviour of Malabar Giant Squirrel, Ratufa indica maxima (Schreber)*. Kerala Forest Research Institute, Research Report, Pp. 55.

Ramakrishnan U, Coss R G and Pelkey N W (1999). Tiger decline caused by the reduction of large ungulate prey: Evidence from a study of leopard diets in southern India. *Biological Conservation* **89:** 113-120.

Ramakrishna, Siddiqui S Z, Sethy P and Dash S (2006). *Faunal Resources of Similipal biosphere reserve, Mayurbhanj, Orissa*. Zoological Survey of India, Kolkata, Conservation Area Series **28**: 1-87 + 8 plates.

Ramankutty N, Gibbs H K, Achard F, DeFries R, Foley J A and Houghton R A (2007). Challenges to estimating carbon emissions from tropical deforestation. *Glob. Change Biol.* **13**: 51–66.

Ramesh T, Snehalatha V, Sankar K and Qureshi Q (2009). Food habits and prey selection of tiger and leopard in Mudumalai tiger reserve, Tamil Nadu, India. *J. Sci. Trans. Environ. Technov* **2**(2): 170-181.

Ranganathan J, Chan K M A, Karanth K U and Smith J L D (2008). Where can tigers persist in future? A landscape-scale, density-based population model for the Indian subcontinent.*Biological Conservation* **141:** 67-77.

Rangarajan M and Shahabuddin G (2006). Displacement and relocation from protected areas: Towards a biological and historical synthesis. *Conserv Soc* **43**: 359-378.

Ranjitsinh M K (1987). Deer, Antelopes and Bovines. In: Israel, S. and T. Sinclair (editors), *Indian Wildlife*,1st Edition,APA Productions (HK) Ltd, Pp.99-100.

Rao K M and Satpathy B (2003). Modal- a unique tasar ecorace. *Indian Silk*, September 2003, Pp 15-17.

Rathore B M S (1996). Joint management options for protected areas: Challenges and opportunities. In *People and protected areas: towards participatory conservation in India,* (eds.) Kothari A, Singh N and Suri S, Sage Publications, New Delhi, Pp. 85-92.

Ravindranath N H, Chaturvedi R K, Murthy I K (2008). Forest conservation, afforestation and reforestation in India: Implications for forest carbon stocks. *Current Science* **216**(2): 216-222.

Ravindranath N H, Murthy I K, Sudha P, Ramprasad V, Nagendra M D V, Sahana C A, Khan H and Srivathsa K G (2007). Methodological issues in forestry mitigation projects: A case study of Kolar district. *Mitigation and Adaptation Strategies for Global Change*, **12**(6): 1077-1098.

Ravindranath N H and Ostwald M (2008). *Carbon inventory methods: Handbook for greenhouse gas inventory, carbon mitigation and round wood production projects.* Springer Verlag, Pp. 304.

Ravindranath N H and Somashekar B S (1995). Potential and economics of forestry options for carbon sequestration in India. *Biomass and Bioenergy* **8**(5): 323-336.

Ravindranath N H, Sudha P and Sandhya R (2001). Forestry for sustainable biomass production and carbon sequestration in India. *Mitigation and Adaptation Strategies for Global Change* **6**: 233-256.

Rice C G (1991). The status of Four-horned Antelope *Tetracerus quadricornis. Journal of the Bombay Natural History Society* 88(1): 63-66.

Richard P (1965). *The World of the Tiger*. ASIN: B0007DU2IU.

Ritland K, Newton C and Marshall H D (2001). Inheritance and population structure of the white-phased 'Kermode' black bear. *Curr Biol* **11**: 1468-1472.

Rodgers W A (1991). *Techniques for wildlife census in India: A field manual.* Wildlife Institute of India, Dehradun, India.

Rodgers W A and Panwar H S (1988). *Planning a wildlife protected area network in India.* Wildlife Institute of India, Dehradun, India.

Roonwal M L and Mohnot S M (1977). *Primates of South Asia. Ecology, Sociobiology, and Behaviour.* Harvard University Press, Cambridge, Massachusetts and London, England, Pp.421 + xviii.

Rout S D (2004). *Medicinal plants of Similipal biosphere reserve*. Unpublished PhD thesis, TM Bhagalpur University, India.

Rout S D and Swain D (2006). The giant squirrel (*Ratufa indica*) in Similipal tiger reserve, Orissa, India. *Tigerpaper* **33**(4): 24-27.

Royle J A and Nichols J D (2003). Estimating abundance from repeated presence-absence data or point counts. *Ecology* **84:** 777-790.

Russel L B and Major M H (1957). Radiation-induced presumed somatic mutations in the house mouse. *Genetics* **42**: 161-175.

Russel W L (1951). X-ray induced mutations in mice. *Cold Spring Harbor Symp Quant Biol* **16**: 327-335.

Sachan S K S, Patra J K and Thatoi H N (2013). Indigenous knowledge of ethnic tribes for utilization of wild mushrooms as food and medicine in Similipal biosphere reserve, Odisha, India. *International Journal of Agricultural Technology* **9**(2): 403-416.

Sachan S K S and Thatoi H N (2017). Biodiversity and biotechnological potential of wild mushroom from Similipal biosphere reserve, Mayurbhanj, Odisha. In *Facets of Similipal* (Editors, Thatoi H N, Sahu, H K, Bastia A K, Biswal A K, Patra A K and Mohapatra C), North Orissa University, Baripada, India, pp. 1-13.

Sagar S R and Singh L A K (2001). Elephant population estimation during 1989 monsoon in Similipal tiger reserve. In *Elephants in Similipal (history, status, issues, technique, and biological notes on elephants)*, (compiled by S S Srivastava and L A K Singh), Similipal Tiger Reserve, Baripada, India, pp. 55-67.

Sagar V, Kaelin C B, Natesh M, Reddy A, Mohapatra R K, Chhattani H, Biswas S, Bhatt S, Paul S, Jhala Y V, Verma M M, Pandav B, Mondol S, Barsh G S, Swain D and Ramakrishnan U (2021). High frequency of an otherwise rare phenotype in a small and isolated tiger population. Under communication with *Proc Natl Acad Sci USA*.

Sahoo D, Sahu H K, Dutta S K and Upadhyay H S (2012). Sighting of Red-necked Keelback (*Rhabdophis subminiatus*) in Similipal Tiger Reserve, Orissa, India. *Tigerpaper* **39**(1): 31–32.

Sahu H K and Swain D (2007). Status of mugger *Crocodylus palustris* in Similipal tiger reserve, Orissa, India. *J. Bombay Nat. Soc.* **104**(2): 214-217.

Sandhyarani S, Murthy K S R and Pullaiah T (2007). Tree Flora in Eastern Ghats of Southern Peninsular India. *Research Journal of Botany* **2**: 176-185

Sankhala K (2005). *Tiger! The story of the Indian tiger*. Natraj Publishers, Dehradun.

Santosh J A and Sukumar R (1995). Some solutions to problems in estimation of elephant density. In *A Week with Elephants: Proceedings of the International Seminar on the Conservation of Asian Elephant (June1993)* (eds J C Daniel and H S Datye). Oxford University Press, Bombay, pp. 394-404.

Saranya K, Reddy C S, Rao P P and Jha C S (2014). Decadal time scale monitoring of Forest Fires in Similipal Biosphere Reserve, India using Remote sensing and GIS. *Environmental Monitoring and Assessment*. April 2014 DOI: 10.1007/s10661-014-3619-7.

Sarkar M P, Brahmachary R L and Dutta J (1991). Short chain free fatty acids as putative pheromones in the marking fluid of tiger. *J Indian Chem Soc* **68**: 255-256.

Sarkar M P, Brahmachary R L and Dutta J (1994). Scent marking in the tiger. *Advances in the Biosciences* **93**: 339-344.

Sathaye J, Makundi W and Andrasko K (1995). A comprehensive mitigation assessment process (COMAP) for the evaluation of forestry mitigation options. *Biomass and Bioenergy* **8**(5): 345-356.

Saxena H O and Brahmam M (1989). *The flora of Similipahar (Similipal), Odisha: With particular reference to the potential economic plants*. Regional Research Laboratory, Bhubaneswar, India, Pp. 231+x.

Saxena H O and Brahmam M (1994-1996). *The flora of Odisha*. Vol.I-IV. Odisha Forest Development Corporation Ltd., Bhubaneswar, India, pp. 2918+LXIV.

Schaller G B (1967). *The deer and the tiger*. Chicago, University of Chicago Press.

Schlager G and Dickie M M (1966). Spontaneous mutation rates at five coat-color loci in mice. *Science* **151**: 205-206.

Schlager G and Dickie M M (1967). Spontaneous mutations and mutation rates in the house mouse. *Genetics* **57**: 319-330.

Seidensticker J, Christie S and Jackson P (1999). Introducing the tiger. In *Riding the tiger: Tiger conservation in human-dominated landscapes*, (eds.) Seidensticker J, Christie S and Jackson P, Cambridge University Press, London, Pp.1-3.

Seidensticker J, Sunquist M E and McDougal C (1990). Leopards living at the edge of Royal Chitwan National Park, Nepal. In *Conservation in Developing Countries: Problems and Prospects,* (J C Daniel and J S Serrao, editors.), Bombay Natural History Society, Bombay. Pages 415-423.

Seidensticker J, Yoganand K and Johnsingh A J T (2011). Sloth bears living in seasonally dry tropical and moist broad leaf forests and their conservation. In *The Ecology and Conservation of Seasonally Dry Forests in Asia*, (W McShea, S J Davies, and N Bhumpakphan, editors), Smithsonian Institution Scholarly Press, Pages 217-236.

Senapati A (2016). Odisha loses 24th elephants in 8 years on railway tracks. *Down to Earth*, Thursday 22 November 2018.

Senapati N and Sahu N K (eds.) (1967). *Odisha district gazetteers, Mayurbhanj.* Printed by the Superintendent, Odisha Government Press, Cuttack.

Seryodkin I V, Goodrich J M, Kostyrya A V, Schleyer B O, Smirnov E N, Kerley L L and Miquelle D G (2005). Relationship between tigers, brown bears, and Himalayan black bears. In Miquelle D G, E N Smirnov and J M Goodrich, editors, *Tigers of Sikhote-Alin Zapovednik: ecology and conservation.* PSP, Vladivostok, Russia, (in Russian), p. 156-163.

Shibahara S, Tomita V, Sakakura T, Nager C, Chaudhuri B and Muller R (1986). Cloning and expression of cDNA-enocoding mouse tyrosinase. *Nucleic Acids Res* **14**: 2413-2427.

Short J (1983). Density and seasonal movements of forest elephant (*Loxodonta Africana cyclotis*, Matschie) in Bia National Park, Ghana. *African Journal of Ecology* **21**: 175-184.

Simberloff D (1998). Flagships, umbrellas and keystones: Is single-species management passé in the landscape era? *Biological Conservation* **83**: 247 - 257.

Singh A P and Chalisgaonkar R (2006). Restoration of corridors to facilitate the movement of wild Asian elephants in Rajaji-Corbett elephant range, India. Roorkee, India.

Singh K P and Kumar K (2012). A note on the lichens from Similipal biosphere reserve, Odisha, India. *Indian Journal of Forestry*, **35**(*3*): 383-390.

Singh K P and Singh J S (1988). Certain structural and functional aspects of dry tropical forests and savannah. *Int. Jourl. Ecol. and Envi. Sci.* **14**: 31-45.

Singh L A K (1997). Wildlife Research in Similipal. In *Similipal: A natural habitat of unique biodiversity*, (Eds.) P C Tripathy and S N Patro, Odisha Environmental Society, Bhubaneswar, India, Pp. 105-112.

Singh L A K (1999). *Tracking tigers: A pocket book for forest guards.* WWF Tiger Conservation Programme, pp. 36.

Singh L A K (1999a). *Born black: The melanistic tigers in India.* WWF-India, New Delhi, Pp. 66.

Singh, L A K (1999b). A profile of Indian Crocodiles. *ENVIS* (*Wildlife and Protected Areas*) **2**(1): 1-4.

Singh L A K (2000). Interpreting visual signs of the Indian Crocodile. *Crocodile Specialist Group Newsletter* **19**(1): 7-9.

Singh L A K (2015). Cascade approach for round-the-year tourism around wildlife landscape. *Tourism Odisha- Voice of the Travel Trade, Hotel and Restaurant Association of Odisha*, June 2015: 14-16.

Singh L A K and Swain D (2003). The four-horned antelope or chousingha (*Tetracerus quadricornis*) in Similipal. *Zoos' Print Journal* **18**(*9*): 1197-98.

Singh U S (2019). Planting trees in large numbers will not mitigate carbon emission to the desired level but conservation of old forests will do. *International Journal of Scientific and Engineering Research* **10**(*7*): 758-773.

Singh V B (1978). The elephant in U.P. (India). A resurvey of its status after 10 years. *Journal of Bombay Natural History Society* **75**(1): 71-82.

Srivastava B P (1979). Status of tiger in India. *In International symposium on tiger, India (February 22 – 24, 1979)*, Project Tiger, Government of India, New Delhi, Pp.11-16.

Srivastava S S (2001). *Management plan of Similipal tiger reserve* (year: 2001-11). Similipal Tiger Reserve, Baripada, India, Pp.157.

Srivastava S S and Singh L A K (1998). Profile of tourists in Similipal. *Zoos' Print*, XIII(4): 12-15.

Srivastava S S and Singh L A K (2003). Status of wild dog (*Cuon alpinus*) in Similipal and possible impacts of 'village dogs'. *Zoos' Print* **18**(*10*): 18-21

Stephen M (2004). *Tiger*. Richmond Hill., Ont.: Firefly Books.

Sugiyama Y (1964). Group composition, population density and some sociological observations of Hanuman langurs (*Presbytis entellus*). *Primates* (Inuyama) **5**: 7-37.

Sukumar R (1985). *Ecology of the Asian elephant (Elephas maximus) and its interaction with man in South India*. Vols. I and II. Ph.D. thesis. Indian Institute of Science, Bangalore.

Sukumar R (1989). *The Asian elephant: Ecology and management*. Cambridge University Press, Cambridge.

Sukumar R, Surendra V K, Santosh J A and Ramakrishnan U (1991). Methods of estimating elephant numbers: A review. In *Censusing elephants in forest: Proceedings of an International Workshop, Southern India, 2-10 January, 1991* (eds. U Ramakrishnan, J A Santosh and R Sukumar), Asian Elephant Conservation Centre, Bangalore, India, pp. 2-29.

Sumner F B (1921). Desert and lava-dwelling mice and the problem of protective coloration in mammals. *J Mammal* **2**: 75-86.

Sumner F B (1929a). The analysis of a concrete case of intergradation between two subspecies. *Proc Natl Acad Sci USA* **15**: 110-120.

Sumner F B (1929b). The analysis of a concrete case of intergradation between two subspecies. II. Additional data and interpretations. *Proc Natl Acad Sci USA* **15**: 481-493.

Sunquist M, Karanth K U and Sunquist F (1999). Ecology, behaviour and resilience of the tiger and its conservation needs. Introducing the tiger. In *Riding the tiger: Tiger conservation in human-dominated landscapes*, J S Seidensticker, S Christie and P Jackson (Eds), Cambridge University Press, London, pp. 5-18.

Suri S (1996). People's involvement in protected areas: Experiences from abroad and lessons for India. In *People and protected areas: towards participatory conservation in India*, (eds.) Kothari A, Singh N and Suri S, Sage Publications, New Delhi, Pp. 247-60.

Swain D (1993). Accidental death of a tusker. *The Indian Forester* **119**(10): 871.

Swain D (1996). Status of elephants in Kuldiha sanctuary, Odisha. *The Indian Forester* **122**(10): 927-932.

Swain D (1997). Participatory biodiversity management in Mayurbhanj. *Wastelands News* **12**(4): 41-43.

Swain D (2001). Joint protected area management: A proposal for the management of sanctuaries and national parks in India. *Int J Sustain Dev World Ecol* **8**: 257-65.

Swain D (2004). *Asian elephants: Past, present and future*, International Book Distributors, Dehradun, India.

Swain D (2005). Similipal biosphere reserve. In *'Proceedings of Indo-US workshop on conservation of biodiversity in Similipal biosphere reserve (14-15 November, 2005)'*. Pp. 38-41. (North Orissa University: Baripada, India).

Swain D (2009). Role of India's tiger reserves in the protection of origin/catchment of water courses: A case study of the Similipal tiger reserve. *XIII World Forestry Congress*, Buenos Aires, Argentina, 18 – 23 October 2009.

Swain D (2014). State of forest in Similipal tiger reserve, Odisha, India: A case study. *The Indian Forester*, **140**(*11*): 1049-1055

Swain D and Behura B K (2013). Melanistic tigers of Similipal tiger reserve, India: Bane or boon? *Tigerpaper* **40**(*1*): 1-4.

Swain D and Behura B K (2013). Human impact on tropical deciduous forests: A case study of Indian sal (*Shorea robusta*) forests of Similipal biosphere reserve. *The Indian Foreste* **139**(*11*): 988-994.

Swain D and Morris P J (2014). Occupancy estimation and modeling in monitoring tigers with implications for conservation and management. *Environment Conservation Journal* **15**(1 and 2): 21-33.

Swain D and Nanda F B (1997). Study of plant biodiversity in a newly established preservation plot inside Similipal National Park, Odisha. In *Similipal: A natural habitat of unique biodiversity*, (Eds.) P C Tripathy and S N Patro, Odisha Environmental Society, Bhubaneswar, India, Pp. 46-59.

Swain D and Patnaik S K (2002). Elephants of Odisha: Conservation issues and management options. *The Indian Forester* **128**(2): 145-154.

Swain D, Rout S D and Behura B K (2008). A study of the population structure of two species of non-human primates in the Similipal tiger reserve, Mayurbhanj, Orissa, India. *The Indian Forester* **134**(*10*): 1296-1308.

Swain D, Sahu H K and Parida B K (2005). Censusing tiger and leopard in Similipal tiger reserve-2004, Odisha, India. *Tigerpaper* **32**: 22-25.

Swain D and Singh L A K (2003). Musth in female Asian elephant. *Zoos' Print Journal* **18**(9): 1202.

Swain D and Singh L A K (2004). Death feigning behaviour by a mouse deer. *The Indian Forester* **130**(7): 829-830.

Thatoi H N (2014). *Assessment of nutritional and medicinal potential of indigenous mushrooms from Similipal biosphere reserve and development of their propagation and massive cultivation technique*. Project report submitted to Department of Science and Technology, Government of Odisha, Bhubaneswar, India.

Thomas L and Karanth K U (2002). Statistical concepts: Estimating absolute densities of prey species using line transect sampling. In *Monitoring tigers and their Prey* (eds. K U Karanth and J D Nichols), Centre for Wildlife Studies, India, Pp. 87-110.

Threon E, Hawkins K, Bermingham E, Ricklefs R E and Mundy N I (2001). The molecular basis of an avian plumage polymorphism in the wild: A melanocortin-1- receptor point mutation is perfectly associated with the melanic plumage morph of the bananaquit, *Coereba flaveola*. *Curr Biol* **11**: 550-557.

Tiger Task Force Report (1972). *Projct Tiger: A planning proposal for preservation of tiger* (*Panthara tigris tigris* Linn) *in India*. Indian Board for Wild Life, New Delhi.

Tiger Task Force Report (2005). *The report of the tiger task force: Joining the dots*. Government of India, New Delhi.

Tripathy P C and Patro S N (Eds). *Similipal: A natural habitat of unique biodiversity*, Orissa Environmental Society, Bhubaneswar, India.

Turkalo A (1996). Studying forest elephants by direct observation in the Dzanga clearing: An update. *Pachyderm* **22**: 59-60.

Turkalo A and Fay J M (1995). Studying forest elephants by direct observation: preliminary results from the Dzangha clearing, Central African Republic. *Pachyderm* **20**: 45-54.

UNESCO (2012). World Network of Biosphere Reserves (WNBR). Available at: http: //www.unesco.org/new/en/natural-sciences/environment/ecological-sciences/biosphere-reserves/world-network-wnbr/sphere-reserves/world-network-wnbr/

Vancuylenberg B W B (1977). Feeding behaviour of the Asiatic elephant in south east Sri Lanka in relation to conservation. *Biol Conserv* **12**: 33-54.

Varman K S, Ramakrishnan U and Sukumar R (1995). Direct and indirect methods of counting elephants: A comparison of results from Mudumalai sanctuary. In *A Week with Elephants: Proceedings of the International Seminar on the Conservation of Asian Elephant* (*June1993*) (eds J C Daniel and H S Datye). Oxford University Press, Bombay, Pp. 331-339.

Vasu N (1911). *The archaeological survey of Mayurbhanj*, Vol. I, Published by Mayurbhanj state, Odisha, India.

Verma M, Negandhi D, Khanna C, Edgaonkar A, David A, Kadekodi G, Costanza R, Gopal R, Bonal B S, Yadav S P and Kumar S (2017). Making the hidden visible: Economic valuation of tiger reserves in India. *Ecosystem Services* **26**: 236-234.

Verma M, Negandhi D, Khanna C, Edgaonkar A, David A, Kadekodi G, Costanza R and Singh R (2015). *Economic valuation of tiger reserves in India: A value+ approach*. Indian Institute of Forest Management, Bhopal, India.

Verma M, Tiwari C, Anand S, Edgaonkar A, David A, Kadekodi G, Ninan K N, Sharma P, Panda P and Thatey Z (2019). *Economic valuation of tiger reserves in India: Phase II*. Indian Institute of Forest Management, Bhopal, India.

Vijayakumar V, Vyas R and Choudhury B C (1999). Status of mugger and its conservation problems in Gujarat. *ENVIS* (*Wildlife and Protected Areas*) **2**(1): 69-76.

Wada A, Okumoto M and Tsudzuki M (1929). Tawny: A novel light coat color mutation found in a wild population of *Mus musculus molossinus*, a new allele at the melanocortin 1 receptor (Mc1r) locus. *Exp Animals* **48**: 73-78.

Wade R (1987). The management of common property resources: Finding a cooperation solution. *Res Obs* **2**(2): 219-34.

Wani M and Kothari A (2007). Conservation and people's livelihood rights in India: Final report of a research project under the UNESCO small grants programme. Available at http: //www.kalpavriksh.org

Wang S W and Macdonald D W (2006). Livestock predation by carnivores in Jigme Singye Wangchuck National Park, Bhutan. *Biological Conservation* **129:** 558-565.

Whitaker R and Andrews H V (2003). Crocodile conservation, Western Asia region: An update. *J. Bombay nat. Hist. Soc.* **100** (2 and 3): 432—445.

White L and Edwards A (eds.) (2000). *Conservation research in the African rain forests: A technical handbook*. Wildlife Conservation Society, New York.

WII (2016). ECO-friendly measures to mitigate impacts of linear infrastructure on wildlife. Wildlife Institute of India, Dehradun, India.

WII (2020). *Protected Areas of India.* Envis Centre, Wildlife Institute of India, wiienvis.nic.in. (visited on 04.04.2020.)

Wilson D E and Reeder D M (eds) (2005). *Mammal species of the world. A taxonomic and geographic reference* (3rd ed). Johns Hopkins University Press. Available at http: //www.press.jhu.edu

World Ecotourism Summit (2002). World Ecotourism Summit (Québec, Canada, 19 – 22 May 2002) Final Report.

Wright B and Mohanty B (2010). *Simlipal tiger reserve; assessment of recent elephant poaching and protection initiatives.* Report submitted to the National Tiger Conservation Authority. Available at http: //projecttiger.nic.in/whtsnew/Simlipal per cent 20Report_June2010_Final2.pdf.

Wright S (1917a). Color inheritance in mammals-I. *J Hered* **8**: 224-235.

Wright S (1917b). Color inheritance in mammals-II. The mouse. *J Hered* **8**: 373-378.

Wright S (1917c). Color inheritance in mammals-III. The rat. *J Hered* **8**: 426-430.

Wright S (1917d). Color inheritance in mammals-V. The guinea pig. *J Hered* **8**: 476-480.

Zacharias M A and Roff J C (2001). Use of focal species in marine conservation and management: a review and critique. *Aquatic Conservation: Marine and Freshwater Ecosystems* **11**: 59 – 76.

Appendices

7.3.04
Mahendra

Appendix-I: List of Plants Recorded in Similipal Biosphere Reserve.

Sl. No.	Local Name	Scientific Name	Family
	PTERIDOPHYTA		
1.		*Huperzia hamiltonii* (Spreng.) Trevis.	Huperziaceae
2.		*Huperzia squarrosa* (G.Forst.) Trevis	Huperziaceae
3.		*Palhinhaea cernua* (L.) Franco. & Vasc.	Lycopodiaceae
4.	Mrityusanjivani	*Selaginella bryopteris* (L.) Baker.	Selaginellaceae
5.		*Selaginella chrysorrhizos* Spring	Selaginellaceae
6.		*Selaginella cillaris* (Retz.) Spring.	Selaginellaceae
7.		*Selaginella indica* (Milde) Tryon	Selaginellaceae
8.		*Selaginella repanda* (Desv. & Poir.) Spring.	Selaginellaceae
9.		*Equisetum ramosissimum* Desf.	Equisetaceae
10.		*Botrychium daucifolium* Wall.	Botrychiaceae
11.		*Angiopteris evecta* (Forst.) Hoffm.	Angiopteridaceae
12.		*Angiopteris helferiana* Presl.	Angiopteridaceae
13.		*Dicranopteris linearis* (Burm. f.) Underwood.	Dicranopteridaceae
14.		*Colysis digitata* (Baker) Ching	Polypodiaceae
15.		*Colysis pedunculata* (Hook. & Grev.) Ching.	Polypodiaceae
16.		*Lepisorus contortus* (Christ.) Ching	Polypodiaceae
17.		*Lepisorus nudus* (Hook.) Ching.	Polypodiaceae
18.		*Leptochilus decurrens* (Bl.) Copel.	Polypodiaceae
19.		*Microsorum membarnaceum* (D. Don) Ching	Polypodiaceae
20.		*Microsorium punctatum* (L.) Copel.	Polypodiaceae
21.		*Paraleptochilus decurrens (Bl.) Copel.*	Polypodiaceae
22.		*Pyrrosia adnascens (Sw.) Ching.*	Polypodiaceae
23.		*Pyrrosia beddomeana* (Giesenh.) Ching	Polypodiaceae
24.		*Pyrrosia lanceolata* (L.) Farw.	Polypodiaceae
25.		*Pyrrosia mollis* (Kunze.) Ching	Polypodiaceae
26.		*Pyrrosia nuda* (Giesenh.) Ching	Polypodiaceae
27.		*Pyrrosia stigmosa (Sw.) Ching.*	Polypodiaceae
28.	Garur-panki	*Drynaria quercifolia (L.) J.Sm.*	Drynariaceae
29.	Mahajal	*Lygodium flexuosum* (L.) Sw.	Lygodiaceae
30.		*Lygodium microphyllum* (Cav.) R. Br.	Lygodiaceae
31.		*Cheilanthes anceps* Blanf.	Cheilanthaceae
32.		*Cheilanthes tenuifolia* (Burm.f.) Sw.	Cheilanthaceae
33.		*Coniogramme fraxinea* (D. Don) Fee	Pteridaceae
34.		*Parahemiotis cordifolia* Roxb.	Pteridaceae
35.		*Pteris arisanensis* L.	Pteridaceae
36.		*Pteris biauriata* L. Sub sps. *biauriata*	Pteridaceae
37.		*Pteris biauriata* L. sub sps. *fornicata*	Pteridaceae
38.		*Pteris cretica* L.	Pteripdaceae
39.		*Pteris ensiformis* Burm. f.	Pteridaceae
40.		*Pteris longipes* D. Don	Pteridaceae
41.		*Pteris pellucida* Presl.	Pteripdaceae
42.		*Pteris guadriaurita* Retz.	Pteripdaceae
43.		*Pteris vittata* L.	Pteripdaceae
44.		*Adiantum cappillus-veneris* L.	Adiantaceae
45.		*Adiantum incisum* Forsk.	Adiantaceae

Sl. No.	Local Name	Scientific Name	Family
46.		*Adiantum philippense* L.	Adiantaceae
47.		*Adiantum proliferum* Roxb.	Adiantaceae
48.		*Coniogramme fraxinea* (D.Don) Fee ex Diels.	Hemionitidaceae
49.		*Hemionitis arifolia* (Burm.) Moore.	Hemionitidaceae
50.		*Antrophyum henryi* Hieron	Antrophyaceae
51.		*Antrophyum reticulatum* (Forst.) Kaulf.	Antrophyaceae
52.		*Ceratopteris thalictroedes* (L.) Brongn.	Parkeriaceae
53.		*Marsilea minuta* L.	Marsileaceae
54.		*Gonocormus prolifer* (Blume) Prantl.	Hymenophyllaceae
55.		*Mecodium exsertum* (Wall. Ex Hook.) Copel.	Hymenophyllaceae
56.		*Alsophila spinulosa* Wall. ex Hook	Cyatheaceae
57.		*Cyathea gigantea* (Wall. ex Hook.) Holtt.	Cyatheaceae
58.		*Microlepia speluncae* (L.) Moore var.	Dennstaedtiaceae
59.		*Microlepia strigosa* (Thunb.) C. Presl.	Dennstaedtiaceae
60.		*Pteridium revolutum* (Blume) Nakai	Dennstaedtiaceae
61.		*Pteridium aquilinum* (L.) Kuhn.	Pteridiaceae
62.		*Sphenomeris chinensis* (L.) Maxon.	Lindsaeaceae
63.		*Ampelopteris prolifera* (Retz.) Copel.	Thelypteridaceae
64.		*Amphineuron opulentum* (Kaulf.) Holtt.	Thelypteridaceae
65.		*Amphineuron terminans* (Hook.) Holtt.	Thelypteridaceae
66.		*Christella dentata* (Forsk.) Brownsey & Jermy.	Thelypteridaceae
67.		*Christella parasitica* (L.) Lev.	Thelypteridaceae
68.		*Cyclosorus parasiticus* (L.) Tard	Thelypteridaceae
69.		*Pronephrium nudatum* (Roxb. ex Griff.) Holtt.	Thelypteridaceae
70.		*Pseudocyclosorus falciloba* (Hook.) Ching.	Thelypteridaceae
71.		*Sphaerostephanos unitus* (L.) Holtt.	Thelypteridaceae
72.		*Thelypteris falciloba* (Hook.) Ching	Thelypteridaceae
73.		*Thelypteris nudatum* (Roxb.) Morto	Thelypteridaceae
74.		*Thelypteris tenera* (Roxb.) C.V. Morton	Thelypteridaceae
75.		*Thelypteris terminans* (J. Sm.) Tagawa & K. Iwats.	Thelypteridaceae
76.		*Trigonospora ciliata* (Benth.) Holtt.	Thelypteridaceae
77.		*Asplenium indicum* Sledge.	Aspleniaceae
78.		*Asplenium laciniatum* D.Don var.	Aspleniaceae
79.		*Asplenium nidus* L.	Aspleniaceae
80.		*Asplenium perakense* Mathew & Christ.	Aspleniaceae
81.		*Asplenium simonsianum* Hook.	Aspleniaceae
82.		*Asplenium unilaterale* Lam.	Aspleniaceae
83.		*Anisocampium cumingianum* Presl.	Athyriaceae
84.		*Athyrium cunningianum* (Presl) Milde	Athyriaceae
85.		*Athyrium falcatum* Bedd.	Athyriaceae
86.		*Diplazium dilatatum* Bl.	Athyriaceae
87.		*Diplazium esculentum* (Retz.) Sw. ex Schred.	Athyriaceae
88.		*Diplazium polypodiodes* Bl. Syn. - *Diplazium asperum* Bl.	Anthyriaceae
89.		*Diplazium spectabile* (Wall. ex Merr.) Ching	Anthyriaceae
90.		*Diplazium subsinuatum* (Wall. ex Hook. & Grev.) Tagawa	Anthyriaceae
91.		*Archinodes aristata* (Frost.f.) Trindale.	Dryopteridaceae
92.		*Dryopteris cochleata* (D.Don) C.Chr.	Dryopteridaceae

Sl. No.	Local Name	Scientific Name	Family
93.		*Dryopteris sparsa* (D.Don) Kuntze.	Dryopteridaceae
94.		*Blechnum orientale* L.	Blechnaceae
95.		*Salvinia cucullata* Roxb. ex Bory.	Salviniaceae
	GYMNOSPERMS		
96.		*Gnetum ula* Brong.	Gnetaceae
97.	Orguna/Odasamari	*Cycas circinalis* L.	Cycadaceae
98.		*Cryptomeria japonica* Don.	Coniferae
99.		*Pinus insularis* Endl.	Coniferae
	ANGIOSPERMS		
100.		*Clematis gouriana* Roxb. ex DC.	Ranunculaceae
101.		*Clematis roylei* Rehder.	Ranunculaceae
102.		*Clematis smilacifolia* Wall.	Ranunculaceae
103.		*Clematis wightiana* Wall.	Ranunculaceae
104.		*Naravelia zeylanica* (L.) DC.	Ranunculaceae
105.		*Ranunculus pensylvanicus* L.f.	Ranunculaceae
106.		*Thalictrum foliolosum* DC.	Ranunculaceae
107.	Rai	*Dillenia aurea* Sm.	Dilleniaceae
108.	Oou	*Dillenia indica* L.	Dilleniaceae
109	Rai	*Dillenia pentagyna* Roxb.	Dilleniaceae
110	Champa	*Magnolia champaca* (L.) Baill. *ex* Pierre Syn. - *Michelia champaca* L.	Magnoliaceae
111.		*Alphonsea lutea* (Roxb.) Hook.f. & Thomson	Annonaceae
112.		*Alphonsea ventricosa* (Roxb.) Hook.f. & Thomson	Annonaceae
113.	Badhiala	*Annona reticulata* L.	Annonaceae
114.	Ata	*Annona squamosa* L.	Annonaceae
115.		*Artabotrys hexapetalus* (L.f.) Bhandari.	Annonaceae
116.		*Desmos chinensis* Lour.	Annonaceae
117.		*Miliusa tomentosa* (Roxb.) J.Sinclair	Annonaceae
118.	Gandha palasa	*Miliusa velutina* (A.DC.) Hook.f. & Thomson	Annonaceae
119.	Champati	*Polyalthia cerasoides* (Roxb.) Bedd.	Annonaceae
120.		*Polyalthia simiarum* (Buch.-Ham. ex Hook. f. & Thomson) Benth. ex Hook. f. & Thomson	Annonaceae
121.		*Polyalthia suberosa* (Roxb.) Thwaites	Annonaceae
122.		*Uvaria hamiltonii* Hook.f. &Thoms.	Annonaceae
123.	Okanbindi	*Cissampelos pareira* L.	Menispermaceae
124.		*Cocculus hirsutus* (L.) W.Theob.	Menispermaceae
125.		*Stephania japonica* (Thunb.) Miers.	Menispermaceae
126.	Guluchi	*Tinospora glabra* (Burm.f.) Merr.	Menispermaceae
127.		*Tinospora sinensis* (Lour.) Merr. Syn. - *Tinospora cordifolia* (Willd.) Miers *ex* Hook.f. & Thoms.	Menispermaceae
128.	Kain	*Nymphaea pubescens* Willd.	Nymphaeaceae
129.	Laphar phula	*Argemone mexicana* L.	Papaveraceae
130.	Sorisa	*Brassica juncea* (L.) Czern.	Brassicaceae
131.	Sorisa	*Brassica rapa* L.	Brassicaceae
132.	Mula	*Raphanus sativus* L.	Brassicaceae
133.	Asadua	*Capparis zeylanica* L.	Capparaceae
134.	Anasorisa	*Cleome viscosa* L.	Capparaceae

Sl. No.	Local Name	Scientific Name	Family
135.	Madana mastaka	*Hybanthus enneaspermus* (L.) F.Muell	Violaceae
136.		*Viola betonicifolia* Sm.	Violaceae
137.	Kunkuma	*Bixa orellana* L.	Bixaceae
138.	Kapasia/Pithaalu	*Cochlospermum religiosum* (L.) Alston.	Bixaceae
139.	Khakada	*Casearia tomentosa* Roxb. Syn. - *Casearia elliptica* Willd.	Salicaceae
140.	Kokra	*Casearia graveolens* Dalz.	Salicaceae
141.		*Casearia rubescens* Dalz.	Salicaceae
142.	Baincha	*Flacourtia indica* (Burm. f.) Merr.	Salicaceae
143.	Baincha	*Flacourtia jangomas* (Lour.) Raeusch.	Salicaceae
144.	Dahanimari	*Homalium napaulense* (DC.) Benth.	Salicaceae
145.		*Homalium tomentosum* (Vent.) Benth.	Salicaceae
146.		*Pittosporum wightii* A. K. Mukh.	Pittosporaceae
147.		*Polygala arvensis* Willd.	Polygalaceae
148.		*Polygala crotalarioides* Buch.-Ham. ex DC.	Polygalaceae
149.		*Polygala elongata* Klein. ex Willd.	Polygalaceae
150.		*Polygala erioptera* DC.	Polygalaceae
151.		*Polygala longifolia* Poir.	Polygalaceae
152.		*Polygala furcata* Royle	Polygalaceae
153.		*Salomonia cantoniensis* Lour.	Polygalaceae
154.		*Salomonia ciliate* (L.) DC.	Polygalaceae
155.		*Drymaria cordata* (L.) Willd. ex Schult.	Caryophyllaceae
156.		*Polycarpaea corymbosa* (L.) Lam.	Caryophyllaceae
157.		*Polycarpon prostratum* (Forssk.) Asch. & Schweinf.	Caryophyllaceae
158.	Badbalbalia	*Portulaca oleracea* L.	Portulacaceae
159.		*Hypericum gaitii* Haines.	Hypericaceae
160.		*Hypericum japonicum* Thunb.	Hypericaceae
161.	Sankada	*Garcinia cowa* Roxb. ex Choisy	Clusiaceae
162.	Sitambo	*Garcinia xanthochymus* Hook.f. ex T. Anders.	Clusiaceae
163.	Nageswar	*Mesua ferrea* L.	Calophyllaceae
164.	Chagachha	*Camellia sinensis* (L.) Kuntze	Theaceae
165.	Sal	*Shorea robusta* Gaertn.	Dipterocarpaceae
166.	Simli	*Bombax ceiba* L.	Malvaceae
167.		*Abelmoschus crinitus* Wall.	Malvaceae
168.		*Abelmoschus manihot* (L.) Medic.	Malvaceae
169.		*Abelmoschus moschatus* Medik.	Malvaceae
170.	Pedipedika	*Abutilon indicum* (L.) Sweet	Malvaceae
171.		*Abutilon persicum* (Burm.f.) Merr.	Malvaceae
172.		*Gossypium barbadense* L.	Malvaceae
173.	Piri-Prirka	*Hibiscus aculeatus* Roxb. Syn. - *Hibiscus furcatus* (J.A.Murr.) Kuntze	Malvaceae
174.		*Hibiscus lobatus* (Murray) Kuntze	Malvaceae
175.		*Hibiscus platanifolius* (Willd.) Sweet	Malvaceae
176.	Kapasia	*Kydia calycina* Roxb.	Malvaceae
177.	Bajramuli	*Sida acuta* Burm.f.	Malvaceae
178.		*Sida cordata* (Burm.f.) Borss.	Malvaceae
179.		*Sida mysorensis* Wight & Arn.	Malvaceae
180.		*Sida rhombifolia* L.	Malvaceae

Sl. No.	Local Name	Scientific Name	Family
181.	Bankapasi	*Thespesia lampas* (Cav.) Dalz.	Malvaceae
182.		*Urena lobata L.subsp. lobata*	Malvaceae
183.		*Urena lobata L.subsp. sinuata*	Malvaceae
184.	Samarkhoi	*Byttneria herbacea* Roxb.	Malvaceae
185.	Kodalo	*Firmiana colorata* (Roxb.) R.Br.	Malvaceae
186.		*Guazuma ulmifolia* Lam.	Malvaceae
187.	Mudimudika	*Helicteres isora* L.	Malvaceae
188.	Telpuri	*Melochia corchorifolia* L.	Malvaceae
189.	Kanaka champa	*Pterospermum acerifolium* (L.) Willd.	Malvaceae
190.	Muchukunda	*Pterospermum xylocarpum* (Gaertn.) Sant. &Wagh	Malvaceae
191.	Genduli	*Firmiana simplex* (L.) W.Wight Syn. - *Sterculia urens* Roxb.	Malvaceae
192.		*Sterculia villosa* Roxb.	Malvaceae
193.		*Waltheria indica* L.	Malvaceae
194.	Bananalita	*Corchorus aestuans* L.	Malvaceae
195.	Panasia	*Elaeocarpus tectorius* (Lour.) Poir.	Elaeocarpaceae
196.		*Elaeocarpus stipularis* var. *siamensis* (Craib) Coode Syn. - *Elaeocarpus wallichii* Kurz	Elaeocarpaceae
197.		*Sloanea sterculiacea* (Benth.) Rehder & E.H. Wilson	Elaeocarpaceae
198.		*Grewia abutilifolia* Vent. ex Juss	Malvaceae
199.		*Grewia serrulata* DC. Syn. - *Grewia disperma* Rottl.	Malvaceae
200.	Mirgi chara	*Grewia eriocarpa* Juss. Syn. - *Grewia elastica*	Malvaceae
201.	Sonaranga	*Grewia hirsuta* Vahl	Malvaceae
202.	Burso	*Grewia sapida* Roxb. ex DC.	Malvaceae
203.	Dhaman	*Grewia tiliifolia* Vahl	Malvaceae
204.	Jata	*Triumfetta annua* L.	Malvaceae
205.		*Triumfetta pentandra* A.Rich.	Malvaceae
206.		*Triumfetta pilosa* Roth	Malvaceae
207.		*Triumfetta rhomboidea* Jacq.	Malvaceae
208.	Tisi	*Linum usitatissimum* L.	Linaceae
209.		*Reinwardtia indica* Dumort.	Linaceae
210.		*Aspidopterys tomentosa* var. *hutchinsonii* (Haines) R.C.Srivast.	Malpighiaceae
211.	Banomali	*Hiptage benghalensis* (L.) Kurz	Malpighiaceae
212.		*Biophytum reinwardtii* (Zucc.) Klotz	Oxalidaceae
213.	Ambliti	*Oxalis corniculata* L.	Oxalidaceae
214.		*Oxalis dehradunensis* Raizada	Oxalidaceae
215.		*Acronychia pedunculata* (L.) Miq.	Rutaceae
216.	Bela	*Aegle marmelos* (L.) Corrêa	Rutaceae
217.		*Atalantia monophylla* DC.	Rutaceae
218.	Bheru	*Chloroxylon swietenia* DC.	Rutaceae
219.	Kamala	*Citrus aurantium* L.	Rutaceae
220.	Jambira (Janglilembu)	*Citrus medica* L.	Rutaceae
221.	Agnijala	*Clausena excavata* Burm.f.	Rutaceae
222.	Chauli	*Glycosmis pentaphylla* (Retz.) DC.	Rutaceae
223.		*Micromelum minutum* (G.Forst.) Wight & Arn.	Rutaceae
224.	Bhursunga	*Murraya koenigii* (L.) Sprengel	Rutaceae

Sl. No.	Local Name	Scientific Name	Family
225.	Kamini	*Murraya paniculata* (L.) Jack	Rutaceae
226.	Baintha	*Naringi crenulata* (Roxb.) Nicolson	Rutaceae
227.		*Toddalia asiatica* (L.) Lam.	Rutaceae
228.	Morei / Pihula	*Zanthoxylum rhetsa* (Roxb.) DC.	Rutaceae
229.	Radha champa	*Ochna obtusata DC* var. *obtusata*	Ochnaceae
230.	Bhuin champa	*Ochna obtusata DC var.pumila*	Ochnaceae
231.	Salai	*Boswellia serrata* Roxb. ex Colebr.	Burseraceae
232.	Moi / Katha kusum	*Garuga pinnata* Roxb.	Burseraceae
233.	Rimuli	*Protium serratum* (Wall. ex Colebr.) Engl.	Burseraceae
234.		*Aphanamixis polystachya* (Wall.) R. Parker	Meliaceae
235.	Nimba	*Azadirachta indica* A.Juss.	Meliaceae
236.	Pittamari	*Cipadessa baccifera* (Roth) Miq.	Meliaceae
237.	Mahanimbo	*Melia azedarach* L.	Meliaceae
238.	Rohini	*Soymida febrifuga* (Roxb.) A. Juss.	Meliaceae
239.	Tun	*Toona ciliata* M. Roem.	Meliaceae
240.	Raiphala	*Heynea trijuga* Roxb. ex Sims Syn. - *Trichilia connaroides* (Wight &Arn.) Bentvelzen.	Meliaceae
241.	Mundica	*Walsura trifoliolata* (A.Juss.) Harms	Meliaceae
242.		*Natsiatum herpeticum* Buch.-Ham. ex Arn	Icacinaceae
243.	Bhadai	*Olax scandens* Roxb.	Olacaceae
244.		*Opilia amentacea* Roxb.	Opiliaceae
245.		*Cassine glauca* (Rottb.) Kuntze	Celastraceae
246.	Pengu	*Celastrus paniculatus* Willd.	Celastraceae
247.		*Euonymus glaber* Roxb.	Celastraceae
248.		*Maytenus bailadillana* (V.Naray. & Mooney) D.C.S.Raju & Babu	Celastraceae
249.		*Siphonodon celastrineus* Griff.	Celastraceae
250.		*Gouania tiliifolia* Lam.	Rhamnaceae
251.		*Helinus lanceolatus* Brandis	Rhamnaceae
252.		*Rhamnus napalensis* (Wall.) M.A. Lawson	Rhamnaceae
253.	Rainjhai	*Ventilago denticulata* Willd.	Rhamnaceae
254.	Pitchule	*Ventilago maderaspatana* Gaertn.	Rhamnaceae
255.		*Ziziphus glabrata* B.Heyne ex Roth	Rhamnaceae
256.	Borokoli	*Ziziphus jujuba* Mill. Syn. - *Ziziphus mauritiana* Lam.	Rhamnaceae
257.	Kantakoli	*Ziziphus oenopolia* (L.) Mill.	Rhamnaceae
258.	Chunkoli/Tinkoli	*Ziziphus rugosa* Lam.	Rhamnaceae
259.	Gonti/ Goto	*Ziziphus xylopyrus* (Retz.) Willd.	Rhamnaceae
260.		*Ampelocissus divaricata* (Wall. ex M.A.Lawson) Planch.	Vitaceae
261.	Paninoha	*Ampelocissus latifolia* (Roxb.) Planch.	Vitaceae
262.		*Ampelocissus tomentosa* (B.Heyne & Roth) Planch.	Vitaceae
263.		*Cyphostemma auriculatum* (Roxb.) P.Singh & B.V.Shetty Syn. - *Cayratia auriculata* (Roxb.) Gamble	Vitaceae
264.		*Cayratia pedata* (Lam.) Gagnep.	Vitaceae
265.		*Cissus adnata* Roxb.	Vitaceae
266.		*Cissus assamica* (M.A.Lawson) Craib	Vitaceae
267.		*Cissus quadrangularis* L.	Vitaceae
268.	Tekual, Panibel	*Cissus repanda* (Wight & Arn.) Vahl	Vitaceae
269.	Diboria	*Cissus repens* Lam.	Vitaceae

Sl. No.	Local Name	Scientific Name	Family
270.		*Leea aequata* L.	Vitaceae
271.		*Leea asiatica* (L.) Ridsdale	Vitaceae
272.		*Leea guineensis* G. Don Syn. - *Leea acuminata* Wallich ex Clarke	Vitaceae
273.	Kalad chana	*Leea indica* (Burm. f.) Merr.	Vitaceae
274.	Jibni	*Leea macrophylla* Roxb. ex Hornem.	Vitaceae
275.		*Tetrastigma lanceolarium* (Roxb.) Planch.	Vitaceae
276.	Lahunga/ Mahangakoli	*Lepisanthes rubiginosa* (Roxb.) Leenh. Syn. - *Erioglossum rubiginosum* (Roxb.) Blume	Sapindaceae
277.	Lichu	*Litchi chinensis* Sonn.	Sapindaceae
278.	Kusuma	*Schleichera oleosa* (Lour.) Merr.	Sapindaceae
279.		*Turpinia cochinchinensis* (Lour.) Merr. Syn. - *Turpinia nepalensis* Wall.	Staphyleaceae
280.		*Meliosma pinnata* (Roxb.) Maxim.	Sabiaceae
281.		*Meliosma simplicifolia* (Roxb.) Walp.	Sabiaceae
282.	Kaju	*Anacardium occidentale* L.	Anacardiaceae
283.	Chara	*Buchanania cochinchinensis* (Lour.) M.R.Almeida Syn. - *Buchanania lanzan* Spreng.	Anacardiaceae
284.	Moi	*Lannea coromandelica* (Houtt.) Merr.	Anacardiaceae
285.	Amba	*Mangifera indica* L.	Anacardiaceae
286.		*Nothopegia heyneana* Gamble	Anacardiaceae
287.		*Rhus chinensis* Mill. Syn. - *Rhus semialata* Murray	Anacardiaceae
288.	Bhalia	*Semecarpus anacardium* L.f.	Anacardiaceae
289.	Ambada	*Spondias pinnata* (L.f.) Kurz	Anacardiaceae
290.	Sajana	*Moringa oleifera* Lam.	Moringaceae
291.	Kanchana	*Bauhinia acuminata* L.	Leguminosae
292.		*Bauhinia malabarica* Roxb.	Leguminosae
293.	Barada	*Bauhinia purpurea* L.	Leguminosae
294.	Amta	*Bauhinia semla* Wunderlin	Leguminosae
295.	Siali	*Bauhinia vahlii* Wight & Arn.	Leguminosae
296.	Kanchan	*Bauhinia variegata* L.	Leguminosae
297.		*Caesalpinia cucullata* Roxb.	Leguminosae
298.		*Caesalpinia pulcherrima* (L.) Sw.	Leguminosae
299.	Sunari	*Cassia fistula* L.	Leguminosae
300.		*Senna hirsuta* (L.) H.S.Irwin & Barneby Syn. - Cassia hirsuta L.	Leguminosae
301.		*Chamaecrista nictitans* subsp. *patellaria* (Collad.) H.S.Irwin & Barneby Syn. - *Cassia lechenaultiana* DC.	Leguminosae
302.	Kala chakunda	*Senna occidentalis* Link. Syn. - *Cassia occidentalis* L.	Leguminosae
303.		*Senna sophera* (L.) Roxb. Syn. - *Cassia sophera* L.	Leguminosae
304.	Chakunda	*Senna tora* (L.) Roxb. Syn. - *Cassia tora* L.	Leguminosae
305.	Krushnachuda	*Delonix regia* (Hook.) Raf.	Leguminosae
306.	Ashok	*Saraca asoca* (Roxb.) Willd.	Leguminosae

Sl. No.	Local Name	Scientific Name	Family
307.	Tentuli	*Tamarindus indica* L.	Leguminosae
308.		*Acacia auriculiformis* Benth.	Leguminosae
309.		*Acacia farnesiana* (L.) Willd.	Leguminosae
310.	Ghar-khair	*Acacia lenticularis* Benth.	Leguminosae
311.	Kantasiris	*Acacia leucophloea* (Roxb.) Willd.	Leguminosae
312.	Babul	*Acacia nilotica* (L.) Delile	Leguminosae
313.	Dantari	*Acacia pennata* (L.) Willd.	Leguminosae
314.	Dantari	*Acacia torta* (Roxb.) Craib	Leguminosae
315.	Kala sirisa	*Albizia chinensis* (Osbeck) Merr.	Leguminosae
316.	Sirisa	*Albizia lebbeck* (L.) Benth.	Leguminosae
317.		*Albizia odoratissima* (L.f.) Benth.	Leguminosae
318.	Satpuri	*Albizia procera* (Roxb.) Benth.	Leguminosae
319.	Gilo/Giredi	*Entada rheedii* Spreng.	Leguminosae
320.	kirkichi	*Mimosa himalayana* Gamble	Leguminosae
321.	Lajkuli	*Mimosa pudica* L.	Leguminosae
322.	Chakunda	*Samanea saman* (Jacq.) Merr.	Leguminosae
323.	Kongda, Bankhira	*Xylia xylocarpa* (Roxb.) Taub.	Leguminosae
324.	Kaicho, Runjo	*Abrus precatorius* L.	Leguminosae
325.	Lajuari	*Aeschynomene indica* L.	Leguminosae
326.		*Alysicarpus vaginalis* (L.) DC.	Leguminosae
327.	Ban Kulthia	*Cajanus scarabaeoides* (L.) Thouars Syn. - *Atylosia scarabaeoides* (L.) Benth.	Leguminosae
328.		*Cajanus crassus* (King) Maesen Syn. - *Atylosia volubilis* (Blanco) Gamble	Leguminosae
329.	Palasa	*Butea monosperma* (Lam.) Taub.	Leguminosae
330.		*Spatholobus parviflorus* (DC.) Kuntze Syn. - *Butea parviflora* Roxb.	Leguminosae
331.	Lata palasa	*Butea superba* Roxb.	Leguminosae
332.		*Canavalia gladiata* (Jacq.) DC.	Leguminosae
333.	Buto	*Cicer arietinum* L.	Leguminosae
334.	Aparajita	*Clitoria ternatea* L.	Leguminosae
335.		*Cordariocalyx motorius* (Houtt.) Ohashi	Leguminosae
336.		*Crotalaria acicularis* Benth.	Leguminosae
337.		*Crotalaria albida* Roth.	Leguminosae
338.		*Crotalaria bialata* Schrank	Leguminosae
339.		*Crotalaria calycina* Schrank	Leguminosae
340.		*Crotalaria Montana* Roth	Leguminosae
341.		*Crotalaria mysorensis* Roth	Leguminosae
342.	Jhunjhunka	*Crotalaria pallida* Aiton	Leguminosae
343.		*Crotalaria prostrata* Willd.	Leguminosae
344.		*Crotalaria retusa* L.	Leguminosae
345.	Jhunka	*Crotalaria spectabilis* Roth	Leguminosae
346.		*Dalbergia lanceolaria* L.f.	Leguminosae
347.	Pahadisisu	*Dalbergia latifolia* Roxb.	Leguminosae
348.	Sujuni	*Dalbergia lanceolaria* subsp. *paniculata* (Roxb.) Thoth.	Leguminosae
349.		*Dalbergia pinnata* (Lour.) Prain.	Leguminosae
350.		*Dalbergia volubilis* Roxb.	Leguminosae
351.		*Desmodium benthamii* N.P. Balakr.	Leguminosae

FOREWORD

Both flora and fauna, that are integral to our ecology, are facing the threat of growing human habitation and associated growth factors. It is incumbent, therefore, to create a balance and harmony between ecology and habitations.

Trees serve as lungs in a densely populated space and sequester atmospheric carbon, besides imparting aesthetics to the concrete jungles. This book titled "Flamboyant Blooming Trees" brings forth the beauty of the multi-hued flowers of the trees, that dot our cities and countryside.

The book has been written in a simple and lucid manner keeping scientific jargons to the minimum. A passerby who loves and admires nature's beauty is often at loss of words when it comes to identifying or discussing a tree. This book is precisely meant for such lovers of nature, so that, not only are they able to enjoy the exuberance of the flowering trees, but can also select trees that suit their location for planting. By implication, they can partner with city planners in greening and ameliorating their city ambience. However, some scientific information that has been sprinkled in right measure adds an academic touch and authenticity to the book.

Plant nursery owners and landscapists would also benefit from this book, by being able to showcase to their clients the tree size, shape, bloom time, bloom colour and intense blooming days. It will also help students of botany, agriculture, horticulture and forestry in their academic and scientific pursuits.

The blooms in this book have been captured from different parts of our vast country and contains trees native to India as well as those introduced from other countries. The book therefore demonstrates its pan - tropical world representation. People from other countries with climatic conditions similar to that of India could also benefit.

Dr, Shobha Nagnur a student of Home Science and a passionate gardener has combined creativity & art in designing this pictorial book. A Home Science degree from an Agriculture University has equipped her to embark upon writing this coffee table book by benefiting from access to course work in horticulture and intense exposure to natural and tended gardens. Being a teacher, researcher and extensionist, she blends in this book her love for nature, keenness of academics, the rigour of research and the desire to disseminate knowledge and information. It is a well researched and highly readable book that communicates through well captured photographs and optimal literature.

I congratulate the author for making a valuable contribution to the existing fund of knowledge. Dr. Shobha Nagnur has put in intense efforts that will certainly help to reach out to the society and let the readers appreciate the trees in their own surroundings and promote expansion of their coverage.

Dr. M.B. Chetti
Vice Chancellor
University of Agriculture Sciences,
Dharwad.

PREFACE

This book on trees is second of the triology-climbers, trees and shrubs. It follows my already published book titled "Elegant Flowering Climbers" that was inspired by B.P.Pal's "Beautiful Climbers".

I love sauntering down to friends' homes and visiting gardens. The nature of my work has always complemented my love for travel, letting me blend professional and personal interests. These natural conspiracies provided me ample opportunities to use my camera with gay abandon. I am at home with nature, and love the floral diversity

As I peeped each time intensely through the lens to frame the images suitably, I began to discover interesting and fascinating facts about the trees. For instance, I never had really looked up the Silver oak to notice the bright orange cluster of flowers that resembled a hair brush, nor had I observed the individual flowers on the tentacle like structures of the Umbrella tree. This is how, my passion let me adopt nature as my teacher, and learn to appreciate the variegation of life.

Not just Dharwad, but India became my canvas as I journeyed around and found limitless scope to notice, observe and photograph the trees that dot the country-side and the avenues of the cities. I noticed the similarity that bound together the species adopted across the country, with just hill tracts as an exception. While trees like Neem, Siris, Tamarind and the like are seen along highways and the country-side, Cassias, Bauhinias, Jacaranda, Gulmohar etc. are the favoured choice of the city landscapes on account of their colorful & picturesque flowers. However, flowering being temperature-dependent, the same plants may bloom at different times during different years & regions. It is not surprising, therefore, if someone sees the trees blooming in April- May in South India, blooming during June- July up North India.

Once I had the photographs, a thorough and scientific study was required before I put them together. While I consulted botanists, horticulturists and foresters, I also scanned through literature and browsed the net for identification and additional information. It was then that I chanced upon the book by D.V. Cowen (1950) titled "Flowering Trees and Shrubs of India" from which I have unabashedly borrowed many a thought and articulation. Not finding better words than Cowen, I quote her below verbatim:

> *"Every year from March to May, when the air is at its hottest, the ground is parched and cracked, and a film of dust covers every leaf and branch, one watches in awe that miracle of nature which brings forth from the branches of so many trees such an extravagant abundance of colourful blossoms and clean, polished foliage. From where, one asks, do these trees get sufficient moisture when no rain has fallen for months. The bright new leaves alone would be a happy sight and one's eyes with the incessant glare of an Indian summer gladly rest on them, but with blossoms of every hue as well, it is impossible to be unaware of the galaxy of colours. It is at this time of the year that so many people feel the urge to know more about the trees they live amongst."*

Finding similar, rather identical thoughts echoing within me, I embarked upon this initiative of mine. The book aims to provide the lay people an easily comprehensible content, desirous of knowing more about the trees that surround them, and keen to satisfy their curiosity and experience personal interest in tree planting.

The trees that I have presented in this book follow no fixed rule, and the selection of a particular tree for the inclusion is solely my personal choice and discretion. The only guiding principle being, that the tree hosted attractive flowers and that it caught the fancy of my eyes.

In today's fast paced modern life, where people don't find time even to stand and stare at the boughs, a leisurely reading is always a challenge. Keeping this in mind, minimal text and maximum visual, characterise the content presentation in this book.

It has been for me, both an exciting and learning experience putting together the thoughts in the form of a book. I hope the readers will find the contents sufficient enough to quench their thirst for basic knowledge about the nature around them, and the photographs good enough to kindle a new appreciation for the beauty and calmness that they bring.

Shobha Nagnur

ACKNOWLEDGEMENT

It is my family, that has motivated me into my life's natural academic aspirations. It has been my parents' home with its natural garden, that initially kindled and thereafter sustained my love for nature. I thank my father, Shri V.B. Nagnur and mother, Smt. Shankuntala Nagnur for enabling me this luxury. Sadly, though my mother does not live to see this book, I can however rest content that I could present her my first book on 'Flowering Climbers'. My sisters Shama Udapudi, Sheela Patil and Deepika Kenkere have through their loving motivation and honest critique taught me the value of qualitative output.

Ashok Dalwai has consistently and constantly goaded me into taking up this work, and stick to timelines. A casual enquiry once a while by my son Manthan has always served to wake me out of my occasional procrastination.

My friends Geeta Shirnalli, Ganga Yenagi, Naheeda Anjum, Hemalata S. Nithyashree D.A., Annapoorna K. and Geeta Channal have been the ones, on whom I could always depend upon for field work and research. Many of the species included in the book were introduced to me by them.

My colleagues and younger friends Surekha and Rajeshwari kept me company for out of office late hour work on this book. My students Supriya Patil, Akshata Ramannavar, Veena Chandavari, Shivaleela Patil, Geeta Gadekar, Praveen Hotkar and Shweta Mannikeri have readily offered me the much needed support through their fresh minds & agile fingers.

Arjun Goudar accompanied me willingly in photographing the beautiful flowers and added a professional touch to the photos. The required net search and creativity in designing the book came from Ulappa Mestri - someone always willing to help.

Thanks to all these and several other well wishers, this book is what it is today. I know that thanks is too simple a word, but probably this is the only way to do it formally.

Shobha Nagnur

CONTENTS

INTRODUCTION

The genetic diversity of nature and uniqueness of each species is best captured in the following saying of Leonardo-da-Vinci, one of the greatest minds of the Renaissance period

> *"To such an extent does nature delight and abound in variety that among her trees, there is not one plant to be found which is exactly like another and not only among the plants, among the boughs, the leaves and the fruits you will not find one which is exactly similar to the other."*

The immortal painter's expressions extol the aesthetic value of trees. However, man driven by several economic considerations also values the commercial potential of the flora. The tree species of plants contain multiple utilitarian values, and are therefore raised and nurtured for more reasons than one. In sum, the trees are important from the perspective of economy, environment and health. The manifold reasons as to why the society requires more trees are:

ENVIRONMENTAL BENEFITS

- Trees are the lungs of the planet, absorbing carbon dioxide, releasing oxygen and reducing the effects of green-house gases (GHGs), and thereby helping mitigate global warming. Trees serve as an important sink for absorption of GHGs
- In the city landscapes, trees absorb sound and reduce noise pollution, especially the one caused by motor vehicles
- In a tropical country like India, trees provide the much needed shade and create a cool micro-climate, particularly welcome during the hot summers
- Micro-climates facilitated by shades bring down the local temperature, and help save energy consumption used in different cooling systems like air conditioners. It is all such small steps, that aid in decelerating the pace of GHG emissions - a positive step towards Sustainable development Goals (SDGs), 2030
- Trees improve air quality by acting as natural air filters, removing dust and smoke from the atmosphere
- Trees provide micro-habitats for squirrels, birds, bees, beetles and flies
- A good tree cover improves the canopy cover in forests & common lands, and thereby supports the floral & faunal bio-diversity including wildlife.

HEALTH BENEFITS

- Just living amidst trees can create a happy feeling; thereby improving the physical and mental health
- As taught by Ayurveda system, just walking by the side of the trees or below them lowers blood pressure, slows down heart rate and brings down stress
- There are large number of tree species, whose different parts contain medicinal and therapeutic ingredients and serve as inputs in multiple industries - pharmaceutical (including ayurvedic system of medicine), cosmetic etc.

ECOLOGICAL BENEFITS

- Trees absorb and store rain water; helping ground water recharge
- The roots that spread under the soil, surface bind the soil and check soil erosion, that may occur due to water and wind
- The fallen leaves make excellent compost that enrich the soil, and provide the much needed organic carbon

- ☆ Some tree species are leguminous in nature, and fix atmospheric nitrogen into soil, improving its nutrient status & crop productivity.

ECONOMIC BENEFITS

- ☆ Trees from earlier times have provided wood which has been the first fuel for cooking and still continues to be the primary fuel in some parts of the country
- ☆ Trees provide timber for building construction, furniture manufacture, agricultural & industrial tools, sporting equipments and several household items. Wood pulp is used to make paper, cardboard etc.
- ☆ Trees provide multiple types of fruits & nuts that form an important component of nutrition security.
- ☆ The bark of some trees is used for making cork as well as for extraction of medicines and chemicals.
- ☆ Many trees possess medicinal properties and so have gained importance in pharma and ayurvedic companies, that create jobs and incomes
- ☆ Several tree species are an important source of fodder for domesticated animals.

AESTHETIC BENEFITS

- ☆ Trees are loved by everyone even if they do not serve any other utilitarian purpose, thanks to their aesthetic appeal, that please the eyes and feast the mind. Trees break the monotony of landscape, particularly during the dry summers by display of green foliage and colourful flowers
- ☆ In city landscapes, they sublimate the harsh lines of the buildings and provide cover to screen unsightly views
- ☆ Trees display an endless variety of shapes, forms, textures and vibrant colours. As the season change, so does the appearance of the tree. By constantly changing throughout the course of year, they add beauty to the surroundings
- ☆ Trees bring together people of the community for walking and bird watching. Children enjoy playing amongst the trees and discover their sense of adventure, besides learning from nature

A wide spectrum of benefits, that trees offer to life on earth has been attempted, it is at best an illustrative list, and certainly not exhaustive

Appreciating, that the canvas of trees is too vast to be treated in a coffee table book, the content has been limited to certain trees only. While all angiosperms flower, the book highlights only those tree species, that exhibit beautiful flowers, making it a pre-requisite to learn about the Intense Blooming Days

INTENSE BLOOMING DAYS (IBDs)

The IBDs are those days when a plant blooms 50 per cent or more of its full potential. This period is also one of wow effect, when the tree boasts of its best display of blooms. The IBDs constitute the start of the blooming days and the duration of the blooming days. The IBDs are vulnerable to temperature, and may therefore face either delayed or advanced bloom by a few days. The studies show that:

- Start of IBDs is delayed after a cold winter
- Duration of IBDs is shortened after a cold winter
- Colder than normal weather causes delayed start and shortens the duration of flowering of certain trees

Loss of trees and deforestation has crippled the eco-system. The greatest challenge of the 21st century is to save the environment from disaster beyond the threshold level of restoration.

Sl. No.	Local Name	Scientific Name	Family
352.		*Desmodium dichotomum* (Willd.) DC.	Leguminosae
353.		*Desmodium gangeticum* (L.) DC.	Leguminosae
354.		*Desmodium heterocarpon* (L.) DC.	Leguminosae
355.		*Desmodium laxiflorum* DC.	Leguminosae
356.		*Desmodium microphyllum* (Thunb.) DC.	Leguminosae
357.	Bandhano	*Desmodium oojeinense* (Roxb.) H. Ohashi	Leguminosae
358.	Birkapi	*Phyllodium pulchellum* (L.) Desv. Syn. - *Desmodium pulchellum* (L.) Benth.	Leguminosae
359.		*Dendrolobium triangulare* (Retz.) Schindl. Syn. - *Desmodium triangulare* (Retz.) Merr.	Leguminosae
360.		*Desmodium triflorum* (L.) DC.	Leguminosae
361.	Salparni	*Tadehagi triquetrum* (L.) H.Ohashi Syn. - *Desmodium triquetrum* (L.) DC.	Leguminosae
362.		*Tadehagi pseudotriquetrum* (DC.) Y.C.Yang & P.H.Huang Syn. - *Desmodium triquetrum* subsp. *pseudotriquetrum* (DC.) Prain	Leguminosae
363.		*Desmodium velutinum* (Willd.) DC.	Leguminosae
364.		*Dumasia villosa* DC.	Leguminosae
365.		*Dunbaria rotundifolia* (Lour.) Merr.	Leguminosae
366.	Paldhua	*Erythrina suberosa* Roxb.	Leguminosae
367.		*Flemingia strobilifera* (L.) W.T.Aiton Syn. - *Flemingia bracteata* (Roxb.) Wight	Leguminosae
368.	Rani Kathi	*Flemingia chappar* Benth.	Leguminosae
369.		*Flemingia macrophylla* (Willd.) Merr. Syn. - *Flemingia nana* Roxb.	Leguminosae
370.		*Flemingia paniculata* Benth.	Leguminosae
371.		*Flemingia prostrata* Roxb.	Leguminosae
372.		*Flemingia stricta* Roxb.	Leguminosae
373.		*Galactia longifolia* (Jacq.) Benth.	Leguminosae
374.		*Indigofera atropurpurea* Hornem.	Leguminosae
375.	Giliri	*Indigofera cassioides* DC.	Leguminosae
376.		*Indigofera glabra* L.	Leguminosae
377.		*Indigofera linifolia* (L.f.) Retz.	Leguminosae
378.		*Indigofera prostrate* Willd.	Leguminosae
379.		*Indigofera spicata* Forssk.	Leguminosae
380.	Simba	*Lablab purpureus* (L.) Sweet	Leguminosae
381.	Guadhuni	*Millettia extensa* (Benth.) Baker	Leguminosae
382.		*Endosamara racemosa* (Roxb.) R.Geesink Syn. - *Millettia racemosa* (Roxb.) Benth	Leguminosae
383.	Bado baidonko	*Mucuna nigricans* (Lour.) Steud.	Leguminosae
384.	Baidanka	*Mucuna pruriens* (L.) DC.	Leguminosae
385.	Karanja	*Pongamia pinnata* (L.) Pierre	Leguminosae
386.		*Pseudarthria viscida* (L.) Wight & Arn.	Leguminosae
387.	Piasal/ Bija	*Pterocarpus marsupium* Roxb.	Leguminosae
388.	Bhuinkakharu	*Pueraria tuberose* (Willd.) DC.	Leguminosae
389.		*Sesbania bispinosa* (Jacq.) W.Wight	Leguminosae
390.	Jhilliphula	*Shuteria involucrata* (Wall.) Wight & Arn.	Leguminosae
391.		*Smithia conferta* Sm.	Leguminosae
392.		*Smithia sensitiva* Aiton	Leguminosae

Sl. No.	Local Name	Scientific Name	Family
393.		*Sophora bakeri* Prain	Leguminosae
394.	Gileri	*Tephrosia purpurea* (L.) Pers.	Leguminosae
395.		*Teramnus labialis* (L.f.) Spreng	Leguminosae
396.		*Teramnus mollis* Benth.	Leguminosae
397.	Krushna parni	*Uraria lagopodoides* (L.) DC. Syn. - *Uraria alopecuroides* (Roxb.) Sweet	Leguminosae
398.	Salparni	*Uraria rufescens* (DC.) Schindl.	Leguminosae
399.		*Vigna adenantha* (G.Mey.) Marechal & al.	Leguminosae
400.	Biri	*Vigna mungo* (L.) Hepper	Leguminosae
401.		*Vigna pilosa* (Klein ex Willd.) Baker	Leguminosae
402.	Banamongo	*Vigna radiata* (L.) R.Wilczek	Leguminosae
403.		*Vigna umbellata* (Thunb.) Ohwi & H.Ohashi	Leguminosae
404.	Jhudanga	*Vigna unguiculata* (L.) Walp.	Leguminosae
405.		*Zornia gibbosa* Span.	Leguminosae
406.	New record	*Duchesnea indica* (Jacks.) Focke	Rosaceae
407.		*Prunus ceylanica* (Wight) Miq.	Rosaceae
408.		*Prunus pygeoides* Koehne	Rosaceae
409.		*Rubus ellipticus* Sm.	Rosaceae
410.	Hemsagar	*Bryophyllum pinnatum* (Lam.) Oken Syn. - *Kalanchoe pinnata* (Lam.) Pers.	Crassulaceae
411.		*Drosera burmanni* Vahl	Droseraceae
412.		*Drosera indica* L.	Droseraceae
413.		*Carallia brachiata* (Lour.) Merr.	Rhizophoraceae
414.	Dhaura	*Anogeissus latifolia* (Roxb. ex DC.) Wall. ex Guillem. & Perr.	Combretaceae
415.	Kokundia	*Getonia floribunda* Roxb. Syn. - *Calycopteris floribunda* (Roxb.) Lam. ex Poir.	Combretaceae
416.		*Combretum nanum* Buch.-Ham. ex D.Don	Combretaceae
417.	Atundi	*Combretum album* Pers. Syn. - *Combretum roxburghii* Spreng.	Combretaceae
418.	Asan	*Terminalia tomentosa* Wight & Arn.	Combretaceae
419.	Arjuna	*Terminalia arjuna* (Roxb. ex DC.) Wight & Arn.	Combretaceae
420.	Bahada	*Terminalia bellirica* (Gaertn.) Roxb.	Combretaceae
421.	Harida	*Terminalia chebula* Retz.	Combretaceae
422.		*Corymbia citriodora* (Hook.) K.D.Hill & L.A.S.Johnson Syn. - *Eucalyptus citriodora* Hook.	Myrtaceae
423.		*Eucalyptus tereticornis* Sm.	Myrtaceae
424.		*Corymbia terminalis* (F.Muell.) Hill & Johnson Syn. - *Eucalyptus terminalis*	Myrtaceae
425.	Pijuli	*Psidium guajava* L.	Myrtaceae
426.	Poijamu	*Syzygium nervosum* A.Cunn. ex DC. Syn. - *Syzygium cerasoides* (Roxb.) Raizada	Myrtaceae
427.	Jamu	*Syzygium cumini* (L.) Skeels	Myrtaceae
428.	Bhalu jamu	*Syzygium fruticosum* DC.	Myrtaceae
429.		*Syzygium salicifolium* (Wight) J.Graham Syn. - *Syzygium heyneanum* (Duthie) Gamble	Myrtaceae
430.	Golab jamu	*Syzygium jambos* (L.) Alston	Myrtaceae
431.	Dhobani	*Syzygium praecox* (Roxb.) Rathakr. & N.C.Nair Syn. - *Syzygium roxburghianum* Raizada	Myrtaceae

Sl. No.	Local Name	Scientific Name	Family
432.	Hinjal	*Barringtonia acutangula* (L.) Gaertn.	Lecythidaceae
433.	Kumbhi	*Careya arborea* Roxb.	Lecythidaceae
434.	Koroli	*Melastoma malabathricum* L.	Melastomataceae
435.	Niresho	*Memecylon umbellatum* Burm. f.	Melastomataceae
436.		*Osbeckia chinensis* L.	Melastomataceae
437.		*Osbeckia muralis* Naudin Syn. - *Osbeckia truncata* D.Don ex Wight & Arn.	Melastomataceae
438.		*Osbeckia stellata* Buch.-Ham. ex Ker Gawl.	Melastomataceae
439.		*Sonerila erecta* Jack Syn. - *Sonerila tenera* Royle	Melastomataceae
440.		*Ammannia baccifera* L.	Lythraceae
441.		*Ammannia multiflora* Roxb.	Lythraceae
442.	Sidha	*Lagerstroemia parviflora* Roxb.	Lythraceae
443.	Patoli	*Lagerstroemia speciosa* (L.) Pers. Syn. - *Lagerstroemia reginae* Roxb.	Lythraceae
444.	Manjuati	*Lawsonia inermis* L.	Lythraceae
445.		*Rotala densiflora* (Roth) Koehne	Lythraceae
446.		*Rotala indica* (Willd.) Koehne	Lythraceae
447.		*Rotala rotundifolia* (Buch.-Ham. ex Roxb.) Koehne	Lythraceae
448.	Dhatuki	*Woodfordia fruticosa* (L.) Kurz	Lythraceae
449.	Dalimba	*Punica granatum* L.	Lythraceae
450.		*Ludwigia hyssopifolia* (G.Don) Exell	Onagraceae
451.		*Ludwigia octovalvis* (Jacq.) P.H.Raven	Onagraceae
452.		*Ludwigia perennis* L.	Onagraceae
453.		*Ludwigia prostrata* Roxb.	Onagraceae
454.	Panisingada	*Trapa natans* L.	Trapaceae
455.		*Turnera ulmifolia* L.	Turneraceae
456.	Papaya	*Carica papaya* L.	Caricaceae
457.	Banakundri	*Coccinia grandis* (L.) Voigt	Cucurbitaceae
458.		*Cucumis melo* L.	Cucurbitaceae
459.		*Cucumis sativus* L.	Cucumberaceae
460.	Kakharu	*Cucurbita maxima* Duchesne	Cucumberaceae
461.	Shivalingi	*Diplocyclos palmatus* (L.) C.Jeffrey	Cucurbitaceae
462.	New report	*Gymnopetalum chinense* (Lour.) Merr.	Cucurbitaceae
463.	New report	*Gynostemma pentaphyllum* (Thunb.) Makino	Cucurbitaceae
464.	Lau	*Lagenaria siceraria* (Molina) Standl.	Cucurbitaceae
465.	Janhi	*Luffa acutangula* (L.) Roxb.	Cucurbitaceae
466.	Janhi	*Luffa cylindrica* (L.) M.Roem. Syn. - *Luffa aegyptiaca* Mill.	Cucurbitaceae
467.	Kalara	*Momordica charantia* L.	Cucurbitaceae
468.	Kankud	*Momordica dioica* Roxb. ex Willd.	Cucurbitaceae
469.	Paharikakharu	*Mukia maderaspatana* (L.) M.Roem.	Cucurbitaceae
470.	Lata baigana	*Sechium edule* (Jacq.) Sw.	Cucurbitaceae
471.	Bankunduri	*Solena heterophylla* Lour.	Cucurbitaceae
472.	Jangal kalara	*Thladiantha cordifolia* (Blume) Cogn var. *cordifolia.*	Cucurbitaceae
473.	Mahakala	*Trichosanthes tricuspidata* Lour. Syn. - *Trichosanthes bracteata* (Lam.) Voigt	Cucurbitaceae
474.		*Begonia picta* Sm.	Begoniaceae

Sl. No.	Local Name	Scientific Name	Family
475.	Nagapheni	*Opuntia ficus-indica* (L.) Mill. Syn. - *Opuntia vulgaris* Mill.	Cactaceae
476.	Pita saga	*Glinus oppositifolius* (L.) Aug.DC.	Molluginaceae
477.		*Mollugo pentaphylla* L.	Molluginaceae
478.	Puruni saga	*Trianthema portulacastrum* L.	Aizoaceae
479.		*Bupleurum ramosissimum* Wight & Arn	Apiaceae
480.	Thalkudi	*Centella asiatica* (L.) Urb.	Apiaceae
481.	Dhania	*Coriandrum sativum* L.	Apiaceae
482.	Janglidhania	*Eryngium foetidum* L.	Apiaceae
483.		*Hydrocotyle sibthorpioides* Lam.	Apiaceae
484.		*Oenanthe javanica* (Blume) DC.	Apiaceae
485.	Banshi gopal	*Peucedanum nagpurense* Prain	Apiaceae
486.		*Pimpinella bracteata* Haines	Apiaceae
487.		*Pimpinella heyneana* Wall.	Apiaceae
488.		*Aralia montana* Blume	Aralaceae
489.	Jari	*Schefflera venulosa* (Wight & Arn.) Harms	Araliaceae
490.		*Trevesia palmata* (Roxb. ex Lindl.) Vis.	Araliaceae
491.	Ankula	*Alangium salviifolium* (L.f.) Wangerin	Alangiaceae
492.	Kadamba	*Neolamarckia cadamba* (Roxb.) Bosser Syn. - *Anthocephalus chinensis* (Lam.) A. Rich. ex Walp.	Rubiaceae
493.		*Argostemma verticillatum* Wall.	Rubiaceae
494.		*Argostemma sarmentosum* Wall.	Rubiaceae
495.	Kumar chikni/ Jordaru	*Psydrax dicoccos* Gaertn. Syn. - *Canthium dicoccum* (Gaertn.) Merr.	Rubiaceae
496.		*Canthium glabrum* Blume	Rubiaceae
497.		*Canthium coromandelicum* (Burm.f.) Alston Syn. - *Canthium parviflorum* Lam.	Rubiaceae
498.	Kalei kanta	*Catunaregam spinosa* (Thunb.) Tirveng.	Rubiaceae
499.		*Chassalia curviflora* (Wall.) Thwaites	Rubiaceae
500.		*Coffea benghalensis* B.Heyne ex Schult.	Rubiaceae
501.		*Dentella repens* (L.) J.R.Forst. & G.Forst.	Rubiaceae
502.		*Benkara fasciculata* (Roxb.) Ridsdale Syn. - *Fagerlindia fasciculata* (Roxb.) Tirveng.	Rubiaceae
503.	Bhurudu	*Gardenia gummifera* L.f.	Rubiaceae
504.	Dambaru	*Gardenia latifolia* Aiton	Rubiaceae
505.		*Gardenia resinifera* Roth	Rubiaceae
506.	Haldu / Kaima	*Haldina cordifolia* (Roxb.) Ridsdale Syn. - *Adina cordifolia* (Roxb.) Hook. f.	Rubiaceae
507.	Bhurkunda	*Hymenodictyon orixense* (Roxb.) Mabb.	Rubiaceae
508.	Banodadri	*Hyptianthera stricta* (Roxb. ex Schult.) Wight & Arn.	Rubiaceae
509.	Lohajharia	*Ixora pavetta* Andr.	Rubiaceae
510.		*Ixora undulata* Roxb.	Rubiaceae
511.		*Knoxia sumatrensis* (Retz.) DC.	Rubiaceae
512.		*Kohautia gracilis* (Wall.) DC. Syn. - *Hedyotis gracilis* Wall.	Rubiaceae
513.		*Meyna spinosa* Roxb. ex Link	Rubiaceae
514.		*Mitracarpus hirtus* (L.) DC. Syn. - *Mitracarpus villosus* (Sw.) DC.	Rubiaceae

Sl. No.	Local Name	Scientific Name	Family
515.	Gudikoima	*Mitragyna parvifolia* (Roxb.) Korth.	Rubiaceae
516.	Pindra	*Morinda citrifolia* L.	Rubiaceae
517.	Achhu	*Morinda pubescens* Sm. Syn. - *Morinda coreia* Buch.-Ham.	Rubiaceae
518.		*Morinda umbellata* L.	Rubiaceae
519.		*Mussaenda incana* Wall.	Rubiaceae
520.		*Neanotis montholonii* (Hook.f.) W.H.Lewis	Rubiaceae
521.		*Neanotis tubulosa* (G.Don) Mabb.	Rubiaceae
522.	New record	*Neanotis wightiana* (Wall. ex Wight & Arn.) W.H.Lewis	Rubiaceae
523.		*Oldenlandia affinis* (Roem. & Schult.) DC. Syn. - *Hedyotis affinis* Roem. & Schult.	Rubiaceae
524.		*Oldenlandia auricularia* (L.) K.Schum. *Syn . - Hedyotis auricularia* L.	Rubiaceae
525.	Gharpodia	*Oldenlandia corymbosa* L. Syn. - *Hedyotis corymbosa* (L.) Lam.	Rubiaceae
526.		*Oldenlandia diffusa* (Willd.) Roxb. Syn. - *Hedyotis diffusa* Willd.	Rubiaceae
527.		*Oldenlandia ovatifolia* (Cav.) DC. Syn. - *Hedyotis ovatifolia* Cav.	Rubiaceae
528.		*Oldenlandia pinifolia* (Wall. ex G.Don) Kuntze Syn. - *Hedyotis pinifolia* Wall. ex G.Don	Rubiaceae
529.		*Oldenlandia verticillata* L. Syn. - *Hedyotis verticillata* (L.) Lam.	Rubiaceae
530.		*Oldenlandia vestita* (R.Br. ex G.Don) Drake . Syn. - *Hedyotis vestita* R.Br. ex G.Don	Rubiaceae
531.		*Ophiorrhiza fasciculata* D.Don	Rubiaceae
532.		*Ophiorrhiza rugosa* Wall.	Rubiaceae
533.	Prasaruni	*Paederia foetida* L.	Rubiaceae
534.	Kukuchalia	*Pavetta crassicaulis* Bremek.	Rubiaceae
535.		*Pavetta tomentosa* Roxb. ex Sm.	Rubiaceae
536.		*Psychotria adenophylla* Wall.	Rubiaceae
537.		*Psychotria denticulata* Wall.	Rubiaceae
538.		*Psychotria monticola* Kurz	Rubiaceae
539.		*Richardia scabra* L.	Rubiaceae
540.	Rangachireita	*Rubia cordifolia* L.	Rubiaceae
541.		*Spermacoce articularis* L.f.	Rubiaceae
542.		*Spermacoce hispida* L.	Rubiaceae
543.		*Spermacoce pusilla* Wall.	Rubiaceae
544.		*Spermacoce lasiocarpa* R.Br. ex Wall. Syn. - *Spermacoce ramanii* Sivar. & R.V.Nair	Rubiaceae
545.		*Spermadictyon suaveolens* Roxb.	Rubiaceae
546.	Telkur	*Tamilnadia uliginosa* (Retz.) Tirveng. & Sastre	Rubiaceae
547.	Jajanka	*Tarenna asiatica* (L.) Kuntze ex K.Schum.	Rubiaceae
548.	New record	*Uncaria sessilifructus* Roxb.	Rubiaceae
549.	Tilai	*Wendlandia tinctoria* (Roxb.) DC.	Rubiaceae
550.	Gokhura	*Acanthospermum hispidum* DC.	Asteraceae
551.		*Adenostemma lavenia* (L.) Kuntze	Asteraceae
552.		*Ageratum conyzoides* (L.) L.	Asteraceae

Sl. No.	Local Name	Scientific Name	Family
553.		*Anaphalis adnata* Wall. ex DC.	Asteraceae
554.		*Artemisia japonica* Thunb. Syn. - *Artemisia parviflora* Roxb. ex D.Don	Asteraceae
555.	Magha latenga	*Bidens biternata* (Lour.) Merr. & Sherff	Asteraceae
556.		*Bidens pilosa* L.	Asteraceae
557.		*Blainvillea acmella* (L.) Philipson.	Asteraceae
558.		*Blumea aromatica* DC.	Asteraceae
559.		*Blumea clarkei* Hook.f.	Asteraceae
560.		*Blumea fistulosa* (Roxb.) Kurz	Asteraceae
561.		*Blumea hieracifolia* Hayata	Asteraceae
562.		*Blumea lacera* (Burm.f.) DC.	Asteraceae
563.		*Blumea laciniata* (Wall. ex Roxb.) DC.	Asteraceae
564.		*Blumea lanceolaria* (Roxb.) Druce	Asteraceae
565.		*Blumea membranacea* DC.	Asteraceae
566.		*Blumea axillaris* (Lam.) DC. Syn. - *Blumea mollis* (D.Don) Merr.	Asteraceae
567.		*Blumea oxyodonta* DC.	Asteraceae
568.	Moharenti	*Blumeopsis flava* (DC.) Gagnep.	Asteraceae
569.	Jamjuria	*Caesulia axillaris* Roxb.	Asteraceae
570.	Nakchika	*Centipeda minima* (L.) A.Braun & Asch.	Asteraceae
571.	Bada Pokasunga	*Chromolaena odorata* (L.) R.M.King & H.Rob.	Asteraceae
572.		*Chrysanthellum americanum* (L.) Vatke	Asteraceae
573.		*Erigeron canadensis* L. Syn. - *Conyza canadensis* (L.) Cronquist	Asteraceae
574.		*Conyza japonica* (Thunb.) Less. ex Less.	Asteraceae
575.		*Erigeron trilobus* (Decne.) Boiss. Syn. - *Conyza stricta* Willd.	Asteraceae
576.		*Cyathocline purpurea* (Buch.-Ham. ex D.Don) Kuntze	Asteraceae
577.		*Dichrocephala integrifolia* (L.f.) Kuntze	Asteraceae
578.	Kesdura	*Eclipta prostrata* (L.) L.	Asteraceae
579.	Tatmuli, Mayurchulia	*Elephantopus scaber* L.	Asteraceae
580.	Sarkara	*Emilia sonchifolia* (L.) DC. ex DC.	Asteraceae
581.		*Emilia zeylanica* C.B.Clarke	Asteraceae
582.	Hidimichi	*Enydra fluctuans* DC.	Asteraceae
583.		*Laphangium luteoalbum* (L.) Tzvelev Syn. - *Gnaphalium luteoalbum* L.	Asteraceae
584.		*Gnaphalium pensylvanicum* Willd.	Asteraceae
585.		*Gnaphalium polycaulon* Pers.	Asteraceae
586.	Painjari	*Grangea maderaspatana* (L.) Poir.	Asteraceae
587.		*Gynura aurantiaca* (Blume) Sch.Bip. ex DC. Syn. - *Crassocephallum crepidioides* (Benth.) S. Moore	Asteraceae
588.		*Guizotia abyssinica* (L.f.) Cass.	Asteraceae
589.	Gota Khadaka	*Laggera alata* Nanth.	Asteraceae
590.		*Laggera crispata* (Vahl) Hepper & J.R.I.Wood	Asteraceae
591.		*Launaea acaulis* (Roxb.) Babc. ex Kerr	Asteraceae
592.		*Mikania micrantha* Kunth	Asteraceae
593.		*Sigesbeckia orientalis* L.	Asteraceae

Sl. No.	Local Name	Scientific Name	Family
594.		*Solidago canadensis* L.	Asteraceae
595.		*Sonchus asper* (L.) Hill	Asteraceae
596.		*Sonchus wightianus* DC.	Asteraceae
597.	Bhuikadam, Halamundi	*Sphaeranthus indicus* L.	Asteraceae
598.		*Synedrella nodiflora* (L.) Gaertn.	Asteraceae
599.	Vaishalyakarni	*Tridax procumbens* (L.) L.	Asteraceae
600.		*Baccharoides anthelmintica* (L.) Moench Syn. - *Vernonia anthelmintica* (L.) Willd.	Asteraceae
601.		*Vernonia aspera* (Roxb.) Ham.	Asteraceae
602.		*Cyanthillium cinereum* (L.) H.Rob. Syn. - *Vernonia cinerea* (L.) Less.	Asteraceae
603.		*Vernonia divergens* (DC.) Edgewood	Asteraceae
604.		*Vernonia squarrosa* Dinter ex Merxm.	Asteraceae
605.		*Pentanema cernuum* (Dalzell) Ling Syn. - *Vicoa cernua* Dalzell	Asteraceae
606.	Bana sebati	*Pentanema indicum* (L.) Ling Syn. - *Vicoa indica* (L.) DC.	Asteraceae
607.		Wedelia urticifolia var. wightii DC.	Asteraceae
608.	Parbatjeera	*Wedelia wallichii* Less.	Asteraceae
609.	Gokhru	*Xanthium strumarium* L. Syn. - *Xanthium indicum* DC.	Asteraceae
610.		*Youngia japonica* (L.) DC.	Asteraceae
611.		*Campanula dimorphantha* Schweinf. Syn. - *Campanula bethamii* Wall.	Campanulaceae
612.		*Lobelia alsinoides* Lam.	Lobeliaceae
613.		*Lobelia heyneana* Schult.	Lobeliaceae
614.		*Anagallis arvensis* L.	Primulaceae
615.		*Anagallis pumila* Sw.	Primulaceae
616.		*Androsace umbellata* (Lour.) Merr. Syn. - *Primula umbellata* (Lour.) Bentv.	Primulaceae
617.		*Ardisia depressa* C.B.Clarke	Primulaceae
618.	Tinkoli	*Ardisia solanacea* (Poir.) Roxb.	Primulaceae
619.		*Embelia ribes* Burm.f.	Primulaceae
620.	Baibirango	*Embelia tsjeriam-cottam* (Roem. & Schult.) A.DC.	Primulaceae
621.	Mahulo	*Madhuca longifolia* (J.Koenig ex L.) J.F.Macbr.	Sapotaceae
622.	Jasthimadhu	*Xantolis tomentosa* (Roxb.) Raf.	Sapotaceae
623.	Makarkendu	*Diospyros malabarica* (Desr.) Kostel.	Ebenaceae
624.	Kendu	*Diospyros melanoxylon* Roxb.	Ebenaceae
625.	Halda/ Jallai	*Diospyros montana* Roxb.	Ebenaceae
626.	Kalicha	*Diospyros sylvatica* Roxb.	Ebenaceae
627.		*Diospyros vera* (Lour.) A. Chev. *Syn.- Diospyros ferrea* (Willd.) Bakh.	Ebenaceae
628.		*Styrax serrulatus* Roxb.	Styraceae
629.	Bhaunri	*Symplocos cochinchinensis* var. *laurina* (Retz.) Noot.	Symplococaceae
630.	Lodha	*Symplocos racemosa* Roxb.	Symplococaceae
631.	Masur danta	*Chionanthus ramiflorus* Roxb. Syn. - *Chionanthus intermedius* (Wight) F.Muell.	Oleaceae

Sl. No.	Local Name	Scientific Name	Family
632.	Pochandia	*Chionanthus mala-elengi* (Dennst.) P.S.Green	Oleaceae
633.	Bana malli	*Jasminum arborescens* Roxb.	Oleaceae
634.		*Jasminum flexile* Vahl Syn. - *Jasminum azoricum* L.	Oleaceae
635.		*Jasminum sambac* (L.) Aiton	Oleaceae
636.	Ban malli	*Jasminum scandens* (Retz.) Vahl	Oleaceae
637.		*Ligustrum gamblei* Ramamoorthy	Oleaceae
638.	Gangasiuli	*Nyctanthes arbor-tristis* L.	Oleaceae
639.	Eksira	*Schrebera swietenioides* Roxb.	Oleaceae
640.	Gandhomalati	*Aganosma caryophyllata* G.Don	Apocynaceae
641.	Chhatiana	*Alstonia scholaris* (L.) R. Br.	Apocynaceae
642.		*Alstonia venenata* R.Br.	Apocynaceae
643.		*Catharanthus pusillus* (Murray) G.Don	Apocynaceae
644.	Sadabihari	*Catharanthus roseus* (L.) G.Don	Apocynaceae
645.	Korua/ Khurchi/ Indrajalo	*Holarrhena pubescens* Wall. ex G.Don Syn. - *Holarrhena antidysenterica* (Roth) Wall. ex A.DC.	Apocynaceae
646.	Dudhilata/ Swanlata	*Ichnocarpus frutescens* (L.) W.T.Aiton	Apocynaceae
647.	Kaner	*Nerium oleander* L. Syn. - *Nerium indicum* Mill.	Apocynaceae
648.	Kathachampa	*Plumeria rubra* L.	Apocynaceae
649.	Patal garud	*Rauvolfia serpentina* (L.) Benth. ex Kurz	Apocynaceae
650.	Tagar	*Tabernaemontana divaricata* (L.) R.Br. ex Roem. & Schult.	Apocynaceae
651.	Kaniar	*Cascabela thevetia* (L.) Lippold Syn. - *Thevetia peruviana* (Pers.) K.Schum.	Apocynaceae
652.	Banakaneari	*Vallaris solanacea* (Roth) Kuntze	Apocynaceae
653.	Pahadia kuduchi	*Wrightia arborea* (Dennst.) Mabb.	Apocynaceae
654.		*Wrightia tinctoria* R.Br.	Apocynaceae
655.	Dudhi	*Cryptolepis dubia* (Burm.f.) M.R.Almeida Syn. - *Cryptolepis buchananii* Roem. & Schult.	Apocynaceae
656.	Dudhi/ Antamula/ Sugandhi	*Hemidesmus indicus* (L.) R. Br. ex Schult.	Apocynaceae
657.	Arkakha	*Calotropis gigantea* (L.) Dryand.	Asclepiadaceae
658.		*Caralluma umbellata* Haw.	Asclepiadaceae
659.		*Ceropegia hirsuta* Wight & Arn.	Asclepiadaceae
660.		*Cynanchum callialatum* Buch.-Ham. ex Wight	Asclepiadaceae
661.	New Record	*Dischidia nummularia* R.Br.	Asclepiadaceae
662.		*Gymnema sylvestre* (Retz.) R.Br. ex Sm.	Asclepiadaceae
663.		*Heterostemma tanjorense* Wight & Arn.	Asclepiadaceae
664.		*Holostemma ada-kodien* Schult. Syn. - *Holostemma annulare* (Roxb.) K. Schum.	Asclepiadaceae
665.		*Hoya alexicaca* (Jacq.) Moon Syn. - *Hoya pendula* Wight & Arn.	Asclepiadaceae
666.		*Hoya parasitica* Wall. ex Wight	Asclepiadaceae
667.		*Pentatropis capensis* (L. f.) Bullock	Asclepiadaceae
668.	Utrali/ Uturudi	*Pergularia daemia* (Forssk.) Chiov.	Asclepiadaceae
669.		*Toxocarpus kleinii* Wight & Arn.	Asclepiadaceae
670.		*Dregea volubilis* (L.f.) Benth. ex Hook.f. Syn. - *Wattakaka volubilis* (L. f.) Stapf	Asclepiadaceae
671.	Kochila	*Strychnos nux-vomica* L.	Loganiaceae

Sl. No.	Local Name	Scientific Name	Family
672.	Kakato/ Koyar	*Strychnos potatorum* L.f.	Loganiaceae
673.		*Buddleja asiatica* Lour.	Scrophulariaceae
674.		*Mitreola petiolata* (J.F.Gmel.) Torr. & A.Gray	Loganiaceae
675.		*Mitrasacme indica* Wight	Loganiaceae
676.		*Mitrasacme pygmaea* R. Br.	Loganiaceae
677.		*Canscora diffusa* (Vahl) R.Br. ex Roem. & Schult. Syn. - *Canscora decurrens* Dalzell	Gentianacea
678.		*Canscora alata* (Roth) Wall. Syn. - *Canscora decussata* (Roxb.) Schult. & Schult.f.	Gentianacea
679.		*Centaurium centaurioides* (Roxb.) S.R.Rao & Hemadri	Gentianacea
680.		*Exacum tetragonum* Roxb. Syn. - *Exacum bicolor* Roxb.	Gentianacea
681.		*Exacum tenue* (Blume) Klack.	Gentianacea
682.		*Hoppea dichotoma* Willd.	Gentianacea
683.		*Swertia angustifolia* Buch.-Ham. ex D. Don	Gentianacea
684.		*Nymphoides hydrophylla* (Lour.) Kuntze	Menyanthaceae
685.	Languliya	*Hydrolea zeylanica* (L.) Vahl	Hydroleaceae
686.	Ghanti	*Cordia macleodii* Hook.f. & Thomson	Boraginaceae
687.		*Cordia dichotoma* G.Forst. Syn. - *Cordia wallichii* G.Don	Boraginaceae
688.		*Ehretia acuminata* R.Br.	Boraginaceae
689.	Mosania	*Ehretia laevis* Roxb.	Boraginaceae
690.		*Coldenia procumbens* L.	Boraginaceae
691.		*Cynoglossum zeylanicum* (Vahl) Brand	Boraginaceae
692.		*Heliotropium indicum* L.	Boraginaceae
693.		*Heliotropium strigosum* Willd.	Boraginaceae
694.		*Trichodesma zeylanicum* (Burm.f.) R.Br.	Boraginaceae
695.		*Argyreia bella* Raizada	Convolvulaceae
696.		*Argyreia daltonii* C.B.Clarke	Convolvulaceae
697.	Jada	*Erycibe paniculata*	Convolvulaceae
698.	Bichhamalia	*Evolvulus alsinoides* (L.) L.	Convolvulaceae
699.		*Evolvulus nummularius* (L.) L.	Convolvulaceae
700.		*Hewittia malabarica* (L.) Suresh Syn. - *Hewittia sublobata* (L. f.) Kuntze	Convolvulaceae
701.	Kalam	*Ipomoea aquatica* Forssk.	Convolvulaceae
702.		*Ipomoea barlerioides* (Choisy) Benth. ex C.B. Clarke	Convolvulaceae
703.		*Ipomoea cairica* (L.) Sweet	Convolvulaceae
704.	Amari	*Ipomoea carnea* Jacq.	Convolvulaceae
705.		*Ipomoea eriocarpa* R. Br.	Convolvulaceae
706.		*Ipomoea nil* (L.) Roth	Convolvulaceae
707.	Bileipadi	*Ipomoea pes-tigridis* L.	Convolvulaceae
708.		*Ipomoea quamoclit* L.	Convolvulaceae
709.		*Ipomoea biflora* (L.) Pers. Syn. - *Ipomoea sinensis* (Desr.) Choisy	Convolvulaceae
710.		*Ipomoea turbinata* Lag. ex Choisy	Convolvulaceae
711.		*Merremia emarginata* (Burm. f.) Hallier f.	Convolvulaceae
712.		*Merremia hirta* (L.) Merr.	Convolvulaceae
713.		*Merremia umbellata* (L.) Hallier f.	Convolvulaceae
714.		*Merremia vitifolia* (Burm. f.) Hallier f.	Convolvulaceae

Sl. No.	Local Name	Scientific Name	Family
715.	Nirmuli	*Cuscuta reflexa* Roxb.	Cuscutaceae
716.	Lanka	*Capsicum annuum* L.	Solanaceae
717.	Dutura	*Datura metel* L.	Solanaceae
718.	Bilati	*Lycopersicon esculentum* Mill.	Solanaceae
719.	Tipai	*Physalis minima* L.	Solanaceae
720.		*Solanum erianthum* D. Don	Solanaceae
721.		*Solanum giganteum* Jacq.	Solanaceae
722.	Tamaku,Dukta	*Nicotiana tabacum* L.	Solanaceae
723.	Baigana	*Solanum melongena* L.	Solanaceae
724.	Nunununia	*Solanum americanum* Mill. Syn. - *Solanum nigrum* L.	Solanaceae
725.	Kantua	*Solanum torvum* Sw.	Solanaceae
726.	Bhejibaigan	*Solanum viarum* Duna	Solanaceae
727.		*Solanum violaceum* Ortega	Solanaceae
728.	Bhejibaigan / Ankaranti	*Solanum virginianum* L. Syn. - *Solanum surattense* Burm. f.	Solanaceae
729.		*Adenosma microcephalum* Hook.f.	Plantaginaceae
730.		*Bacopa monnieri* (L.) Wettst.	Plantaginaceae
731.		*Buchnera hispida* Buch.-Ham. ex D.Don	Orobanchaceae
732.		*Centranthera indica* (L.) Gamble	Orobanchaceae
733.		*Centranthera tranquebarica* (Spreng.) Merr.	Orobanchaceae
734.		*Limnophila aquatica* Alston	Plantaginaceae
735.		*Limnophila aromatica* (Lam.) Merr.	Plantaginaceae
736.		*Limnophila connata* (Buch.-Ham. ex D. Don) Hand.-Mazz.	Plantaginaceae
737.		*Limnophila heterophylla* (Roxb.) Benth.	Plantaginaceae
738.		*Limnophila indica* (L.) Druce	Plantaginaceae
739.		*Limnophila rugosa* (Roth) Merr.	Plantaginaceae
740.		*Lindenbergia muraria* (Roxburgh ex D. Don) Brühl	Linderniaceae
741.		*Lindernia anagallis* (Burm.f.) Pennell	Linderniaceae
742.		*Lindernia antipoda* (L.) Alston	Linderniaceae
743.		*Lindernia caespitosa* (Blume) Panigrahi Syn. - *Lindernia pusilla* (Willd.) Bold.	Linderniaceae
744.		*Lindernia ciliata* (Colsm.) Pennell	Linderniaceae
745.		*Lindernia crustacea* (L.) F.Muell.	Linderniaceae
746.		*Lindernia hookeri* Wettst.	Linderniaceae
747.		*Lindernia nummulariifolia* (D.Don) Wettst.	Linderniaceae
748.		*Lindernia viscosa* (Hornem.) Merr.	Linderniaceae
749.		*Mecardonia procumbens* (Mill.) Small	Plantaginaceae
750.	Khelopapada	*Scoparia dulcis* L.	Plantaginaceae
751.		*Striga angustifolia* (D. Don) C.J. Saldanha	Orobanchaceae
752.		*Striga asiatica* (L.) Kuntze	Orobanchaceae
753.		*Torenia cordifolia* Roxb.	Linderniaceae
754.		*Torenia violacea* (Azaola ex Blanco) Pennell	Linderniaceae
755.		*Aeginetia indica* L.	Orobanchaceae
756.		*Utricularia aurea* Lour.	Lentibulariaceae
757.		*Utricularia bifida* L.	Lentibulariaceae
758.		*Utricularia caerulea* L.	Lentibulariaceae
759.		*Utricularia scandens* Benj.	Lentibulariaceae
760.		*Aeschynanthus parasiticus* (Roxb.) Wall.	Gesneriaceae

Sl. No.	Local Name	Scientific Name	Family
761.		*Microchirita hamosa* (R.Br.) Yin Z.Wang Syn. - *Chirita hamosa* R.Br.	Gesneriaceae
762.		*Rhynchoglossum obliquum* Blume	Gesneriaceae
763.	Phanphana	*Oroxylum indicum* (L.) Kurz	Bignoniaceae
764.		*Radermachera xylocarpa* (Roxb.) Roxb. ex K.Schum.	Bignoniaceae
765.		*Spathodea campanulata* P.Beauv.	Bignoniaceae
766.	Parudi / Patuli	*Stereospermum chelonoides* (L.f.) DC.	Bignoniaceae
767.	Rasi	*Sesamum indicum* L. Syn. - *Sesamum orientale* L.	Pedaliaceae
768.	Baghanakhi	*Martynia annua* L.	Martyniaceae
769.		*Andrographis elongata* (Vahl) T.Anderson	Acanthaceae
770.		*Andrographis ovata* (T.Anderson ex Bedd.) Benth. & Hook.f.	Acanthaceae
771.		*Andrographis paniculata* (Burm.f.) Nees	Acanthaceae
772.		*Barleria cristata* L.	Acanthaceae
773.		*Barleria strigosa* Willd.	Acanthaceae
774.		*Barleria gibsonii* Dalzell	Acanthaceae
775.	Daskeranta	*Barleria prionitis* L.	Acanthaceae
776.		*Barleria lupulina* Lindl.	Acanthaceae
777.		*Blepharis maderaspatensis* (L.) B.Heyne ex Roth	Acanthaceae
778.		*Dicliptera bupleuroides* Nees	Acanthaceae
779.		*Dicliptera verticillata* (Forssk.) C.Chr.	Acanthaceae
780.		*Dipteracanthus beddomei* (C.B.Clarke) Santapau	Acanthaceae
781.		*Cardanthera balsamica* (L.f.) Benth. Syn. - *Dipteracanthus cernuus* (Roxb.) Santapu	Acanthaceae
782.		*Dipteracanthus suffruticosus* Voigt	Acanthaceae
783.		*Eranthemum capense* L.	Acanthaceae
784.		*Eranthemum pulchellum* Andrews Syn. - *Eranthemum nervosum* (Vahl) R.Br. ex Roem. & Schult.	Acanthaceae
785.		*Eranthemum purpurascens* Wight ex Nees Syn. - *Eranthemum pulchellum* Andrews	Acanthaceae
786.		*Hemigraphis latebrosa* (Roth) Nees	Acanthaceae
787.	Koilekha	*Hygrophila auriculata* (Schumach.) Heine	Acanthaceae
788.		*Hygrophila salicifolia* (Vahl) Nees	Acanthaceae
789.		*Indoneesiella echioides* (L.) Sreem.	Acanthaceae
790.		*Justicia adhatoda* L.	Acanthaceae
791.		*Justicia betonica* L.	Acanthaceae
792.		*Justicia diffusa* Willd.	Acanthaceae
793.		*Justicia gendarussa* Burm.f.	Acanthaceae
794.		*Justicia japonica* Thunb.	Acanthaceae
795.		*Justicia nilgherrensis* (Nees) Wall. ex C.B.Clarke	Acanthaceae
796.		*Justicia quinqueangularis* K.D.Koenig ex Roxb.	Acanthaceae
797.		*Justicia japonica* Thunb. Syn. - *Justicia simplex* D. Don	Acanthaceae
798.		*Lepidagathis fasciculata* (Retz.) Nees	Acanthaceae
799.		*Lepidagathis incurva* Buch.-Ham. ex D. Don	Acanthaceae
800.		*Lepidagathis purpuricaulis* Nees	Acanthaceae
801.		*Nelsonia canescens* (Lam.) Spreng.	Acanthaceae
802.		*Peristrophe speciosa* Nees	Acanthaceae
803.		*Petalidium barlerioides* (Roth) Nees	Acanthaceae

Sl. No.	Local Name	Scientific Name	Family
804.		*Phaulopsis imbricata* (Forssk.) Sweet	Acanthaceae
805.		*Rhinacanthus nasutus* (L.) Kurz	Acanthaceae
806.		*Ruellia tuberosa* L.	Acanthaceae
807.		*Rungia pectinata* (L.) Nees	Acanthaceae
808.		*Rungia repens* (L.) Nees	Acanthaceae
809.		*Staurogyne glutinosa* Kuntze	Acanthaceae
810.	Painya	*Strobilanthes auriculatus* Nees	Acanthaceae
811.		*Strobilanthes heyneanus* Nees	Acanthaceae
812.		*Strobilanthes lupulinus* Nees	Acanthaceae
813.		*Strobilanthes scaber* Nees	Acanthaceae
814.	Chakrakedar	*Thunbergia fragrans* Roxb.	Acanthaceae
815.		*Callicarpa longifolia* Lam.	Lamiaceae
816.		*Callicarpa macrophylla* Vahl	Lamiaceae
817.		*Callicarpa tomentosa* (L.) L. Syn. - *Callicarpa arborea* Roxb.	Lamiaceae
818.		*Clerodendrum indicum* (L.) Kuntze	Lamiaceae
819.	Samarkand/ Tirkatia	*Rotheca serrata* (L.) Steane & Mabb. Syn. -*Clerodendrum serratum* (L.) Moon	Lamiaceae
820.	Kunti	*Clerodendrum infortunatum* L. Syn. -*Clerodendrum viscosum* Vent.	Lamiaceae
821.	Gambhari	*Gmelina arborea* Roxb.	Lamiaceae
822.	Putus	*Lantana camara* L.	Verbenaceae
823.	Naguari	*Lippia javanica* (Burm.f.) Spreng.	Verbenaceae
824.	Gosingi	*Phyla nodiflora* (L.) Greene	Verbenaceae
825.		*Premna barbata* Wall. ex Schauer Syn. - *Premna calycina* Haines	Lamiaceae
826.		*Premna coriacea* C.B.Clarke	Lamiaceae
827.	Gandhanu	*Premna mollissima* Roth Syn. - *Premna latifolia* Roxb.	Lamiaceae
828.	Gitia	*Premna herbacea* Roxb. Syn. - *Pygmaeopremna herbacea* (Roxb.) Moldenke	Lamiaceae
829.		*Stachytarpheta jamaicensis* (L.) Vahl	Verbenaceae
830.	Poiyan	*Symphorema polyandrum* Wight	Lamiaceae
831.	Sagwan	*Tectona grandis* L.f.	Lamiaceae
832.	Muria	*Vitex glabrata* R.Br.	Lamiaceae
833.		*Vitex leucoxylon* L.f.	Lamiaceae
834.	Begunia, Nirgundi	*Vitex negundo* L.	Lamiaceae
835.	Chadaigudi	*Vitex peduncularis* Wall. ex Schauer	Lamiaceae
836.		*Platostoma hispidum* (L.) A.J. Paton Syn. - *Acrocephalus hispidus* (L.) Nicolson & Sivad.	Lamiaceae
837.		*Ajuga macrosperma* Wall. ex Benth.	Lamiaceae
838.		*Anisochilus carnosus* (L.f.) Wall.	Lamiaceae
839.		*Anisomeles indica* (L.) Kuntze	Lamiaceae
840.		*Colebrookea oppositifolia* Sm.	Lamiaceae
841.		*Pogostemon stellatus* (Lour.) Kuntze Syn. - *Eusteralis stellata* (Lour.) Panigrahi	Lamiaceae
842.		*Gomphostemma parviflorum* Wall. ex Benth.	Lamiaceae
843.	Gangatulsi, Bantulsi	*Hyptis suaveolens* (L.) Poit.	Lamiaceae

Sl. No.	Local Name	Scientific Name	Family
844.		*Leonotis nepetifolia* (L.) R.Br.	Lamiaceae
845.		*Leucas aspera* (Willd.) Link	Lamiaceae
846.	Gaisa	*Leucas zeylanica* (L.) W.T.Aiton Syn. - *Leucas indica* (L.) Sm.	Lamiaceae
847.		*Leucas lanata* Benth.	Lamiaceae
848.		*Leucas decemdentata* (Willd.) Sm. Syn. -*Leucas mollissima* var. *scaberula* Hook.f.	Lamiaceae
849.		*Clinopodium capitellatum* (Benth.) Kuntze Syn. - *Micromeria capitellata* Benth.	Lamiaceae
850.		*Micromeria biflora* (Buch.-Ham. ex D.Don) Benth.	Lamiaceae
851.		*Clinopodium capitellatum* (Benth.) Kuntze Syn. - *Micromeria capitellata* Benth.	Lamiaceae
852.		*Mosla dianthera* (Buch.-Ham. ex Roxb.) Maxim.	Lamiaceae
853.	Bantulsi	*Ocimum americanum* L.	Lamiaceae
854.		*Ocimum basilicum* L.	Lamiaceae
855.	Tulsi	*Ocimum tenuiflorum* L. Syn. - *Ocimum sanctum* L.	Lamiaceae
856.	New record	*Orthosiphon aristatus* (Blume) Miq.	Lamiaceae
857.		*Orthosiphon pallidus* Royle ex Benth.	Lamiaceae
858.	Chandua	*Orthosiphon rubicundus* (D.Don) Benth.	Lamiaceae
859.		*Orthosiphon thymiflorus* (Roth) Sleesen	Lamiaceae
860.		*Platystoma africana* P. Beauv.	Lamiaceae
861.		*Isodon japonicus* (Burm.f.) H.Hara Syn. - *Plectranthus japonicus* (Burm.f.) Koidz.	Lamiaceae
862.		*Plectranthus mollis* (Aiton) Spreng.	Lamiaceae
863.		*Isodon ternifolius* (D.Don) Kudô Syn. - *Plectranthus ternifolius* D.Don	Lamiaceae
864.		*Isodon wightii* (Benth.) H.Hara Syn. - *Plectranthus wightii* Benth.	Lamiaceae
865.		*Pogostemon auricularius* (L.) Hassk.	Lamiaceae
866.		*Pogostemon benghalensis* (Burm.f.) Kuntze	Lamiaceae
867.		*Teucrium viscidum* Blume	Lamiaceae
868.	Kharkharia, Parni	*Boerhavia diffusa* L.	Nyctaginaceae
869.		*Bougainvillea glabra* Choisy	Nyctaginaceae
870.		*Bougainvillea spectabilis* Willd.	Nyctaginaceae
871.	Apamaranga	*Achyranthes aspera* L.	Amaranthaceae
872.		*Achyranthes bidentata* Blume	Amaranthaceae
873.		*Aerva lanata* (L.) Juss.	Amaranthaceae
874.		*Aerva sanguinolenta* (L.) Blume	Amaranthaceae
875.		*Allmania nodiflora* (L.) R.Br. ex Wight	Amaranthaceae
876.		*Alternanthera sessilis* (L.) R.Br. ex DC.	Amaranthaceae
877.	Khada saga	*Amaranthus caudatus* L.	Amaranthaceae
878.	Kanta khada	*Amaranthus spinosus* L.	Amaranthaceae
879.	Nautia	*Amaranthus tricolor* L.	Amaranthaceae
880.		*Amaranthus viridis* L.	Amaranthaceae
881.		*Celosia argentea* L.	Amaranthaceae
882.		*Cyathula prostrata* (L.) Blume	Amaranthaceae
883.		*Gomphrena celosioides* Mart.	Amaranthaceae

Sl. No.	Local Name	Scientific Name	Family
884.		*Pupalia lappacea* (L.) Juss.	Amaranthaceae
885.		*Chenopodium album* L.	Chenopodiaceae
886.		*Persicaria barbata* (L.) H.Hara Syn. - *Polygonum barbatum* L.	Polygonaceae
887.		*Persicaria stagnina* (Buch.-Ham. ex Meisn.) Qaiser Syn. - *Polygonum barbatum* var. *stagninum* (Buch.-Ham. ex Meisn.) Steward	Polygonaceae
888.		*Persicaria chinensis* (L.) H. Gross Syn. - *Polygonum chinense* L.	Polygonaceae
889.		*Persicaria glabra* (Willd.) M.Gómez Syn. - *Polygonum glabrum* Willd.	Polygonaceae
890.		*Persicaria hydropiper* subsp. *microcara* (Danser) Soják	Polygonaceae
891.		*Persicaria dichotoma* (Blume) Masam.	Polygonaceae
892.		*Polygonum plebeium* R.Br.	Polygonaceae
893.		*Persicaria strigosa* (R.Br.) Nakai	Polygonaceae
894.		*Polypleurum wallichii* (R. Br. ex Griff.) Warm.	Podostemaceae
895.	Iswar-kula	*Aristolochia indica* L.	Aristolochiaceae
896.		*Peperomia blanda* (Jacq.) Kunth	Piperaceae
897.		*Peperomia tetraphylla* (G.Forst.) Hook. & Arn.	Piperaceae
898.	Chaikatha	*Piper trioicum* Roxb.	Piperaceae
899.		*Actinodaphne gullavara* (Buch.-Ham. ex Nees) M.R.Almeida Syn. - *Actinodaphne angustifolia* Nees	Lauraceae
900.		*Cassytha filiformis* L.	Lauraceae
901.	Ledha chhali	*Litsea glutinosa* (Lour.) C.B.Rob.	Lauraceae
902.	Paja	*Litsea monopetala* (Roxb.) Pers.	Lauraceae
903.		*Neocinnamomum caudatum* (Nees) Merr.	Lauraceae
904.		*Machilus glaucescens* (Nees) Wight Syn. - *Persea villosa* (Roxb.) Kosterm.	Lauraceae
905.		*Ocotea lancifolia* (Schott) Mez Syn. - *Phoebe lanceolata* (Nees) Nees	Lauraceae
906.		*Phoebe wightii* Meisn.	Lauraceae
907.		*Grevillea robusta* A.Cunn. ex R.Br.	Proteaceae
908.		*Elaeagnus conferta* Roxb.	Elaeagnaceae
909.		*Dendrophthoe falcata* (L.f.) Ettingsh.	Loranthaceae
910.		*Macrosolen cochinchinensis* (Lour.) Tiegh.	Loranthaceae
911.	Malanga	*Scurrula parasitica* L.	Loranthaceae
912.	Malanga	*Viscum articulatum* Burm. f.	Santalaceae
913.	Malanga	*Viscum monoicum* Roxb. ex DC.	Santalaceae
914.	Malanga	*Viscum cruciatum* Sieber ex Boiss. Syn. - *Viscum orientale* Willd.	Santalaceae
915.		*Osyris wightiana* Wall. ex Wight	Santalaceae
916.	Chandan	*Santalum album* L.	Santalaceae
917.		*Acalypha indica* L.	Euphorbiaceae
918.		*Alchornea mollis* (Benth.) Müll.Arg.	Euphorbiaceae
919.	Nunnunia	*Antidesma acidum* Retz.	Phyllanthaceae
920.		*Antidesma montanum* Blume	Phyllanthaceae
921.		*Antidesma bunius* (L.) Spreng.	Phyllanthaceae
922.	Amtua / Nuniari	*Antidesma ghaesembilla* Gaertn.	Phyllanthaceae

Sl. No.	Local Name	Scientific Name	Family
923.	Tabo	*Aporosa octandra* (Buch.-Ham. ex D.Don) Vickery	Phyllanthaceae
924.	Rajkoli	*Baccaurea ramiflora* Lour.	Phyllanthaceae
925.		*Baliospermum solanifolium* (Burm.) Suresh Syn. - *Baliospermum montanum* (Willd.) Müll.Arg.	Euphorbiaceae
926.		*Bischofia javanica* Blume	Phyllanthaceae
927.	Jajana	*Breynia vitis-idaea* (Burm.f.) C.E.C.Fisch.	Phyllanthaceae
928.		*Bridelia glauca* Blume Syn. - *Bridelia pubescens* Kurz	Phyllanthaceae
929.	Kasi	*Bridelia retusa* (L.) A.Juss.	Phyllanthaceae
930.	Nota kasi	*Bridelia stipularis* (L.) Blume	Phyllanthaceae
931.	Karada	*Cleistanthus collinus* (Roxb.) Benth. ex Hook.f.	Phyllanthaceae
932.		*Cleistanthus patulus* (Roxb.) Müll.Arg.	Phyllanthaceae
933.		*Croton bonplandianus* Baill.	Euphorbiaceae
934.	Furudi	*Croton caudatus* Geiseler	Euphorbiaceae
935.	Masud / Putudi	*Croton persimilis* Müll.Arg.	Euphorbiaceae
936.		*Cnesmone javanica* Blume	Euphorbiaceae
937.		*Dimorphocalyx glabellus* Thwaites	Euphorbiaceae
938.	Dudhi	*Euphorbia hirta* L.	Euphorbiaceae
939.	Seju	*Euphorbia nivulia* Buch.-Ham.	Euphorbiaceae
940.		*Euphorbia prostrata* Aiton	Euphorbiaceae
941.		*Euphorbia thymifolia* L.	Euphorbiaceae
942.		*Glochidion lanceolarium* (Roxb.) Voigt	Phyllanthaceae
943.		*Glochidion heyneanum* (Wight & Arn.) Wight Syn. - *Glochidion velutinum* Wight	Phyllanthaceae
944.		*Glochidion zeylanicum* (Gaertn.) A.Juss.	Phyllanthaceae
945.	Pani jhorabi, Jamla	*Homonoia riparia* Lour.	Euphorbiaceae
946.	Banpri jada	*Jatropha curcas* L.	Euphorbiaceae
947.		*Jatropha gossypiifolia* L.	Euphorbiaceae
948.	Kukuda-hadi	*Lasiococca comberi* Haines	Euphorbiaceae
949.		*Macaranga denticulata* (Blume) Müll.Arg.	Euphorbiaceae
950.		*Macaranga peltata* (Roxb.) Müll.Arg.	Euphorbiaceae
951.	Sinduri	*Mallotus philippensis* (Lam.) Müll.Arg.	Euphorbiaceae
952.		*Pachystylidium hirsutum* (Blume) Pax & K.Hoffm.	Euphorbiaceae
953.		*Phyllanthus amarus* Schumach. & Thonn.	Phyllanthaceae
954.		*Phyllanthus debilis* Klein ex Willd.	Phyllanthaceae
955.	Anla	*Phyllanthus emblica* L.	Phyllanthaceae
956.	Bhui-amla/ Bari amla	*Phyllanthus fraternus* G.L.Webster	Phyllanthaceae
957.	Jhar	*Phyllanthus lawii* J.Graham	Phyllanthaceae
958.	Jandaki	*Phyllanthus reticulatus* Poir.	Phyllanthaceae
959.		*Phyllanthus urinaria* L.	Phyllanthaceae
960.		*Phyllanthus virgatus* G.Forst.	Phyllanthaceae
961.	Jada	*Ricinus communis* L.	Euphorbiaceae
962.		*Falconeria insignis* Royle Syn. - *Sapium insigne* (Royle) Trimen	Euphorbiaceae
963.		*Sauropus quadrangularis* (Willd.) Müll.Arg.	Phyllanthaceae
964.		*Microstachys chamaelea* (L.) Müll.Arg. Syn. - *Sebastiania chamaelea* (L.) Müll.Arg.	Euphorbiaceae

Sl. No.	Local Name	Scientific Name	Family
965.		*Flueggea virosa* (Roxb. ex Willd.) Royle Syn. - *Securinega virosa* (Roxb. ex Willd.) Baill.	Phyllanthaceae
966.		*Suregada multiflora* (A.Juss.) Baill.	Euphorbiaceae
967.		*Tragia involucrata* L.	Euphorbiaceae
968.	Panigambhari	*Mallotus nudiflorus* (L.) Kulju & Welzen Syn. - *Trewia nudiflora* var. *tomentosa* Susila & N.P.Balakr.	Euphorbiaceae
969.	New record	*Triadica cochinchinensis* Lour.	Euphorbiaceae
970.		*Celtis tetrandra* Roxb.	Cannabaceae
971.	Charla,Taruda	*Holoptelea integrifolia* Planch.	Ulmaceae
972.	Kharkas	*Trema orientalis* (L.) Blume	Cannabaceae
973.	Jeota, Jautha	*Artocarpus lacucha* Buch.-Ham.	Moraceae
974.		*Artocarpus heterophyllus* Lam.	Moraceae
975.		*Maclura cochinchinensis* (Lour.) Corner Syn. - *Cudrania cochinchinensis* (Lour.) Yakuro Kudo & Masam.	Moraceae
976.	Bara	*Ficus benghalensis* L.	Moraceae
977.		*Ficus benjamina* L.	Moraceae
978.		*Ficus hederacea* Roxb. Syn. - *Ficus fructicosa* Roxb.	Moraceae
979.		*Ficus heterophylla* L.f.	Moraceae
980.		*Ficus hispida* L.f.	Moraceae
981.	Dimiri	*Ficus racemosa* L. Syn. - *Ficus lanceolata* Buch.-Ham. ex Roxb.	Moraceae
982.		*Ficus microcarpa* L.f.	Moraceae
983.		*Ficus nervosa* B.Heyne ex Roth	Moraceae
984.	Kharsara	*Ficus tinctoria* subsp. *gibbosa* (Blume) Corner Syn. - *Ficus cuspidifera* Miq.	Moraceae
985.	Pipalo/ Aswastha	*Ficus religiosa* L.	Moraceae
986.		*Ficus pumila* L. Syn. - *Ficus scandens* Lam.	Moraceae
987.	Potkuli	*Ficus semicordata* Buch.-Ham. ex Sm.	Moraceae
988.		*Ficus virens* Aiton	Moraceae
989.	Tutkoli	*Morus australis* Poir.	Moraceae
990.		*Morus macroura* Miq.	Moraceae
991.	Sara	*Streblus asper* Lour.	Moraceae
992.	Phutkuli	*Streblus taxoides* (Roth) Kurz	Moraceae
993.		*Boehmeria platyphylla* D. Don	Urticaceae
994.		*Elatostema cuneatum* Wight	Urticaceae
995.		*Laportea interrupta* (L.) Chew	Urticaceae
996.		Maoutia puya (Hook.) Wedd.	Urticaceae
997.		*Oreocnide frutescens* (Thunb.) Miq.	Urticaceae
998.		*Pilea microphylla* (L.) Liebm.	Urticaceae
999.		*Pilea scripta* (Buch.-Ham. ex D. Don) Wedd.	Urticaceae
1000.		*Gonostegia pentandra* (Roxb.) Miq. Syn. - *Pouzolzia pentandra* (Roxb.) Benn.	Urticaceae
1001.		*Populus ciliata* Wall. ex Royle	Salicaceae
1002.	Panibegunia	*Salix tetrasperma* Roxb.	Salicaceae
1003.		*Ceratophyllum demersum* L.	Ceratophyliaceae
1004.		*Blyxa aubertii* Rich.	Hydrocharitaceae
1005.		*Ottelia alismoides* (L.) Pers.	Hydrocharitaceae

Sl. No.	Local Name	Scientific Name	Family
1006.		*Vallisneria natans* (Lour.) H.Hara	Hydrocharitaceae
1007.		*Burmannia coelestis* D.Don	Burmanniaceae
1008.	Rasna	*Acampe carinata* (Griff.) Panigrahi	Orchidaceae
1009.		*Acampe ochracea* (Lindl.) Hochr.	Orchidaceae
1010.		*Acampe praemorsa* (Roxb.) Blatt. & McCann	Orchidaceae
1011.		*Acampe rigida* (Buch.-Ham. ex Sm.) P.F.Hunt	Orchidaceae
1012.		*Acanthephippium bicolor* Lindl.	Orchidaceae
1013.		*Acanthephippium sylhetense* Lindl.	Orchidaceae
1014.		*Aerides multiflora* Roxb.	Orchidaceae
1015.		*Aerides odorata* Lour.	Orchidaceae
1016.		*Bulbophyllum cariniflorum* Rchb.f.	Orchidaceae
1017.		*Bulbophyllum crassipes* Hook.f.	Orchidaceae
1018.		*Bulbophyllum macraei* (Lindl.) Rchb.f.	Orchidaceae
1019.		*Bulbophyllum polyrrhizum* Lindl.	Orchidaceae
1020.		*Bulbophyllum triste* Rchb.f.	Orchidaceae
1021.		*Bulbophyllum umbellatum* Lindl.	Orchidaceae
1022.		*Calanthe triplicata* (Willemet) Ames	Orchidaceae
1023.		*Chiloschista parishii* Seidenf.	Orchidaceae
1024.		*Cirrhopetalum panigrahianum* (S.Misra) S.Misra	Orchidaceae
1025.		*Cleisostoma appendiculatum* (Lindl.) Benth. & Hook.f. ex B.D.Jacks	Orchidaceae
1026.		*Crepidium purpureum* (Lindl.) Szlach.	Orchidaceae
1027.		*Cymbidium aloifolium* (L.) Sw.	Orchidaceae
1028.		*Cymbidium bicolor* Lindl.	Orchidaceae
1029.		*Dendrobium aphyllum* (Roxb.) C.E.C.Fisch.	Orchidaceae
1030.		*Dendrobium bicameratum* Lindl.	Orchidaceae
1031.		*Dendrobium cathcartii* Hook.f.	Orchidaceae
1032.		*Dendrobium crepidatum* Lindl. & Paxton	Orchidaceae
1033.		*Dendrobium formosum* Roxb. ex Lindl.	Orchidaceae
1034.		*Dendrobium herbaceum* Lindl.	Orchidaceae
1035.		*Dendrobium macrostachyum* Lindl.	Orchidaceae
1036.		*Dendrobium moschatum* (Buch.-Ham.) Sw.	Orchidaceae
1037.		*Dendrobium peguanum* Lindl.	Orchidaceae
1038.		*Dendrobium regium* Prain	Orchidaceae
1039.		*Dendrobium transparens* Wall. ex Lindl.	Orchidaceae
1040.		*Dienia ophrydis* (J.Koenig) Seidenf.	Orchidaceae
1041.		*Diploprora championi* (Lindl. ex Benth.) Hook. f. ex Schltr.	Orchidaceae
1042.	New record	*Disperis neilgherrensis* Wight	Orchidaceae
1043.	New record	*Epipogium roseum* (D. Don) Lindl.,	Orchidaceae
1044.		*Eria bambusifolia* Lindl.	Orchidaceae
1045		*Eria meghasaniensis* (S.Misra) S.Misra	Orchidaceae
1046.		*Eulophia explanata* Lindl.	Orchidaceae
1047.		*Eulophia graminea* Lindl.	Orchidaceae
1048.	New record	*Eulophia mackinnonii* Duthie	Orchidaceae
1049.		*Eulophia spectabilis* (Dennst.) Suresh	Orchidaceae
1050.		*Flickingeria macraei* (Lindl.) Seidenf.	Orchidaceae
1051.		*Geodorum recurvum* (Roxb.) Alston	Orchidaceae
1052.		*Goodyera fumata* Thwaites	Orchidaceae
1053.		*Goodyera hispida* Lindl.	Orchidaceae

Sl. No.	Local Name	Scientific Name	Family
1054.		*Goodyera procera* (Ker Gawl.) Hook.	Orchidaceae
1055.		*Goodyera thailandica* Seidenf.	Orchidaceae
1056.		*Habenaria commelinifolia* (Roxb.) Wall. ex Lindl.	Orchidaceae
1057.		*Habenaria diphylla* (Nimmo) Dalzell	Orchidaceae
1058.	Dev Sunda	*Habenaria furcifera* Lindl.	Orchidaceae
1059.		*Habenaria gibsonii* var. *foetida* Blatt. & McCann	Orchidaceae
1060.		*Habenaria plantaginea* Lindl.	Orchidaceae
1061.		*Habenaria reniformis* (D.Don) Hook.f.	Orchidaceae
1062.		*Habenaria stenopetala* Lindl.	Orchidaceae
1063.		*Kingidium deliciosum* (Rchb.f.) H.R.Sweet	Orchidaceae
1064.		*Liparis elliptica* Wight	Orchidaceae
1065.		*Liparis nervosa* (Thunb.) Lindl.	Orchidaceae
1066.		*Liparis odorata* (Willd.) Lindl.	Orchidaceae
1067.		*Liparis viridiflora* (Blume) Lindl.	Orchidaceae
1068.		*Luisia brachystachys* (Lindl.) Blume	Orchidaceae
1069.		*Luisia trichorrhiza* (Hook.) Blume	Orchidaceae
1070.		*Luisia tristis* (G.Forst.) Hook.f.	Orchidaceae
1071.		*Luisiopsis inconspicua* (Hook.f.) Sath.Kumar & P.C.S.Kumar	Orchidaceae
1072.		*Micropera pallida* (Roxb.) Lindl.	Orchidaceae
1073.		*Nervilia aragoana* Gaudich.	Orchidaceae
1074.		*Nervilia crociformis* (Zoll. & Moritzi) Seidenf.	Orchidaceae
1075.		*Nervilia infundibulifolia* Blatt. & McCann	Orchidaceae
1076.		*Nervilia plicata* (Andrews) Schltr.	Orchidaceae
1077.		*Oberonia ensiformis* (Sm.) Lindl.	Orchidaceae
1078.		*Oberonia falconeri* Hook.f.	Orchidaceae
1079.		*Oberonia gammiei* King & Pantl.	Orchidaceae
1080.		*Oberonia mucronata* (D.Don) Ormerod & Seidenf.	Orchidaceae
1081.		*Oberonia proudlockii* King & Pantl.	Orchidaceae
1082.		*Oberonia pyrulifera* Lindl.	Orchidaceae
1083.		*Odisha cleistantha* S.Misra	Orchidaceae
1084.		*Pecteilis gigantea* (Sm.) Raf.	Orchidaceae
1085.		*Pelatantheria insectifera* (Rchb.f.) Ridl.	Orchidaceae
1086.		*Peristylus constrictus* (Lindl.) Lindl.	Orchidaceae
1087.		*Peristylus goodyeroides* (D.Don) Lindl.	Orchidaceae
1088.		*Peristylus lawii* Wight	Orchidaceae
1089.		*Peristylus parishii* Rchb.f.	Orchidaceae
1090.		*Peristylus plantagineus* (Lindl.) Lindl.	Orchidaceae
1091.		*Pholidota imbricata* Lindl.	Orchidaceae
1092.		*Rhynchostylis retusa* (L.) Blume	Orchidaceae
1093.		*Seidenfia versicolor* (Lindl.) Marg. & Szlach.	Orchidaceae
1094.		*Smitinandia micrantha* (Lindl.) Holttum	Orchidaceae
1095.		*Spiranthes sinensis* (Pers.) Ames	Orchidaceae
1096.		*Staurochilus ramosus* (Lindl.) Seidenf.	Orchidaceae
1097.		*Tainia penangiana* Hook.f.	Orchidaceae
1098.		*Thunia bracteata* (Roxb.) Schltr.	Orchidaceae
1099.		*Tropidia angulosa* (Lindl.) Blume	Orchidaceae
1100.		*Tropidia pedunculata* Blume	Orchidaceae
1101.		*Vanda tessellata* (Roxb.) Hook. ex G.Don	Orchidaceae
1102.		*Vanda testacea* (Lindl.) Rchb.f.	Orchidaceae

Sl. No.	Local Name	Scientific Name	Family
1103.		*Zeuxine affinis* (Lindl.) Benth. ex Hook.f.	Orchidaceae
1104.	Recently found in SBR	*Zeuxine gracilis* (Breda) Blume	Orchidaceae
1105.		*Zeuxine nervosa* (Wall. ex Lindl.) Benth. ex Trimen	Orchidaceae
1106.	Ban Kadali	*Musa paradisiaca* L.	Musaceae
1107.		*Alpinia malaccensis* (Burm.f.) Roscoe	Zingiberaceae
1108.		*Cheilocostus speciosus* (J.Koenig) C.D.Specht	Zingiberaceae
1109.	Ban haldi, Amada	*Curcuma amada* Roxb.	Zingiberaceae
1110.	Palua	*Curcuma angustifolia* Roxb.	Zingiberaceae
1111.	Ban haldi / Palua	*Curcuma aromatica* Salisb.	Zingiberaceae
1112.		*Curcuma zedoaria* (Christm.) Roscoe	Zingiberaceae
1113.		*Boesenbergia longiflora* (Wall.) Kuntze Syn. - *Curcumorpha longiflora* (Wall.) A.S.Rao & D.M.Verma	Zingiberaceae
1114.		*Globba marantina* L.	Zingiberaceae
1115.		*Globba racemosa* Sm.	Zingiberaceae
1116.		*Hedychium coccineum* Buch.-Ham. ex Sm.	Zingiberaceae
1117.		*Hedychium coronarium* J.Koenig	Zingiberaceae
1118.		*Kaempferia rotunda* L.	Zingiberaceae
1119.		*Zingiber capitatum* Roxb.	Zingiberaceae
1120.		*Zingiber montanum* (J.Koenig) Link ex A.Dietr. Syn. - *Zingiber purpureum* Roscoe	Zingiberaceae
1121.	Ban-ada	*Zingiber rubens* Roxb.	Zingiberaceae
1122.		*Canna indica* L.	Cannaceae
1123.		*Ananas comosus* (L.) Merr.	Bromeliaceae
1124.		*Crinum amoenum* Ker Gawl. ex Roxb.	Amaryllidaceae
1125.	Kondai	*Crinum viviparum* (Lam.) R.Ansari & V.J.Nair	Amaryllidaceae
1126.		*Pancratium triflorum* Roxb.	Amaryllidaceae
1127.	Sisal	*Agave sisalana* Perrine	Asparagaceae
1128.	Talmuli	*Curculigo orchioides* Gaertn.	Hypoxidaceae
1129.		*Molineria trichocarpa* (Wight) N.P.Balakr. Syn. - *Curculigo trichocarpa* (Wight) Bennet & Raizada	Hypoxidaceae
1130.		*Tacca leontopetaloides* (L.) Kuntze	Taccaceae
1131.		*Dioscorea alata* L.	Dioscoreaceae
1132		*Dioscorea belophylla* (Prain) Voigt ex Haines	Dioscoreaceae
1133		*Dioscorea bulbifera* L.	Dioscoreaceae
1134.	Kanta alu	*Dioscorea glabra* Roxb.	Dioscoreaceae
1135.	Suta alu	*Dioscorea hamiltonii* Hook.f.	Dioscoreaceae
1136.	Bainya	*Dioscorea hispida* Dennst.	Dioscoreaceae
1137.	Pani alu	*Dioscorea oppositifolia* L.	Dioscoreaceae
1138.	Karabha alu	*Dioscorea pentaphylla* L.	Dioscoreaceae
1139.	Kosa alu /Saiga alu	*Dioscorea pubera* Blume	Dioscoreaceae
1140.		*Dioscorea tomentosa* J.Koenig ex Spreng.	Dioscoreaceae
1141.	Tunga alu	*Dioscorea wallichii* Hook.f.	Dioscoreaceae
1142.	Satawari	*Asparagus racemosus* Willd.	Asparagaceae
1143.	Jhinka	*Chlorophytum arundinaceum* Baker	Asparagaceae
1144.		*Chlorophytum tuberosum* (Roxb.) Baker	Asparagaceae
1145.		*Disporum cantoniense* (Lour.) Merr.	Asparagaceae
1146.		*Dracaena terniflora* Roxb.	Asparagaceae
1147.		*Drimia indica* (Roxb.) Jessop	Asparagaceae
1148.		*Gloriosa superba* L.	Colchicaceae

Sl. No.	Local Name	Scientific Name	Family
1149.		*Iphigenia indica* (L.) A.Gray ex Kunth	Colchicaceae
1150.	Ramdantani	*Smilax perfoliata* Lour.	Smilaceae
1151.	Ramdantani / Muturi	*Smilax zeylanica* L.	Smilaceae
1152.		*Monochoria vaginalis* (Burm.f.) C.Presl	Pontederiaceae
1153.		*Dictyospermum ovalifolium* Wight	Commelinaceae
1154.		*Commelina appendiculata* C.B.Clarke	Commelinaceae
1155.		*Commelina benghalensis* L.	Commelinaceae
1156.		*Commelina diffusa* Burm.f.	Commelinaceae
1157.		*Commelina erecta* L.	Commelinaceae
1158.		*Commelina paludosa* Blume	Commelinaceae
1159.		*Commelina suffruticosa* Blume	Commelinaceae
1160.		*Cyanotis arachnoidea* C.B.Clarke	Commelinaceae
1161.		*Cyanotis cristata* (L.) D.Don	Commelinaceae
1162.		*Cyanotis fasciculata* (B.Heyne ex Roth) Schult. & Schult.f.	Commelinaceae
1163.		*Cyanotis tuberosa* (Roxb.) Schult. & Schult.f.	Commelinaceae
1164.		*Floscopa scandens* Lour.	Commelinaceae
1165.		*Murdannia edulis* (Stokes) Faden	Commelinaceae
1166.		*Murdannia japonica* (Thunb.) Faden	Commelinaceae
1167.		*Murdannia nudiflora* (L.) Brenan	Commelinaceae
1168.		*Murdannia spirata* (L.) G.Brückn.	Commelinaceae
1169.		*Murdannia vaginata* (L.) G.Brückn.	Commelinaceae
1170.		*Cyanotis axillaris* (L.) D.Don ex Sweet Syn. - *Tonningia axillaris* (L.) Kuntze	Commelinaceae
1171.		*Juncus prismatocarpus* R.Br.	Juncaceae
1172.	Gauri-bet	*Calamus latifolius* Roxb.	Arecaceae
1173.		*Calamus viminalis* Willd.	Arecaceae
1174.		*Caryota urens* L.	Arecaceae
1175.	Bhui Khejuri	*Phoenix acaulis* Roxb.	Arecaceae
1176.		*Alocasia fornicata* (Roxb.) Schott	Araceae
1177.		*Alocasia macrorrhizos* (L.) G.Don	Araceae
1178.		*Amorphophallus paeoniifolius* (Dennst.) Nicolson	Araceae
1179.		*Arisaema tortuosum* (Wall.) Schott	Araceae
1180.	Saru	*Colocasia esculenta* (L.) Schott	Araceae
1181.	Kanta saru	*Lasia spinosa* (L.) Thwaites	Araceae
1182.		*Amorphophallus margaritifer* (Roxb.) Kunth	Araceae
1183.		*Pothos scandens* L.	Araceae
1184.		*Rhaphidophora glauca* (Wall.) Schott	Araceae
1185.		*Remusatia vivipara* (Roxb.) Schott	Araceae
1186.		*Rhaphidophora decursiva* (Roxb.) Schott	Araceae
1187.		*Scindapsus officinalis* (Roxb.) Schott	Araceae
1188.		*Theriophonum minutum* (Willd.) Baill.	Araceae
1189.		*Butomopsis latifolia* (D.Don) Kunth	Butomaceae
1190.		*Aponogeton natans* (L.) Engl. & K.Krause	Aponogetonaceae
1191.		*Paepalanthus melaleucus* (Bong.) Kunth	Eriocaulaceae
1192.		*Eriocaulon quinquangulare* L.	Eriocaulaceae
1193.		*Eriocaulon ritchieanum* Ruhland	Eriocaulaceae
1194.		*Eriocaulon ritchieanum* Ruhland	Eriocaulaceae
1195.		*Eriocaulon truncatum* Buch.-Ham. ex Mart.	Eriocaulaceae
1196.		*Eriocaulon xeranthemum* Mart.	Eriocaulaceae

Sl. No.	Local Name	Scientific Name	Family
1197.		*Carex baccans* Nees	Cyperaceae
1198.		*Carex filicina* Nees	Cyperaceae
1199.		*Carex phacota* Spreng.	Cyperaceae
1200.		*Carex stramentitia* Boott ex Boeckeler	Cyperaceae
1201.		*Kyllinga brevifolia* Rottb.	Cyperaceae
1202.		*Cyperus compactus* Retz.	Cyperaceae
1203.		*Cyperus cuspidatus* Kunth	Cyperaceae
1204.		*Cyperus cyperoides* (L.) Kuntze	Cyperaceae
1205.		*Cyperus distans* L.f.	Cyperaceae
1206.		*Pycreus flavidus* (Retz.) T.Koyama	Cyperaceae
1207.		*Cyperus haspan* L.	Cyperaceae
1208.		*Cyperus iria* L.	Cyperaceae
1209.		*Cyperus malaccensis* Lam.	Cyperaceae
1210.		*Cyperus niveus* Retz.	Cyperaceae
1211.		*Cyperus nutans* Vahl	Cyperaceae
1212.		*Cyperus pilosus* Vahl	Cyperaceae
1213.		*Cyperus procerus* Rottb.	Cyperaceae
1214.		*Cyperus diffusus* Vahl	Cyperaceae
1215.		*Pycreus pumilus* (L.) Nees	Cyperaceae
1216.		*Cyperus rotundus* L.	Cyperaceae
1217.		*Pycreus sanguinolentus* (Vahl) Nees	Cyperaceae
1218.		*Kyllinga odorata* Vahl	Cyperaceae
1219.		*Cyperus tenuispica* Steud.	Cyperaceae
1220.		*Eleocharis congesta* D.Don	Cyperaceae
1221.		*Eleocharis retroflexa* (Poir.) Urb.	Cyperaceae
1222.		*Fimbristylis aestivalis* Vahl	Cyperaceae
1223.		*Fimbristylis bisumbellata* (Forssk.) Bubani	Cyperaceae
1224.		*Fimbristylis dichotoma* (L.) Vahl	Cyperaceae
1225.		*Fimbristylis littoralis* Gaudich.	Cyperaceae
1226.		*Fimbristylis quinquangularis* (Vahl) Kunth	Cyperaceae
1227.		*Fimbristylis schoenoides* (Retz.) Vahl	Cyperaceae
1228.		*Fuirena ciliaris* (L.) Roxb.	Cyperaceae
1229.		*Courtoisina cyperoides* (Roxb.) Soják	Cyperaceae
1230.		*Lipocarpha chinensis* (Osbeck) J.Kern	Cyperaceae
1231.		*Lipocarpha gracilis* (Rich. ex Pers.) Nees	Cyperaceae
1232.		*Schoenoplectiella articulata* (L.) Lye	Cyperaceae
1233.		*Actinoscirpus grossus* (L.f.) Goetgh. & D.A.Simpson	Cyperaceae
1234.		*Schoenoplectiella juncoides* (Roxb.) Lye	Cyperaceae
1235.		*Schoenoplectiella mucronata* (L.) J.Jung & H.K.Choi	Cyperaceae
1236.		*Lipocarpha squarrosa* (L.) Goetgh.	Cyperaceae
1237.		*Scleria lithosperma* (L.) Sw.	Cyperaceae
1238.		*Scleria terrestris* (L.) Fassett	Cyperaceae
1239.		*Scleria tessellata* Willd.	Cyperaceae
1240.		*Alloteropsis cimicina* (L.) Stapf	Poaceae
1241.		*Alloteropsis semialata* (R.Br.) Hitchc.	Poaceae
1242.		*Andropogon ascinodis* C.B.Clarke	Poaceae
1243.		*Apluda mutica* L.	Poaceae
1244.		*Apocopis courtallumensis* (Steud.) Henrard	Poaceae
1245.		*Apocopis paleacea* (Trin.) Hochr.	Poaceae

Sl. No.	Local Name	Scientific Name	Family
1246. A		*Aristida setacea* Retz.	
1247.		*Arthraxon lanceolatus* (Roxb.) Hochst.	Poaceae
1248.		*Arthraxon lancifolius* (Trin.) Hochst.	Poaceae
1249.		*Arthraxon hispidus* (Thunb.) Makino Syn. - *Arthraxon quartinianus* (A.Rich.) Nash	Poaceae
1250. A		*Arundinella bengalensis* (Spreng.) Druce	Poaceae
1251.		*Arundinella setosa* Trin.	Poaceae
1252.		*Arundo donax* L.	Poaceae
1253.		*Bambusa nutans* Wall. ex Munro	Poaceae
1254.		*Bambusa tulda* Roxb.	Poaceae
1255.		*Bothriochloa bladhii* (Retz.) S.T.Blake	Poaceae
1256.		*Capillipedium parviflorum* (R.Br.) Stapf Syn. - *Bothriochloa parviflora* (R.Br.) Ohwi	Poaceae
1257.		*Bothriochloa pertusa* (L.) A.Camus	Poaceae
1258.		*Brachiaria distachya* (L.) Stapf	Poaceae
1259.		*Brachiaria ramosa* (L.) Stapf	Poaceae
1260.		*Capillipedium assimile* (Steud.) A.Camus	Poaceae
1261.		*Centotheca lappacea* (L.) Desv.	Poaceae
1262.		*Chionachne gigantea* (J.Koenig) Veldkamp Syn. - *Chionachne koenigii* (Spreng.) Thwaites	Poaceae
1263.	Guguchia	*Chrysopogon aciculatus* (Retz.) Trin.	Poaceae
1264.		*Chrysopogon verticillatus* (Roxb.) Trin. ex Steud.	Poaceae
1265.		*Coelachne simpliciuscula* (Steud.) Munro ex Benth.	Poaceae
1266.	New record	*Coelorachis khasiana* (Hack.) Stapf ex Bor	Poaceae
1267.		*Coix aquatica* Roxb.	Poaceae
1268.		*Chionachne gigantea* (J.Koenig) Veldkamp Syn. - *Coix gigantea* J.Koenig	Poaceae
1269.		*Coix lacryma-jobi* L.	Poaceae
1270.		*Cymbopogon caesius* (Hook. & Arn.) Stapf	Poaceae
1271.	Dhantari	*Cymbopogon flexuosus* (Nees ex Steud.) W.Watson	Poaceae
1272.	Dhanantri	*Cymbopogon martini* (Roxb.) W.Watson	Poaceae
1273.	Duba	*Cynodon dactylon* (L.) Pers.	Poaceae
1274.		*Cyrtococcum oxyphyllum* (Steud.) Stapf	Poaceae
1275.		*Cyrtococcum patens* (L.) A.Camus	Poaceae
1276.		*Dactyloctenium aegyptium* (L.) Willd.	Poaceae
1277.	Saliabanso	*Dendrocalamus strictus* (Roxb.) Nees	Poaceae
1278.		*Desmostachya bipinnata* (L.) Stapf	Poaceae
1279.		*Dichanthium annulatum* (Forssk.) Stapf	Poaceae
1280.		*Dichanthium caricosum* (L.) A.Camus	Poaceae
1281.		*Digitaria abludens* (Roem. & Schult.) Veldkamp	Poaceae
1282.		*Digitaria bicornis* (Lam.) Roem. & Schult.	Poaceae
1283.		*Digitaria ciliaris* (Retz.) Koeler	Poaceae
1284.		*Digitaria longiflora* (Retz.) Pers.	Poaceae
1285.		*Digitaria stricta* Roth	Poaceae
1286.		*Echinochloa colona* (L.) Link	Poaceae
1287.		*Eleusine indica* (L.) Gaertn.	Poaceae
1288.		*Eragrostiella nardoides* (Trin.) Bor	Poaceae
1289.		*Eragrostis atrovirens* (Desf.) Trin. ex Steud.	Poaceae
1290.		*Eragrostis coarctata* Stapf	Poaceae

Sl. No.	Local Name	Scientific Name	Family
1291.		*Eragrostis gangetica* (Roxb.) Steud.	Poaceae
1292.		*Eragrostis amabilis* (L.) Wight & Arn. Syn. - *Eragrostis tenella* (L.) P.Beauv. ex Roem. & Schult.	Poaceae
1293.		*Eragrostis unioloides* (Retz.) Nees ex Steud.	Poaceae
1294.	New record	*Eulalia fastigiata* (Nees ex Steud.) Haines	Poaceae
1295.	Banga-serum	*Eulalia trispicata* (Schult.) Henrard	Poaceae
1296.	Bubai,Sabai	*Eulaliopsis binata* (Retz.) C.E.Hubb.	Poaceae
1297.		*Garnotia tenella* (Arn. ex Miq.) Janowski	Poaceae
1298.		*Hackelochloa granularis* (L.) Kuntze	Poaceae
1299.	Sukla	*Heteropogon contortus* (L.) P.Beauv. ex Roem. & Schult.	Poaceae
1300.		*Hygroryza aristata* (Retz.) Nees ex Wight & Arn.	Poaceae
1301.		*Ichnanthus pallens* var. *major* (Nees) Stieber Syn. - *Ichnanthus vicinus* (Bailey) Merr.	Poaceae
1302.	Chiru, Juna	*Imperata cylindrica* (L.) Raeusch.	Poaceae
1303.		*Isachne albens* Trin.	Poaceae
1304.		*Isachne globosa* (Thunb.) Kuntze	Poaceae
1305.		*Isachne globosa* (Thunb.) Kuntze Syn. - *Isachne miliacea* Roth	Poaceae
1306.		*Ischaemum polystachyum* J.Presl Syn. - *Ischaemum hirtum* Hack.	Poaceae
1307.		*Polytrias indica* (Houtt.) Veldkamp Syn. - *Ischaemum indicum* (Houtt.) Merr.	Poaceae
1308.		*Ischaemum rugosum* Salisb.	Poaceae
1309.		*Iseilema laxum* Hack.	Poaceae
1310.		*Leersia hexandra* Sw.	Poaceae
1311.		*Leptochloa chinensis* (L.) Nees	Poaceae
1312.		*Microstegium ciliatum* (Trin.) A.Camus	Poaceae
1313.		*Mnesithea laevis* (Retz.) Kunth	Poaceae
1314.		*Oplismenus burmanni* (Retz.) P.Beauv.	Poaceae
1315.		*Oplismenus compositus* (L.) P.Beauv.	Poaceae
1316.		*Oryza meyeriana* (Zoll. & Moritzi) Baill.	Poaceae
1317.		*Oryza rufipogon* Griff.	Poaceae
1318.	Dhana	*Oryza sativa* L.	Poaceae
1319.		*Oryza officinalis* Wall. ex Watt	Poaceae
1320.		*Panicum brevifolium* L.	Poaceae
1321.		*Panicum notatum* Retz.	Poaceae
1322.	Gundlu / Gundla	*Panicum sumatrense* Roth	Poaceae
1323.		*Panicum curviflorum* Hornem. Syn. - *Panicum trypheron* Schult.	Poaceae
1324.		*Paspalidium flavidum* (Retz.) A.Camus	Poaceae
1325.		*Paspalum conjugatum* P.J.Bergius	Poaceae
1326.	Kodo	*Paspalum scrobiculatum* L.	Poaceae
1327.		*Pennisetum pedicellatum* Trin.	Poaceae
1328.		*Pennisetum polystachyon* Schult.	Poaceae
1329.		*Pennisetum purpureum* Schumach.	Poaceae
1330.		*Perotis indica* (L.) Kuntze	Poaceae
1331.		*Phragmites karka* (Retz.) Trin. ex Steud.	Poaceae
1332.		*Pogonatherum paniceum* (Lam.) Hack.	Poaceae

Sl. No.	Local Name	Scientific Name	Family
1333.		*Pogonatherum rufobarbatum* Griff.	Poaceae
1334.		*Eulalia contorta* (Brongn.) Kuntze Syn. - *Pseudopogonatherum contortum* (Brongn.) A.Camus	Poaceae
1335.		*Pseudosorghum fasciculare* (Roxb.) A.Camus	Poaceae
1336.		*Rottboellia cochinchinensis* (Lour.) Clayton	Poaceae
1337.		*Saccharum fallax* Balansa	Poaceae
1338.		*Saccharum narenga* (Nees ex Steud.) Hack.	Poaceae
1339.	Puyal	*Saccharum spontaneum* L.	Poaceae
1340.		*Sacciolepis indica* (L.) Chase	Poaceae
1341.		*Sacciolepis interrupta* (Willd.) Stapf	Poaceae
1342.		*Schizachyrium brevifolium* (Sw.) Buse	Poaceae
1343.		*Sehima nervosum* (Rottler) Stapf	Poaceae
1344.		*Setaria intermedia* Roem. & Schult.	Poaceae
1345.		*Setaria palmifolia* (J.Koenig) Stapf	Poaceae
1346.		*Setaria pumila* (Poir.) Roem. & Schult.	Poaceae
1347.	Juna	*Sorghum bicolor* (L.) Moench	Poaceae
1348.		*Sorghum nitidum* (Vahl) Pers.	Poaceae
1349.		*Sporobolus diandrus* (Retz.) P.Beauv.	Poaceae
1350.		*Sporobolus indicus* var. *purpureosuffusus* (Ohwi) T.Koyama	Poaceae
1351.		*Sporobolus wallichii* Munro ex Thwaites	Poaceae
1352.		*Themeda arundinacea* (Roxb.) A.Camus	Poaceae
1353.		*Themeda caudata* (Nees ex Hook. & Arn.) A.Camus	Poaceae
1354.		*Themeda laxa* (Andersson) A.Camus	Poaceae
1355.		*Themeda quadrivalvis* (L.) Kuntze	Poaceae
1356.		*Themeda triandra* Forssk.	Poaceae
1357.		*Themeda villosa* (Lam.) A.Camus	Poaceae
1358.	Phuljhadu	*Thysanolaena latifolia* (Roxb. ex Hornem.) Honda	Poaceae
1359.	Maka	*Zea mays* L.	Poaceae

Appendix-II: List of Lichens Recorded in Similipal Biosphere Reserve (Singh and Kumar, 2012).

Family-Arthoniaceae

1. *Arthonia medusula* (Pers.) Nyl.
2. *Arthonia translucens* Stirt.
3. *Arthonia tumidula* (Ach.) Ach.
4. *Arthothelium albescens* Patw. & Makhija
5. *Arthothelium confertum* (A.L.Sm) Makhija & Patw.
6. *Arthothelium erumpens* Müll. Arg.
7. *Arthothelium pycnocarpoides* Müll. Arg.
8. *Herpothallon isidiatum* Jagdeesh & al
9. *Tylophoron protruden7s* Nyl.
10. Cryptothecia multipuntata Jagdeesh & al.
11. *Cryptothecia bengalensis* Jagadeesh & al.
12. *Cryptothecia effusa* (Müll. Arg.) R. Sant.
13. *Cryptothecia involuta* Stirt.
14. *Cryptothecia subtecta* Stirt.

Family-Arthopyreniaceae

15. *Mycomicrothelia conothelina* (Nyl.) Hawksw.

Family-Caliciaceae

16. *Buellia alboatrior* (Nyl.) Zahlbr
17. *Buellia stellulata* (Taylor) Mudd.
18. *Hafellia curatellae* (Malme) Marbach
19. *Dirinaria applanata* D.D. Awasthi.
20. *Dirinaria consimilis* (Stirt.) D.D. Awasthi.

Family-Chrysothricaceae

21. Chrysothrix candelaris (L.) Laundon

Family-Collemataceae

22. Leptogium austroamericanum (Malme) Dodge

Family-Ectolechiaceae

23. *Calopedia puiggarii* (Müll. Arg.) Vezda
24. *Sporopodium argillaceum* (Müll. Arg.) Zahlbr.

Family-Graphidaceae

25. *Diorygma hieroglyphicum* (Pers.) Kalb & al.
26. *Diorygma junghuhnii* (Mont. & Bosch.) Kalb. & al.
27. *Diorygma megasporrum* Kalb. & al.
28. *Diorygma pruinnosum* (Eschw.) Kalb. Staiger
29. *Diorygma radiatum* (Awasthi & S.R.Singh) Kr. P. Singh & Swarnalatha
30. *Dyplolabia afzelii* (Ach.) A. Massal
31. *Glyphis cicatricosa* Ach.
32. *Glyphis duriuscula* Stirt.
33. *Glyphis scyphulifera* (Ach.) Staiger
34. *Graphis caesiella* Vain
35. *Graphis cincta* (Pers) Aptroot
36. *Graphis distincta* Makhija & Adaw
37. *Graphis furcata* Fee
38. *Graphis glaucescence* Fee
39. *Graphis hiascens* (Fee) Archer
40. *Graphis japonica* (Mull. Arg) A.W. Archer
41. *Graphis agarekarii* Patw. & C.R.Kulk.
42. *Graphis albidofarinaceae* Adaw. & Makhija
43. *Graphis albissima* Müll. Arg
44. *Graphis chloroalba* Makhija & Adaw.
45. *Graphis filiformis* Adaw. & Makhija
46. *Graphis handelii* Zahlbr
47. *Graphis insulana* (Mull. Arg) Lucking
48. *Graphis librata* C. Knight
49. *Graphis perticosa* (Kremp.) A.W. Archer
50. *Graphis pinicola* Zahlbr
51. *Graphis platycarpa* Eschw
52. *Graphis pyrrhocheiloides* Zahlbr
53. *Graphis scripta* (L.) Ach.
54. *Graphis streblocarpa* (Bel.) Nyl.
55. *Graphis subasahinae* Nagarkar & Patw.
56. *Graphis tenella* Ach.
57. *Leiorruma exaltatum* (Mont. &V.D. Bosch) Müll. Arg.
58. *Pallidogramme chrysenteron* (Mont.) Staiger & al.
59. *Phaeographis brasiliensis* (A.Massal.) Kalb. & Matthes-Leicht
60. *Platythecium grammitis* (Fée) Staiger
61. *Sarcographa tricosa* (Ach.) Mull. Arg.

Family-Haematommataceae

62. *Haematomma wattii* (Stirt.) Zahlbr.

Family-Lecanoraceae

63. *Lecanora cinerofusca var. cinerofusca* H.Magn
64. *Lecanora iseana* Räsänen

65. *Lecidella enteroleucella* (Nyl.) Hertel
66. *Ramboldia russula* (Ach.) Kalb., Lumbsch & Elix

Family-Lecideaceae

67. *Lecidea granifera* (Ach.) Vain.
68. *Lecidea lapicida* (Ach.) Ach.
69. *Lecidea plana* (J. Lahm.) Nyl.
70. *Lecidea secernes* H. Magn

Family-Letrouitiaceae

71. *Letrouitia leprolyta* (Nyl.) Hafellner
72. *Letrouitia domingenesis* Hafellner & Bellem
73. *Letrouitia transgrassa* (Malme) Hafellner & Bellem

Family: Monoblastaceae

74. *Anisomeridium terminatum* (Nyl.) R.C. Harris

Family-Parmeliaceae

75. *Bulbothrix isidiza* (Nyl.) Hale
76. *Myelochroa xantholepis* (Mant. & Bosch.) Elix
77. *Parmotrema andium* (Müll. Arg.) Hale
78. *Parmotrema ravum* (Krog & Swinscow) Sèrus
79. *Parmotrema tinctorum* (Despr. ex. Nyl.) Hale

Family-Pertusariaceae

80. *Pertusaria kodaikanalensis* Choisy
81. *Pertusaria melastomela* Nyl.
82. *Pertusaria multipuncta* (Turner) Nyl
83. *Pertusaria quassiae* (Fee) Nyl.

Family-Physciaceae

84. *Heterodermia pseudospeciosa* (Kurok) Culb
85. *Heterodermia obscurata* (Nyl.) Trevis
86. *Heterodermia albidiflava* (Kurok.) D.D. Awasthi
87. Heterodermia diademata (Taylor) D.D. Awasthi
88. *Pyxine coccifera* (Fee) Nyl.
89. *Rinodina oxydata* (A. Massal.) Massal.

Family-Pilocarpaceae

90. *Byssoloma tricholomum* (Mont.) Zahlbr.

Family-Pyrenulaceae

91. *Anthracothecium thwaitesii* (Leight) Müll. Arg.
92. *Anthracothecium variolosium* (Pers.) Müll.Arg.
93. *Pyrenula acutalis* R.C Harris
94. *Pyrenula anomela* (Ach.) Vain.
95. *Pyrenula citriformis* R.C. Harris
96. *Pyrenula introducta* (Nyl.) Zahlr.
97. *Pyrenula leucotrypa* (Nyl.) Upreti.
98. *Pyrenula mamillana* (Ach.) Trevis.
99. *Pyrenula nitens* Fee
100. *Pyrenula sublaevigata* (Patw. & Makhija) Upreti
101. *Pyrenula subnitida* (Nyl.) Müll. Arg.
102. *Pyrenula thelomopha* Tuck.

Family-Ramalinaceae

103. *Bacidia alutacea* (Kremp.) Zahlbr.
104. *Bacidia convexula* (Müll. Arg.) Zahlbr.
105. Bacidia millegrana (Taylor) Zahlbr.
106. *Bacidia medialis* (Tuck.ex Nyl.) Zahlbr.
107. *Bacidia phaeolomoides* (Müll. Arg.) Zahlbr.
108. *Bacidiospora psorina* (Nyl.ex. Hue) Kalb.

Family-Roccellaceae

109. *Opegrapha rufescens* Pers

Family-Stictidaceae

110. *Conotrema lumbricoides* Sipman

Family-Strigulaceae

111. *Strigula complanata* (Fée) Mont.

Family-Teloschistaceae

112. *Caloplaca aurantia* (Pers.) Hellborn
113. *Caloplaca bassiae* (Willd.ex. Ach.) Zahlbr.
114. *Caloplaca encephalartii* (Kremp) Zahlbr.
115. *Caloplaca herbidella* (Nyl.ex. Hue) Magn.
116. *Caloplaca vitellinula* (Nyl) H. Olivier

Family-Theolotremataceae

117. *Chapsa pseudophlyctic* (Nyl.)A. Frisch
118. *Diploschistes muscorum* spp. bartelttii Lumbsch.
119. *Myriotrema norstictideum* (Patw. & Nagarkar) D.D. Awasthi
120. *Ocellularia lankaensis* Hale
121. *Reimnitzia sentensis* (Tuck.) Kalb.
122. *Thelotrema albo-olivaceum* Vain.

Family-Trypetheliaceae

123. *Bathelium benghalense* Müll. Arg.
124. *Bathelium madreporiforme* (Eschw.) Trevis
125. *Bathelium tuberculosum* (Makhija & Patw) Harris
126. *Laurera cumingii* (Mont) Zahlbr.
127. *Laurera keralensis* Upreti & A. Singh

128. *Laurera kundarensis* Upreti &A.Singh
129. *Laurera vezdae* Makhija & Patw
130. *Laurera aurantiaca* Makhija & Patw.
131. *Megalotremis biocellata* Aptroot
132. *Polymeridium proponens* (Nyl.) R.C. Harris
133. *Pseudopyrenula pupula* (Ach.) Müll. Arg.
134. *Pseudopyrenula subnudata* Müll. Arg.
135. *Pseudopyrenula subvelata* Müll. Arg.
136. *Trypethelium eluterae* Spreng
137. *Trypethelium endosulphureum* Makhija & Patw.
138. *Trypethelium platystomum* Mont.
139. *Trypethelium tropicum* (Ach.) Müll. Arg.

Family-Verrucariaceae

140. *Staurothele drummondii* (Tuck.) Tuck.
141. *Verrucaria rupestris* Schrad

Appendix-III: List of Bryophytes Recorded in Similipal Biosphere Reserve (Alam *et al.,* 2013).

CLASS: JUNGERMANNIOPSIDA Stotler and Crand.-Stotl.,

ORDER - METZGERIALES R. M. R. M. SCHUST. EMEND SCHLJAK.

Metzgeriaceae H. Klingg`r.

Metzgeria Raddi

1. *Metzgeria himalayensis* Kashyap

Pelliaceae H. Klinggr.

Pellia Raddi

2. *Pellia epiphylla* (L.) Corda

Pallaviciniaceae Mig.

Pallavicinia S. Gray

3. *Pallavicinia lyellii* (Hook.) L. Carruth.

ORDER-JUNGERMANNIALES H. KLINGGR.

Lophocoleaceae Vanden Berghen

Lophocolea (Dumort.) Dumort.

4. *Lophocolea bidentata* (L.) Dumort.

Heteroscyphus Schiffn.

5. *Heteroscyphus argutus* (Nees) Schiffn.

CLASS: MARCHANTIOPSIDA Cronquist, Takht. and W. Zimm.,

ORDER – MARCHANTIALES LIMPR.

Targioniaceae Dumort.

Targionia L.

6. *Targionia hypophylla* L.
7. *Targionia indica* Udar et A. Gupta

Cythodiaceae K. Müll.

Cyathodium Kunze

8. *Cyathodium cavernarum* Kunze

Aytoniaceae Cavers

Asterella P. Beauv.

9. *Asterella blumeana* (Nees) Kachroo

Plagiochasma Lehm. et Lindenb.

10. *Plagiochasma appendiculatum* Lehm. et Lindenb.
11. *Plagiochasma intermedium* Lindenb. et Gottsche. in G. L. and N.

Reboulia Raddi

12. *Reboulia hemisphaerica* (L.) Raddi

Dumortieraceae D. G. Long

Dumortiera Nees

13. *Dumortiera hirsuta* (Sw.) Nees in Reinw. Bl. et Nees

Marchantiaceae Lindl.

Marchantia L. emend Raddi

14. *Marchantia polymorpha* L.
15. *Marchantia subintegra* Mitt.
16. *Marchantia palmata* Nees

Conocephalaceae Müll. Frib. ex Grolle,

Conocephalum Wigg.

17. *Conocephalum conicum* (L.) Lindb.

Ricciaceae Reichenb.

Riccia L.

18 *Riccia crystalina* L.

19 *Riccia fluitans* L.

Ricciocarpos Corda

20. *Ricciocarpous natans* L.

Aneuraceae H. Klinggr.

Riccardia S. Gray

21. *Riccardia levieri* Schiffn.

CLASS –ANTHOCEROTOPSIDA DE BARY EX JANCZ. CORR. PROSK.

ORDER – ANTHOCEROTALES LIMPR. IN COHN

Anthocerotaceae (Gray) Dumort. corr. Trevis. emend. Hassel

Anthoceros L.

22. *Anthoceros punctatus* L.

CLASS BRYOPSIDA ROTHM.

ORDER POTTIALES M. FLEISCH.

Pottiaceae Schimp.

Anoectangium Schwagr.

23. *Anoectangium stracheyanum* Mitt.

Barbula Hedw.,

24. *Barbula arcuata* Griff.

    Hyophila Brid.

25. *Hyophila involuta* (Hook.) Jaeg.

ORDER-POLYTRICHALES M. FLEISCH.

Polytrichaceae Schwägr

    Polytrichum Hedw.

26. *Polytrichum juniperum* Hedw.

ORDER-BRYALES LIMPR.

Bryaceae Schwägr.

    Bryum Hedw.

27 *Bryum argenteum* Hedw.

ORDER: FUNARIALES M. FLEISCH.

Funariaceae Schwägr

    Funaria Hedw.

28. *Funaria hygrometrica* Hedw.

ORDER: DICRANALES H. PHILIB EX M. FLEISCH.

Fissidentaceae Schimp.

    Fissidens Hedw.

29 *Fissidens bryoides* Hedw.

ORDER- HYPNALES W. R. BUCK ET VITT

Sematophyllaceae Broth.

    Foreauella Dix. et P. Vard.

30. *Foreauella orthotheca* (Schwaegr.) Dix. et P. Vard.

Leskeaceae Schimp

    Herpetineuron (C. Muell.) Card.

31. *Herpetineuron toccoae* (Sull. et Lesq.) Card.

Neckeraceae Schimp.

    Neckeropsis Reichdt.

32. *Neckeropsis exserta* (Schwaegr.) Broth.

    Pinnatella Fleisch.

33. *Pinnatella calcutensis* Fleisch.

Appencix-IV: List of Corticolous Algae Recorded in Similipal Biosphere Reserve (Bhakta *et al.*, 2014).

1. *Gloeocapsa lignicola* Rabenhorst
2. *Myxosarcina spectabilis* sensu Vashista
3. *Gloeocapsa alpina* (Nägeli) Brand
4. *Chroococcidiopsis indica* Desikachary
5. *Chroococcidiopsis kashaypi* Friedmann
6. *Aphanocapsa testacea* Nägeli
7. *Aphanocapsa grevillei* (Berkeley) Rabenhorst
8. *Nostoc punctiforme* (Kütz) Hariot
9. *Nostoc commune* Vaucher ex Born. et Flah.
10. *N. piscinale* Kützing ex Born. et Flah.
11. *N. parmelioides* Kützing ex Born. et Flah.
12. *Phormidium rubritericola* Gardner
13. *P. holdenii* (Forti) Anagnostidis
14. *Stigonema turfaccum* Cooke
15. *Tolypothrix byssoidea* (Berk.) Kirchner
16. *Scytonema mirabile* (Dillw.) Born
17. *S. pseudopunctatum* Skuja
18. *Fischerella muscicola* (Thuret) Gom
19. *Desmococcus olivaceus* (Persoon ex Acharius) J.R.Laundon

Appendix-VA: List of Medicinal Plants recorded in Similipal Biosphere Reserve.

Species	Part(s) Used	Medicinal Use(s)
Abrus precartorius L.	Root, leaf, seed	Substitute of rice liquor; cures bodypain, skin diseases
Achyranthes aspera L.	Whole plant, leaf	Used in rheumatism, snake bites, urinary infections
Alstonia scholaris (L.) R.Br.	Root, bark	Cures leprosy, skin diseases, malaria, diarrhoea, asthma
Amaranthus spinosus L.	Whole plant	Used as febrifuge, astringent; cures dyspepsia
Andrographis paniculata (Burm.f.) Wall.ex Nees	Whole plant, leaf	Cures dysentery, worm infection; used in liver tonic
Argemone maxicana L.	Root, seed	Used in skin diseases
Argyreia speciosa Sweet	Root, stem, leaf	Used as carminative, cardiotonic, expectorant, nerve tonic
Azadiracta indica L.	Leaf, flower, seed	Used as antibacterial, antiviral; cures skin infections
Bacopa monieri (L.) Wetts	Whole plant	Used as digestive, anti inflammatory, diuretic; cures asthma, skin diseases
Barleria cristata L.	Root, leaf	Used for cough, inflammations
Barleria strigosa Wild	Root, leaf	Used for cough, cold, inflammations.
Barringtonia acutangula (L.) Gaertn.	Root, leaf, seed	Cures diarrhoea; used as tonic, anthelmintic, expectorant.
Bauhinia acuminata L.	Bark, leaf,	Used to treat leprosy, asthma, malaria, and piles.
Bauhinia purpurea L.	Root, bark	Used in diarrhoea, worm infection..
Bauhinia vahlii Wt.& Arn	Root, bark	Used as aphrodisiac, tonic, cures dysentery.
Bauhinia variegata L.	Root, bark	Used in scrofula, diarrhoea, ulcers, leprosy
Bidens pillosa L.	Root, leaf, seed	Used in eye and eardrops; cures asthma
Bixa orellana L.	Root, bark, leaves, fruit	Cures antipyretic, snakebite, gonorrhea, jaundice
Blechnum oriental L.	Rhizome	Used in urinary disorders, anthelmintic; cures boils
Buchanania lanzan Spreng	Bark, leaves	Cures mouth sores, chronic constipation
Butea monosperma (Lam.) Taub	Root, bark, leaf, flower, seed	Useful in night blindness, elephantiasis; cures piles, tumours, menstrual disorder
Callicarpa arborea Roxb.	Bark	Applied to cutaneous ailments
Callicarpa macrophylla Vahl	Root, leaf	Used in stomach disorders, rheumatic joints
Calotropis gigantea (L.) R.Br.	Bark	Used for leprosy.
Canscora decussata (Roxb.) J.A. & J.H. Schultes	Leaf	Used in insanity, epilepsy, nervous debility.
Canscora diffusa (Vahl)R.Br. ex R. & S.	Whole plant	Useful in nervous, debility,mental diseases.
Canthium dicoccum (Gaertn.)Teys & Benn.	Bark	Employed as febrifuge, fractures, dysentery
Careya arborea Roxb.	Bark, seed	Used in indigestion, flatulence.
Cassia auriculata L.	Bark, seed	Used in eye troubles, diabetes, chylus urine
Cassia fistula L.	Leaf	Taken to purify blood; used as purgative, laxative for habitual constipation.
Cassia occidentalis L.	Root, leaf, seed	Used in whooping cough, skin diseases, gastric disorder.
Cassine glauca (Rottb.) Kuntze	Bark	Given in pneumonia, cholera, epilepsy, hysteria.
Catharanthus roseus (L.) G. Don	Leaf	Used in diabetes, menorrhagia, wasp stings.
Catunaregam spinosa (Thunb.) Tirveng	Bark, fruit, root	Useful in fever, diarrhoea, dysentery, gout, leprosy, asthma.
Celastrus paniculata Willd.	Bark, seed	Taken in stomach pain, swellings
Celosia cristata L.	Flowers, seed	Used in diarrhoea, cough, dysentery.
Centipeda minima (L.) A.Br. & Asch.	Whole plant, seed	Used in ophthalmia, vermifuge.
Cheilocostus speciosus (J.Koenig) C.D.Specht	Root, leaf	Used as tonic, anthelmintic; used in liver disorder
Chlorophytum arundinaceum Baker	Root	Used as tonic.
Cissampelos pareira L.	Root	Used in dyspepsia, dropsy, urinary troubles

Species	Part(s) Used	Medicinal Use(s)
Citrus medica L.	Fruit	Used in dysentery
Clausena excavata Burm.	Root, flower, leaf	Used for colic
Clausena heptaphylla (Roxb.) Wight & Arn.	Leaf	Used for colic
Clematis gouriana Roxb. ex DC.	Leaf, stem	Used as vesicant, poisonous
Clerodendron serratum (L.) Moon	Root, leaf, seed	Used in rheumatism, dyspepsia, dropsy, febrifuge, cephalagia.
Clerodendron viscosum Vent.	Leaf, flower	Used to treat tumour, leucoderma, burning sensation of skin.
Clitoria ternatea L.	Root	Taken in diuretic, eye diseases, diptheria.
Coccinia grandis (L.) Voight	Leaf, root, whole plant	Cure filarial swellings, diabetes, gonorrhoea.
Commelina bengalensis L.	Whole plant	Considered demulcent, emollient, laxative, refrigerant; used in leprosy
Combretum roxburghii Spreng.	Root, bark, leaf	Used as blood purifier; cure loose motion, sore
Corchorus aestuans L.	Seed	Used to treat stomach pain, pneumonia
Crotalaria albida Heyne ex Roth	Root	Used as purgative.
Croton roxburghii Balak	Root, bark	Cure sore throat, dysentery, snakebite, antifertility
Cyathocline purpurea (Buch.-Ham. ex D.Don) Kuntze	Leaf	Used in tonic, anthelmintic, cold
Cyperus rotundas L.	Tubers	Used in stomach, bowel complaints
Dalbergia latifolia Roxb.	Whole plant	Cure dyspepsia, diarrhoea, leprosy.
Datura metel L.	Leaf, flower	Used for cough, asthma, rheumatism, gastric disorder; used as tonic
Derris indica (Lam.) Bennet	Oil	Applied on leprosy, skin diseases, fever, rheumatism, scabies.
Desmodium gyrans (L.f.) DC.	Root	Used as antidote to snakebite, scropion sting.
Desmodium oojeinense (Roxb.) Ohashi	Root, seed	Cure orchitis.
Dicliptera bupleuroides Nees	Whole plant	Used as tonic.
Dillenia aurea Sm.	Bark	Recommended for thrush and weak gum.
Diospyros malabarica (Desr.) Kastel	Bark, leaf, flower, fruit	Useful in diarrhoea, dysentery, leprosy, skin diseases, vaginal disorder.
Diospyros melanoxylon Roxb.	Leaf, flower, bark	Used in urinary and skin diseases, diarrhoea, dyspepsia
Elephantopus scaber L.	Root, leaf	Used as an emollient in dysuria, diarrhoea, dysentery, stomach pain, ulcer
Enydra fluctuans Lour.	Leaf	Used in nervous affections, headache, eye, hookworm.
Erycibe paniculata Roxb.	Bark	Given in cholera.
Eryngium foetidum L.	Root	Properties show stomachic.
Euphorbia hirta L.	Leaf	Cure cough, asthma, colic, dysentery, genito-urinary tract.
Exacum bicolor Roxb.	Whole plant	Used as tonic, febrifuge.
Flcourtia langomas (Lour.) Raeusch.	Leaf, fruit	Used in diarrhoea, diphoretic, liver complaints.
Flemingia chappar Buch.-Ham. ex Benth.	Root, leaf	Used in epilepsy, hysteria, vermifuge.
Flemingia macrophylla (Willd) Prain ex Merr.	Root	Used for ulcers, swellings.
Flemingia strobilifera (L.) R. Br.	Root, leaf	Used in epilepsy, hysteria.
Floscopa scandens Lour.	Whole plant	Used in fractured bones, sore eyes.
Gardenia gummifera L.f.	Gum	Used in nervous disorder, diarrhoea.
Gloriosa superba L.	Tubers	Used in treatment of gout, rheumatism, skin diseases.
Gmelina arborea Roxb.	Leaf	Cure gonorrhoea, cough, ulcer.
Gravillea robusta A. Cunn. Ex. R. Br.	Bark	Contains gum, tannin.
Grewia hirsute Vahl.	Fruit	Used in diarrhoea, dysentery.
Haldina cordifolia (Roxb.) Ridsdale	Bark, root	Useful in diarrhoea, dysentery, *pitta*, wounds, ulcers, fever.
Helicteres isora L.	Bark, fruit	Used in diarrhoea, dyspepsia, gastric troubles.

Species	Part(s) Used	Medicinal Use(s)
Hibiscus mutabilis L.	Leaf, flower	Used for cough, menorrhagia, dysuria; given in pectoral and pulmonary complaints.
Hiptage benghalensis (L.) Kurz.	Leaf	Used in chronic rheumatism, asthma, cutaneous diseases; used for scabies.
Holarrhena pubescens (Buch.-Ham.) Wall. ex G. Don.	Bark, seed	Used in amoebic dysentery, diarrhoea, malaria fever.
Hygrophila auriculata (Schum.) Heine	Root, leaf	Used in jaundice, dropsy, rheumatism, urinogenital diseases.
Hypericum japonicum Thunb. Ex Murr.	Leaf	Used in asthma, dysentery, styptic.
Ichnocarpus frutescens R.Br.	Root, leaf	Used in leprosy, skin diseases.
Indigofera cassioides Rottler ex DC.	Root	Given for pneumonias; used for cough.
Indigofera glabra L.	Leaf	Used as tonic, emollient in external applications.
Iphigenia indica (L.) A. Gray ex Kunth	Bulb	Cure convulsion, colic.
Jasminum scandens Vahl.	Root	Used for ringworm.
Jatropha gossypifolia L.	Shoot, leaf	Used in dental diseases, carbuncles.
Justicia adhatoda L.	Leaf	Used in cold, cough, asthma.
Justicia betonica L.	Whole plant	Applications for boils, swellings; used in diarrhoea.
Kydia calysina Roxb.	Leaf	Cure rheumatism, lumbago.
Lagerstroemia parviflora Roxb.	Bark, leaf, fruit	Used in diabetes.
Lagerstroemia reginae Roxb.	Leaf, fruits	Used in diabetes.
Lawsonia inermis L.	Leaf, root	Used against sore throat, headache,
Lantana camara L.	Leaf	Used for itch, antiseptic for wounds; given in tetanus, rheumatism, malaria.
Leea asiatica (L.) Ridsdale	Tuber	Used against guinea worms.
Leucas indica (L.) R.Br. ex Vatke	Leaf	Used as sedative, stomachic, vermifuge, old sores, dermatosis.
Leucas mollissima Wall. ex Benth.	Whole plant	Used for diuretic, digestion, antiseptic.
Litchi chinensis Sonner.	Seed	Used as anodyne in neuralgic disorders, orchitis
Litsea glutinosa (Lour.) Robins	Bark, root	Used in diarrhoea, dysentery, emmenagogue
Litsea monoptela (Roxb.) Pers	Bark, seed	Used in diarrhoea, sprain, bruishes; used for rheumatism
Lygodium microphyllum (Cav.) R.Br.	Leaf	Used in dysentery, skin diseases, swellings
Madhuca longifolia (J. Koenig ex L.) J.F. Macbr	Bark, flowers, seed	Used against inflammations, sprains; useful in strangury, verminosis
Mangifera indica L.	Bud	Cure chronic dysentery.
Melaleuca leucadendron L.	Leaf	Used in chronic laryngitis, bronchitis.
Melastoma malabatharicum L.	Bark, leaf	Used for skin troubles.
Merrimia umbellate (L.) Hall.f.	Whole plant, seed	Used for fistulae, tumours, cutaneous diseases
Mimosa pudica L.	Leaf, root	Used in iron deficiency, birth control, sexual weakness
Mitragyna parviflora (Roxb.) Korth.	Bark, root, leaf	Useful in gastopathy, colic, fever, wounds, dyspepsia, leprosy, fever
Millettia extensa (Benth.) Baker	Root	Applied to sores of cattle to kill vermins.
Morinda pubescens Sm.	Root, leaf	Useful in haemorrhages, dysentery, boils, gout, inflammations
Morus australis Poir.	Fruit, bark, leaf	Used for sore throat, dyspepsia, and melancholia
Mosla dianthera (Buch. -Ham. ex Roxb.) Maxim.	Whole plant	Useful in hyperthyreosis
Mucuna pruriens (L.) DC.	Root	Used for diseases of nervous system, renal affections, dropsy, elephantiasis
Murraya paniculata (L.) Jack	Leaf, root, bark	Used for diarrhoea, dysentery, vomiting, renal pain
Mussaenda incana Wall	Root	Cure coughs, chewed as an appetizer
Ocimum americanum L.	Leaf, seed	Useful in anorexia, leprosy, malaria and fever
Ocimum gratissimum L.	Oil, seed	Used for earache, toothache, colic, headache, neuralgia
Oxalis corniculata L.	Whole plant	Useful in dyspepsia, scurvy, ulcers, cardiopathy, anemia

Species	Part(s) Used	Medicinal Use(s)
Paederia foetida L.	Leaf, root	Used in bowel troubles, diuretic, piles, liver pain
Parthenium hysterophorus L.	Whole plant, root	Used as tonic, febrifuge, emmenagogue, analgesis, dysentery
Peperomia tetraphylla (Forst) Hook & Arn.	Leaf	Used in kidney diseases
Phobe lanceolata Nees	Berry	Applied to sores
Phyllanthus emblica L.	Fruit	Useful in diabetes, cough, asthma, dysentery, haemorrhages, menorrhagia
Phyllanthus fraternus Webster	Whole plant, root	Used for stomach trouble, dyspepsia, colic, dropsy, jaundice
Piper trioicum Roxb.	Fruit	Used for dyspepsia, vomiting, colic, and cold
Plumbago zeylanicum L.	Root	Used for dyspepsia, piles, diarrhoea, and skin diseases
Polyalthia longifolia (Sonn.) Thw.	Bark	Useful in fever, skin diseases, diabetes
Polygonum strigosum R.Br.	Leaf	Used as fish poison
Prunus ceylanica (Wight) Miq.	Bark	Used for stomachache
Ranunculus pennsylvanicus L.f.	Whole plant	Used to raise blisters, vesicant, antibacterial activity
Rauvolfia serpentina (L.) Benth.	Root, leaf	Cure blood pressure, opacities of the cornea of eye, snakebite
Reinwardtia indica Dum.	Whole plant	Used in paralysis, wounds
Reinacanthus nasuta (L.) Kurz.	Root, leaf	Used in eczema, ringworm; used against cancer, antiseptic
Rhus semialata Murray	Gallnut	Used in colic, diarrhoea, dysentery, and haemorrhage
Rungia parviflora Nees	Leaf	Applied to contusions, febrifuge
Salix tetrasperma Roxb.	Leaf, bark	Given in rheumatism, epilepsy, venereal diseases, piles, and swellings
Saraca asoca (Roxb.) de Wilde	Bark, leaf, flower, seed	Useful in fever, colic, ulcers, menorrhagia, pimples
Semecarpus anacardium L.f.	Fruit	Useful in beriberi, cancer, sciatica, cough, asthma, diabetes
Setaria palmifolia (Koenig) Stapf	Whole plant	Used in irregular menstruation
Shorea robusta Gaertn. f.	Bark, leaf, fruit, resin	Useful in ulcers, wounds, gonorrhoea, leucorrhoea
Sida acuta Burm.f.	Leaf, root	Applied to testicular swellings, elephantiasis; used in nervous and urinary disorder
Sida cordata (Burm.f) Borssum.	Whole plant	Used for gonorrhoea, hyperdiuresis, diarrhoea.
Sida rhombifolia L.	Root, stem	Used in dermatopathy; tuberculosis, leucorrhoea, dipsia.
Sigesbelkia orientalis L.	Whole plant	Cure ulcers, sores, parasitic infections, renal colic
Smilax perfoliata Lour.	Root	Used for rheumatism, urinary complaints, dysentery, skin diseases
Solanum erianthum D.Don.	Root, leaf	Used for urinary troubles; given in vaginal discharges.
Solanum virginianum L.	Whole plant	Useful in colic, dyspepsia, leprosy, skin diseases, cardiac disorder.
Stachytarpheta jamaicensis (L.)Vahl.	Leaf, bark	Used for intestinal worms, veneral diseases, ulcers, dropsy, diarrhoea
Stephania japonica (Thunb.) Miers	Root	Used for fever, diarrhoea, dyspepsia, urinary diseases
Streblus asper Lour.	Bark, leaf	Used as antiseptic
Strobilanthus auriculatus Nees	Leaf	Used externally as a febrifuge
Syzygium cumini (L.) Skeels	Bark, leaf, fruits	Useful in diabetes, fever, gastropathy
Tephrosia purpurea (L.) Pers.	Leaf	Used in bronchitis, bilious, boils, pimples, bleeding piles
Terminalia tomentosa Heyne ex Roth.	Bark	Used as styptic, cardiotonic, purgative
Terminalia arjuna (Roxb. ex DC)	Bark	Useful in fractures, ulcers, diabetes, cardiopathy
Terminalia bellirica (Gaertn.) Roxb.	Bark, fruits	Useful in cough, bronchitis, dropsy, dyspepsia, diarrhoea, dysentery
Terminalia chebula Retz.	Mature & immature fruits	Useful in ulcers, inflammations, gastopathy, leprosy
Toddalia asiatica (L.)	Root, bark	Used in diarrhoea, debility, antipyretic.
Trema orientalis (L.)Bl.	Root	Given in diarrhoea, blood in urine, epilepsy

Species	Part(s) Used	Medicinal Use(s)
Tridax procumbens L.	Leaf	Used to check bleeding of wounds; bronchial catarrh, dysentery, diarrhoea.
Triumfetta rhomboidea Jacq.	Root, bark, flowers	Used for diarrhoea, dysentery, intestinal ulcers, leprosy
Vernonia anthelmintica (L.) Willd.	Fruits	Useful in inflammations, cough, asthma, leprosy, colic, fever
Vernonia aspera Buch.-Ham.	Root	Used in articular rheumatism
Vernonia cinerea (L.) Less	Whole plant	Given in diarrhoea, stomachache, cough, cold, fever, and rheumatism.
Vitex negundo L.	Leaf	Used in rheumatism, dyspepsia, piles, and dysentery
Woodfordia fruticosa (L.) Kurz	Flowers	Useful in leprosy, haemorrhages, menorrhagia, diabetes, ulcers
Xantolis tomentosa (Roxb.)Raf.	Fruit, bark	Given in cholera, cough, cold
Ziziphus rugosa L.	Bark, flowers	Given in diarrhoea, menorrhagia.

Appendix-VB: List of Mushrooms Recorded from Similipal Biosphere Reserve.

Sl.No.	Mushroom Species	Habitat	Host/ Substratum	Edibility	Types (Toxic/Non toxic)
1	*Agaricus* sp.	Mixed stands	Soil	Non-edible	Unknown
2	*Agaricus* sp.	Decaying Paddy straw soil	Paddy straw soil	Non-edible	Unknown
3	*Amanita pseudoprinceps* Y.Y. Cui, Q. Cai& Zhu L. Yang	Decaying Sal leaf mixed soil	Soil	Edible	Non toxic
4	*Amanita vaginata* (Bull.) Lam	Mixed stands	Soil	Non-edible	Non toxic
5	*Amanita* sp.1	Mixed stands	Soil	Non edible	Unknown
6	*Amanita* sp.2	Mixed stands	Soil	Non-edible	Unknown
7	*Amanita* sp.3	Decaying Sal leaf mixed soil	Soil	Edible	Non toxic
8	*Astraeus asiaticus* Phosri, M.P. Martín & Watling	Decaying Sal leaf mixed soil	Soil	Edible	Non toxic
9	*Astreatus hygrometricus* (Pers.) Morgan	Mixed stands	Soil	Edible	Non toxic
10	*Auricularia* sp.	Dung	Dung	Non-edible	Unknown
11	*Boletus edulis* Bull.	Mycorrhizal	Soil	Non-Edible	Non toxic
12	*Boletus* sp.1	Mixed stands	Soil	Non-edible	Toxic
13	*Boletus* sp.2	Mixed stands	Soil	Non-edible	Toxic
14	*Calvatia* sp.	Grass field	Soil	Edible	Non toxic
15	*Cantharellus lateritius (Berk.) Singer*	Mixed stands	Soil	Non edible	Unknown
16	*Chlorophyllum* sp.	Grass field	Soil	Non-edible	Unknown
17	*Coprinus comatus* (O.F. Müll.) Pers.	Grass lands	Soil	Non-edible	Unknown
18	*Coprinus disseminatus* (Pers.) J.E. Lange	Decaying trunk of plant	Plant	Non-edible	Unknown
19	*Coprinus fimetarius* (Schaeff.) Fr.	Elephant dung	Dung	Non-edible	Unknown
20	*Coprinus plicatilis* (Curtis) Fr.	Mixed stands	Soil	Non edible	Unknown
21	*Crepidotus applanatus* (Pers.) P. Kumm.	Dung	Dung	Non-edible	Unknown
22	*Daldinia concentrica* (Bolton) Ces. & De Not.	Decaying trunk of plant	Plant	Non-edible	Unknown
23	*Entoloma* sp.	Mycorrhizal	Soil	Non edible	Unknown
24	*Filoboletus manipularis* (Berk.) Singer	Decaying branch of plant	Plant	Non-edible	Unknown
25	*Ganoderma applanatum* (Pers.) Pat.	Decaying plant	Plant	Non-edible	Unknown
26	*Ganoderma chalceum* (Cooke) Steyaert	Decaying trunk of plant	Plant	Non-edible	Non toxic
27	*Ganoderma lipsience* (Batsch.)	Decaying plant	Plant	Non-edible	Unknown
28	*Ganoderma sessiliforme* Murrill, Bull.	Decaying trunk of plant	Plant	Non-edible	Unknown
29	*Ganoderma tsugae* Wasser, Zmitr. & M. Didukh	Decaying trunk of plant	Plant	Non-edible	Unknown
30	*Ganoderma* sp.	Decaying trunk of plant	Plant	Non-edible	Unknown
31	*Geastrum* sp.	Decaying Sal leaf mixed soil	Soil	Edible	Non toxic
32	*Gymnopilus* sp.	Decaying branch of plant	Plant	Non-edible	Unknown
33	*Lactarius* sp.1	Mycorrhizal	Soil	Non-edible	Unknown
34	*Lactarius* sp.2	Mixed stands	Soil	Non edible	Unknown
35	*Lentinus tuber-regium* (Fr.) Fr.	Decaying trunk of plant	Plant	Edible	Non toxic
36	*Lentinus* sp1	Decaying trunk of plant	Plant	Edible	Non toxic
37	*Lentinus* sp.2	Decaying trunk of plant	Plant	Edible	Non toxic
38	*Lentinus* sp.3	Decaying trunk of plant	Plant	Edible	Non toxic

Sl.No.	Mushroom Species	Habitat	Host/ Substratum	Edibility	Types (Toxic/Non toxic)
39	*Lenzytes* sp.	Decaying plant	Plant	Non-edible	Unknown
40	*Leucoagaricus* sp.	Mixed stands	Soil	Non-edible	Unknown
41	*Lycoperdon* sp.	Decaying Sal leaf mixed soil	Soil	Edible	Non toxic
42	*Marasmius* sp.	Decaying plant leaf	Plant Leaf	Non-edible	Unknown
43	*Microporus xanthopus* (Fr.) Kuntze	Branch of the plant	Plant	Non-edible	Unknown
44	*Oxyporus populinus* (Schumach.) Donk	Decaying trunk of plant	Plant	Non-edible	Unknown
45	*Panaeolina* sp.	Elephant dung	Dung	Non-edible	Unknown
46	*Panaeolus acuminatus* (Schaeff.) Quél.	Grass lands	Soil	Non-edible	Unknown
47	*Phallus indusiatus* (Vent.) Desv.	Forest soil	Soil	Non-edible	Nontoxic
48	*Phellinus* sp.	Decaying plant	Plant	Non-edible	Unknown
49	*Pleurotus* sp.1	Decaying trunk of plant	Plant	Edible	Non toxic
50	*Pleurotus* sp.2	Decaying trunk of plant	Plant	Edible	Non toxic
51	*Polyporus* sp.	Decaying trunk of plant	Plant	Non-edible	Unknown
52	*Psathyrella candolleana* (Fr.) Maire	Mixed stands	Soil	Non-edible	Unknown
53	*Psathyrella* sp.	Mixed stands	Soil	Non-edible	Unknown
54	*Pycnoporus cinnabarinus* (Jacq.) P. Karst.	Branch of the plant	Plant	Non-edible	Unknown
55	*Russula albonigra* (Krombh.) Fr.	Mycorrhizal	Soil	Edible	Non toxic
56	*Russula cyanoxantha* (Schaeff.) Fr.	Decaying Sal leaf mixed soil	Soil	Edible	Non toxic
57	*Russula delica* Fr.	Mixed stands	Soil	Edible	Non toxic
58	*Russula densifolia* Secr. ex Gillet	Decaying Sal leaf mixed soil	Soil	Edible	Non toxic
59	*Russula pseudodelica* J.E. Lange	Mycorrhizal	Soil	Edible	Non toxic
60	*Russula violeipes* Quél.	Decaying Sal leaf mixed soil	Soil	Edible	Non toxic
61	*Russula* sp.1	Mycorrhizal	Soil	edible	Unknown
62	*Russula* sp.2	Mixed stands	Soil	edible	Unknown
63	*Russula* sp. 3	Mixed stands	Soil	Edible	Non toxic
64	*Russula* sp.4	Mixed stands	Soil	edible	Unknown
65	*Russula* sp.5	Decaying Sal leaf mixed soil	Soil	Edible	Non toxic
66	*Russula* sp.6	Decaying Sal leaf mixed soil	Soil	Edible	Non toxic
67	*Russula* sp.7	Decaying Sal leaf mixed soil	Soil	Edible	Non toxic
68	*Russula* sp.8	Decaying Sal leaf mixed soil	Soil	Edible	Non toxic
69	*Russula* sp .9	Mixed stands	Soil	Edible	Non toxic
70	*Scleroderma* sp.	Mixed stands	Soil	Non edible	Unknown
71	*Sparasis* sp.1	Decaying branch of plant	Plant	Edible	Unknown
72	*Termitomyces clypeatus* R. Heim	Termite nest	Soil	Edible	Non toxic
73	*Termitomyces heimii* Natarajan	Termite nest	Soil	Edible	Non toxic
74	*Termitomyces microcarpus* (Berk. & Broome) R. Heim	Grass land and Termite nest	Soil	Edible	Non toxic
75	*Termitomyces* sp.	Soil with underground termite nest	Soil	Edible	Non toxic
76	*Trametes* sp.1	Decaying trunk of plant	Plant	Non-edible	Unknown
77	*Tricholoma* sp.	Mycorrhizal	Soil	Non edible	Unknown
78	*Tyromyces* sp.	Decaying trunk of plant	Plant	Non-edible	Unknown
79	*Tuber* sp.	Decaying Sal leaf mixed soil	Soil	Edible	Non toxic
80	*Volvariella volvacea* (Bull.) Singer	Decaying paddy straw	Paddy straw	Edible	Non toxic
81	*Xylaria polymorpha* (Pers.) Grev.	Decaying trunk of plant	Plant	Non-edible	Unknown

Appendix-VI: List of Mammals Recorded in Simplipal Biosphere Reserve.

1. Asian Elephant (Odia: Hati) *Elephas maximus* Linnaeus, 1758
2. Madras treeshrew (Odia: Bana chuchundra) *Anathana ellioti* (Waterhouse, 1850)
3. Rhesus monkey (Odia: Pati Mankada) *Macaca mulatta* (Zimmermann, 1780)
4. Northern plain's langur (Odia: Hanu Mankada) *Semnopithecus entellus* (Dufresne, 1797)
5. Five-striped squirrel (Odia: Gunduchi musa) *Funambulus pennantii* Wroughton, 1905
6. Indian giant squirrel (Odia: Belera Musa) *Ratufa indica* (Erxleben,1777)
7. Giant Indian flying squirrel (Odia: Olei Budi) *Petaurista philippensis* (Elliot, 1839)
8. Lesser bandicoot (Odia: No specific name) *Bandicota bengalensis* (Grayand Hardwicke, 1833)
9. Large bandicoot (Odia: No specific name) *Bandicota indica* (Bechstein, 1800)
10. Blanford's rat (Odia: No specific name) *Madromys blanfordi* (Thomas, 1881)
11. Little Indian field mouse (Odia: Musa) *Mus booduga* (Gray, 1837)
12. House mouse (Odia: Chutia Musa) *Mus musculus* Linnaeus, 1758
13. House rat (Odia: Musa) *Rattus rattus* (Linnaeus, 1758)
14. Indian long-tailed tree mouse (Odia: No specific name) *Vandeleuria oleracea* (Bennett, 1832)
15. Antelope rat (Odia: No specific name) *Tatera indica* (Hardwicke, 1807)
16. Indian crested porcupine (Odia: Jhinka) *Hystrix indica* Kerr, 1792
17. Indian hare (Odia: Thekua) Lepus nigricollis F. Cuvier, 1823
18. Gray musk shrew (Odia: Chuchundra) *Suncus murinus* (Linnaeus, 1766)
19. White-toothed pigmy shrew (Odia: No specific name) *Suncus etruscus* (Savi, 1822)
20. Short-nosed fruit bat (Odia: Badudi) *Cynopterus sphinx* (Vahl, 1797)
21. Indian flying fox (Odia: Badudi) *Pteropus giganteus* (Brunnich, 1782)
22. Indian fulvous bat (Odia: Badudi) *Rousettus leschenaulti* (Desmarest, 1820)
23. Rufous horseshoe bat (Odia: Badudi) *Rhinolophus rouxii* Temminck, 1835
24. Sri Lanka leaf-nosed bat (Odia: Badudi) *Hipposideros lankadiva* Kelaart, 1850
25. Lesser rat-tailed bat (Odia: Badudi) *Rhinopoma hardwickii* Gray, 1831
26. Pouch bearing bat (Odia: Badudi) *Saccolaimus saccolaimus* (Temminck, 1838)
27. Kelaart's pipistrelle (Odia: Chemini) *Pipistrellus ceylonicus* (Kelaart, 1852)
28. Indian pipistrelle (Odia: Chemini) *Pipistrellus coromandra* (Gray, 1838)
29. Indian pigmy pipistrelle (Odia: Chemini) *Pipistrellus tenuis* (Temminck, 1840)
30. Greater yellow bat (Odia: Haladia Badudi) *Scotophilus heathii* (Horsfield, 1831)
31. Lesser yellow bat (Odia: Haladia Badudi) *Scotophilus kuhlii* Leach, 1821
32. Indian pangolin (Odia: Bajrakapta) *Manis crassicaudata* Gray, 1827
33. Jungle cat (Odia: Bana Bhua, Katas) *Felis chaus* Schreber, 1777
34. Leopard cat (Odia: Chitta Biradi) *Prionailurus bengalensis* (Kerr, 1792)
35. Rusty spotted cat (Odia: No specific name) *Prionailurus rubiginosus* (I. Geoffroy S-H, 1831)
36. Fishing cat (Odia: Machharankana Biradi) *Prionailurus viverrinus* (Bennett, 1833)
37. Leopard (Odia: Kalara-patria Bagha) *Panthera pardus* (Linnaeus, 1758)
38. Royal Bengal tiger (Odia: Mahabala Bagha) *Panthera tigris* (Linnaeus, 1758)
39. Common palm civet (Odia: Salia Patani) *Paradoxurus hermaphrodites* (Pallas, 1777)
40. Small Indian civet (Odia: Salia Patani) *Viverricula indica* (Desmarest, 1804)
41. Indian grey mongoose (Odia: Bada Neula) *Urva edwardsii* (I. Geoffroy S-H, 1818)
42. Small Asian mongoose (Odia: Sana Neula) *Urva auropunctatus* (Hodgson, 1836)
43. Ruddy mongoose (Odia: Kala Langudia Neula) *Urva smithii* Gray, 1837
44. Stripe-necked mongoose (Odia: No specific name) *Urva vitticollis* Bennet, 1835
45. Striped Hyena (Odia: Hundala, Heta Bagha) *Hyaena hyaena* (Linnaeus,1758)

46. Golden jackal (Odia: Bilua, Siala) *Canis aureus* Linnaeus, 1758
47. Gray wolf (Odia: Gadhia) *Canis lupus* Linnaeus, 1758
48. Wild dog/ Dhole (Odia: Balia Kukura, Bana Kukura) *Cuon alpinus* (Pallas, 1811)
49. Bengal fox (Odia: Koki Siali) *Vulpes bengalensis* (Shaw, 1800)
50. Sloth bear (Odia: Bhalu) *Melursus ursinus* (Shaw, 1791)
51. Honey badger or Ratel (Odia: Gada Bhalu) *Mellivora capensis* (Schreber, 1776)
52. Asian small-clawed otter (Odia: Pani Odha) *Aonyx cinereus* (Illiger, 1815)
53. Smooth-coated otter (Odia: Pani Odha) *Lutrogale perspicillata* (I. Geoffroy S-H, 1826)
54. Wild boar (Odia: Barha) *Sus scrofa* Linnaeus, 1758
55. Indian mouse deer/ Indian chevrotain (Odia: Gurandi) *Moschiola indica* (Gray, 1852)
56. Spotted deer/ Chital (Odia: Harina, Mruga) *Axis axis* (Erxleben, 1777)
57. Sambar (Odia: Sambar) *Rusa unicolor* Kerr, 1792
58. Northern red muntjac or Barking Deer (Odia: Kutura) *Muntiacus vaginalis* (Boddaert, 1785)
59. Gaur/ Indian bison (Odia: Gayala) *Bos gaurus* Smith, 1827
60. Four-horned antelope (Odia: Chausingha) *Tetracerus quadricornis* (Blainville, 1816)

Appendix-VII: List of Birds Recorded in Simplipal Biosphere Reserve (Manoj V Nair *Pers. Com.*).

Sl. No.	Common Name	Scientific Name
	PODICIPEDIDAE	
1	Great Crested Grebe	*Podiceps cristatus*
2	Little Grebe	*Tachybaptus ruficollis*
3	Black-necked Grebe	*Podiceps nigricollis*
	PHALACROCORACIDAE	
4	Little Cormorant	*Phalacrocorax niger*
5	Indian Cormorant	*Phalacrocorax fuscicollis*
6	Great Cormorant	*Phalacrocorax carbo*
	PHASIANIDAE	
7	Grey Francolin	*Francolinus pondicerianus*
8	Black Francolin	*Francolinus francolinus*
9	Common Peafowl	*Pavo cristatus*
10	Painted Spurfowl	*Galloperdix lunulata*
11	Red Spurfowl	*Galloperdix spadicea*
12	Red Junglefowl	*Gallus gallus*
13	Jungle Bush Quail	*Perdicula asiatica*
14	Rain Quail	*Coturnix coromandelica*
15	Blue-breasted Quail	*Coturnix chinensis*
16	Painted Bush Quail	*Crythro rsinyricha*
	DENDROCYGNIDAE	
17	Lesser Whistling Duck	*Dendrocygna javanica*
18	Fulvous Whistling Duck	*Dendrocygna bicolor*
	ANATIDAE	
19	Cotton Pygmy-goose	*Nettapus coromandelianus*
20	Common Teal	*Anas crecca*
21	Garganey	*Anas querquedula*
22	Spot-billed Duck	*Anas poecilorhyncha*
23	Gadwall	*Anas strepera*
24	Eurasian Wigeon	*Anas Penelope*
25	Northern Pintail	*Anas acuta*
26	Northern Shoveler	*Anas clypeata*
27	Common Pochard	*Aythya ferina*
28	Ferruginous Pochard	*Aythya nyroca*
29	Tufted Duck	*Aythya fuligula*
30	Red Crested Pochard	*Netta rufina*
31	Rudy Shelduck	*Tadorna ferruginea*
32	Baer's Pochard	*Aythya baeri*
	TURNICIDAE	
33	Barred Buttonquail	*Turnix suscitator*
34	Small Buttonquail	*Turnix sylvaticus*
35	Yellow Legged Buttonquail	*Turnix tanki*
	PICIDAE	
36	Eurasian Wryneck	*Jynx torquilla*
37	Speckled Piculet	*Picumnus innominatus*
38	Brown-capped Pygmy Woodpecker	*Dencdrocopos nanus*
39	Fulvous-breasted Woodpecker	*Dendrocopus macei*
40	Yellow-crowned Woodpecker	*Dendrocopos mahrattensis*
41	Rufous Woodpecker	*Celeus brachyurus*
42	Lesser Yellownape	*Picus chlorolophus*
43	Greater Yellownape	*Picus flavinucha*
44	Streak-throated Woodpecker	*Picus xanthopygaeus*
45	Grey-headed Woodpecker	*Picus canus*
46	Black-rumped Flameback	*Dinopium benghalense*
47	Greater Flameback	*Chrysocolaptes lucidus*
48	Heart Spotted Woodpecker	*Hemicircus canente*
49#	White Bellied Woodpecker	*Drycopus javensis*
	MEGALAIMIDAE	
50	Coppersmith Barbet	*Megalaima haemacephala*
51	Brown-headed Barbet	*Megalaima zeylanica*
52	Lineated Barbet	Megalaima lineate
53	Blue-throated Barbet	*Megalaima asiatica*
	BUCEROTIDAE	
54	Indian Grey Hornbill	*Ocyceros birostris*
55	Oriental Pied Hornbill	*Anthracoceros albirostris*
56	Malabar Pied Hornbill	*Anthracoceros coronatus*
	UPUPIDAE	
57	Common Hoopoe	*Upupa epops*
	TROGONIDAE	
58	Malabar Trogon	*Harpactes fasciatus*
	CORACIIDAE	
59	Indian Roller	*Coracias benghalensis*
	ALCEDINIDAE	
60	Common Kingfisher	*Alcedo atthis*
61	White-throated Kingfisher	*Halcyon smyrnensis*
62	Stork-billed Kingfisher	*Halcyon capensis*
63	Pied Kingfisher	*Ceryle rudis*
64	Blue-eared Kingfisher	*Alcedo meninting*
	MEROPIDAE	
65	Green Bee-eater	*Merops orientalis*
66	Blue-tailed Bee-eater	*Merops philippinus*
67	Blue-bearded Bee-eater	*Nyctyornis athertoni*
68	Chestnut-headed Bee-eater	*Merops leschenaultia*

Sl. No.	Common Name	Scientific Name
	CUCULIDAE	
69	Asian Koel	*Eudynamys scolopacea*
70	Pied Cuckoo	*Clamator jacobinus*
71	Chestnut-winged Cuckoo	*Clamator coromandus*
72	Common Hawk Cuckoo	*Hierococcyx varius*
73	Large Hawk Cuckoo	*Hierococcyx sparverioides*
74	Indian Cuckoo	*Cuculus micropterus*
75	Eurasian Cuckoo	*Cuculus canorus*
76*	Oriental Cuckoo	*Cuculus saturates*
77	Lesser Cuckoo	*Cuculus poliocephalus*
78	Grey-bellied Cuckoo	*Cacomantis passerines*
79	Plaintive Cuckoo	*Cacomantis merulinus*
80	Banded Bay Cuckoo	*Cacomantis sonneratii*
81	Drongo Cuckoo	*Surniculus lugubris*
82	Sirkeer Malkoha	*Phaenicophaeus leshcenaultii*
83	Blue-faced Malkoha	*Phaenicophaeus viridirostris*
84	Green Billed Malkoha	*Phaenicophaeus tristis*
	CENTROPIDAE	
85	Greater Coucal	*Centropus sinensis*
86#	Lesser Coucal	*Centropus bengalensis*
	PSITTACIDAE	
87	Rose-ringed Parakeet	*Psittacula krameri*
88	Alexandrine Parakeet	*Psittacula eupatria*
89	Plum-headed Parakeet	*Psittacula cyanocephala*
90	Vernal Hanging Parrot	*Loriculus vernalis*
	APODIDAE	
91	Asian Palm Swift	*Cypsiurus balasiensis*
92	Alpine Swift	*Apus melba*
93	White-rumped Needletail	*Zoonavena sylvatica*
94	Fork-tailed Swift	*Apus pacificus*
95	House Little Swift	*Apus affinis*
	HEMIPROCNIDAE	
96	Crested Tree Swift	*Hemiprocne coronate*
	TYTONIDAE	
97	Barn Owl	*Tyto alba*
98	Grass Owl	*Tyto capensis*
	STRIGIDAE	
99	Collared Scops Owl	*Otus bakkamoena*
100	Oriental Scops Owl	*Otus sunia*
101	Spotted Owlet	*Athene brama*
102	Jungle Owlet	*Glaucidium radiatum*
103	Brown Hawk Owl	*Ninox scutulata*
104	Short-eared Owl	*Asio flammeus*
105	Eurasian Eagle Owl	*Bubo bubo*
106	Spot-bellied Eagle Owl	*Bubo nipalensis*

Sl. No.	Common Name	Scientific Name
107	Brown Fish Owl	*Ketupa zeylonenis*
108	Brown Wood Owl	*Strix leptogrammica*
109	Mottled Wood Owl	*Strix ocellata*
	CAPRIMULGIDAE	
110	Grey Nightjar	*Caprimulgus indicus*
111	Indian Nightjar	*Caprimulgus asiaticus*
112	Savanna Nightjar	*Caprimulgus affinis*
113	Large-tailed Nightjar	*Caprimulgus macrurus*
	COLUMBIDAE	
114	Rock Pigeon	*Columba livia*
115	Pale-capped Pigeon	*Columba punicea*
116	Green Imperial Pigeon	*Ducula aenea*
117	Mountain Imperial Pigeon	*Ducula badia*
118	Oriental Turtle Dove	*Streptopelia orientalis*
119	Laughing Dove	*Streptopelia senegalensis*
120	Spotted Dove	*Streptopelia chinensis*
121	Red Collared Dove	*Streptopelia tranquebarica*
122	Eurasian Collared Dove	*Streptopelia decaocto*
123	Emerald Dove	*Chalcophaps indica*
124	Yellow-footed Green pigeon	*Treron phoenicoptera*
125	Thick-billed Green Pigeon	*Treron curvirostra*
126	Orange-breasted Green Pigeon	*Treron bicincta*
127	Pompadour Green Pigeon	*Treron pompadora*
	PTEROCLIDIDAE	
128	Chestnut-bellied Sandgrouse	*Pterocles exustus*
	RALLIDAE	
129	Purple Swamphen	*Porphyrio porphyrio*
130	Water Cock	*Gallicrex cinerea*
131	Common Moorhen	*Gallinula chloropus*
132	White-breasted Waterhen	*Amaurornis phoenicurus*
133	Brown Crake	*Amaurornis akool*
134	Ruddy-breasted Crake	*Porzana fusca*
135	Slaty-breasted Rail	*Gallirallus striatus*
136	Common Coot	*Fulica atra*
	ROSTRATULIDAE	
137	Greater Painted-Snipe	*Rostratula bengalensis*
	SCOLOPACIDAE	
138	Common Snipe	*Gallinago gallinago*
139	Common Redshank	*Tringa tetanus*
140	Common Greenshank	*Tringa nebularia*
141	Green Sandpiper	*Tringa ochropus*
142	Common Sandpiper	*Tringa hypoleucos*
143	Wood Sandpiper	*Tringa glareola*

Sl. No.	Common Name	Scientific Name
144	Little Stint	*Calidris minuta*
145	Temminck's Stint	*Calidris temminckii*
	BURHINIDAE	
146	Eurasian Thick-knee	*Burhinus oedicnemus*
	RECURVIROSTRIDAE	
147	Black-winged Stilt	*Himantopus himantopus*
	JACANIDAE	
148	Bronze-winged Jacana	*Metopidius indicus*
149	Pheasant-tailed Jacana	*Hydrophasianus chirurgus*
	GLAREOLIDAE	
150	Oriental Pratincole	*Glareola maldivarum*
151	Small Pratincole	*Glareola lactea*
	CHARADRIIDAE	
152	Pacific Golden Plover	*Pluvialis fulva*
153	Grey Plover	*Pluvialis squatarola*
154	Little Ringed Plover	*Charadrius dubius*
155	River Lapwing	*Vanellus duvaucelii*
156	Red-wattled Lapwing	*Vanellus indicus*
157	Yellow-wattled Lapwing	*Vanellus malabaricus*
158	Kentish Plover	*Charadrius alexandrinus*
159	Grey-headed Lapwing	*Vanellus cinerens*
	LARIDAE	
160	Brown-headed Gull	*Larus brunnicephalus*
161	Gull-billed Tern	*Gelochelidon nilotica*
162	River Tern	*Sterna aurantia*
163	Black bellied Tern	*Sterna acuticauda*
164	Whiskered Tern	*Chlidonias hybrid*
	ACCIPITRIDAE	
165	Jerdon's Baza	*Aviceda jerdoni*
166	Black Baza	*Aviceda leuphotes*
167	Osprey	*Pandion haliaetus*
168	Black-shouldered Kite	*Elanus caeruleus*
169	Black Kite	*Milvus migrans*
170	Black-eared Kite	*Milvus lineatus*
171	Brahminy Kite	*Haliastur Indus*
172#	Long Billed Vulure	*Gyps indicus*
173	Cinereous Vulture	*Aegypius monachus*
174#	Red-headed Vulture	*Sarcogyps calvus*
175#	Egyptian Vulture	*Neophron percnopterus*
176#	White Rumped Vulture	*Gyps bengalensis*
177	Short-toed Snake Eagle	*Circaetus gallicus*
178	Crested Serpent Eagle	*Spilornis cheela*
179#	Black Eagle	*Ictinaetus malayensis*
180	Eurasian Marsh Harrier	*Circus aeruginosus*
181	Pallid Harrier	*Circus macrourus*
182	Montagu's Harrier	*Circus pygargus*
183	Pied Harrier	*Circus melanoleucos*
184	Shikra	*Accipiter badius*

Sl. No.	Common Name	Scientific Name
185	Besra	*Accipiter virgatus*
186	Eurasian Sparrowhawk	*Accipiter nisus*
187	Crested Goshawk	*Accipiter trivirgatus*
188	Oriental Honey Buzzard	*Pernis ptilorhyncus*
189	White-eyed Buzzard	*Butastur teesa*
190	Common Buzzard	*Buteo buteo*
191	Greater Spotted Eagle	*Aquila clanga*
192	Indian Spotted Eagle	*Aquila hastata*
193	Tawny Eagle	*Aquila rapax*
194	Steppe Eagle	*Aquila nipalensis*
195	Bonelli's Eagle	*Hieraaetus fasciatus*
196	Booted Eagle	*Hieraaetus pennatus*
197#	Rufous-bellied Eagle	*Hieraaetus kienerii*
198	Changeable Hawk Eagle	*Spizaetus cirrhatus*
199	Mountain Hawk Eagle	*Spizaetus nipalensis*
200	Pallas Fish Eagle	*Hliaeetus leucoryphus*
201	Grey Headed Fish Eagle	*Ichthyophaga ichthyaetus*
	FALCONIDAE	
202	Collared Falconet	*Microhierax caerulescens*
203	Common Kestrel	*Falco tinnunculus*
204	Lessr Kestrel	*Falco naumanni*
205	Peregrine Falcon	*Falco peregrinus*
206	Laggar Falcon	*Falco jugger*
207	Eurasian Hobby	*Falco subbuteo*
	ANHINGIDAE	
208	Darter	*Anhinga melanogaster*
	ARDEIDAE	
209	Little Heron	*Butorides striatus*
210	Pond Heron	*Ardeola grayii*
211	Purple Heron	*Ardea purpurea*
212	Grey Heron	*Ardea cinerea*
213	Cattle Egret	*Bubulcus ibis*
214	Little Egret	*Egretta garzetta*
215	Intermediate Egret	*Mesophoyx intermedia*
216	Great Egret	*Casmerodius albus*
217	Black Crowned Night Heron	*Nycticorax nycticorax*
218	Black Bittern	*Dupetor flavicollis*
219	Yellow Bittern	*Ixobrychus sinensis*
220	Cinnamon Bittern	*Ixobrychus cinnamomeus*
	CICONIDAE	
221	Asian Openbill	*Anastomus oscitans*
222	Wooly-necked Stork	*Ciconia episcopus*
223	Lesser Adjutant	*Leptoptilos javanicus*
224	Painted Stork	*Mycteria leucocephala*
	THRESKIORNITHIDAE	
225	Black-headed White Ibis	*Threskiornis melanocephalus*
226	Black Ibis	*Pseudibis papillosa*

Sl. No.	Common Name	Scientific Name
	CAMPEPHAGIDAE	
227	Rosy Minivet	*Pericrocotus roseus*
228	Scarlet Minivet	*Pericrocotus roseus*
229	Long-tailed Minivet	*Pericrocotus ethologus*
230	Small Minivet	*Pericrocotus cinnamomeus*
231	Large Cuckoo-shrike	*Coracina novaehollandiae*
232	Large Woodshrike	*Tephrodornis gularis*
233	Common Woodshrike	*Tephrodornis pondicerianus*
234	Black headed Cuckoo-shrike	*Coracina melanoptera*
235	Black-winged Cuckoo-shrike	*Coracina melaschistos*
236	Bar winged Flycatcher-shrike	*Hemipus picatus*
	LANIDAE	
237	Long-tailed Shrike	*Lanius schach*
238	Grey-backed Shrike	*Lanius tephronotus*
239	Bay-backed Shrike	*Lanius vittatus*
240	Brown Shrike	*Lanius cristatus*
	CORVIDAE	
241	Rufous Treepie	*Dendrocitta vagabunda*
242	Grey Treepie	*Dendrocitta formosae*
243	House Crow	*Corvus splendens*
244	Large-billed Crow	*Corvus macrorhyncos*
	DICRURIDAE	
245	Black Drongo	*Dicrurus macrocercus*
246	White-bellied Drongo	*Dicrurus caerulescens*
247	Bronzed Drongo	*Dicrurus aeneus*
248	Ashy Drongo	*Dicrurus leucophaeus*
249	Spangled Drongo	*Dicrurus hottentottus*
250	Greater Racket Tailed Drongo	*Dicrurus paradiseus*
	ORIOLIDAE	
251	Indian Golden Oriole	*Oriolus oriolus*
252	Black-hooded Oriole	*Oriolus xanthornus*
253	Black-naped Oriole	*Oriolus chinensis*
	ARTAMIDAE	
254	Ashy Wood Swallow	*Artamus fuscus*
	IRENIDAE	
255	Common Iora	*Aegithina tiphia*
256	Blue-winged Leafbird	*Chloropsis cochinchinensis*
257	Golden-fronted Leafbird	*Chloropsis aurifrons*
258	Asian Fairy Bluebird	*Irene puella*
	PITTIDAE	
259	Indian Pitta	*Pitta brachyuran*
	MUSCICAPIDAE	
260	Asian Paradise-flycatcher	*Terpsiphone paradise*
261	Black-naped Monarch	*Hypothymis azurea*

Sl. No.	Common Name	Scientific Name
262	White-browed Fantail	*Rhipidura aureola*
263	White-throated Fantail	*Rhipidura albicollis*
264	Oriental Magpie Robin	*Copsychus saularis*
265	White-rumped Shama	*Copsychus malabaricus*
266	Indian Robin	*Saxicoloides fulicata*
267	Bluethroat	*Luscinia svecica*
268	Black Redstart	*Phoenicurus ochruros*
269	Desert Wheatear	*Oenanthe deserti*
270	Variable Wheatear	*Oenanthe picata*
271	Common Stonechat	*Saxicola torquata*
272	White-tailed Stonechat	*Saxicola leucura*
273	Pied Bushchat	*Saxicola caprata*
274	Blue Rock Thrush	*Monticola solitaries*
275	Blue-capped Rock Thrush	*Monticola cinclorhynchus*
276	Malabar Whistling Thrush	*Myophonus horsfieldii*
277	Pied Thrush	*Zoothera wardii*
278	Tickell's Thrush	*Turdus unicolor*
279	Orange headed thrush	*Zoothera citrine*
280	Scaly Thrush	*Zoothera dauma*
281*	Black-breasted Thrush	*Turdus dissimilis*
282	Dark-throated Thrush	*Turdus ruficollis*
283	Eurasian Blackbird	*Turdud merula*
284	Red-throated flycatcher	*Ficedula parva*
285	Tickell's Blue Flycatcher	*Cyornis tickelliae*
286	Asian Brown Flycatcher	*Muscicapa dauurica*
287	Brown-breasted Flycatcher	*Muscicapa muttui*
288	Ultramarine Flycatcher	*Ficedula superciliaris*
289	Little Pied Flycatcher	*Ficedula westermanni*
290	Verditer Flycatcher	*Eumyias thalassina*
291	Pale-chinned Flycatcher	*Cyornis poliogenys*
292	Blue-throated flycatcher	*Cyornis rubeculoides*
293	Grey-headed Canary Flycatcher	*Culicicapa ceylonensis*
	PARIDAE	
294	Great Tit	*Parus major*
295	Indian Yellow Tit	*Parus aplonotus*
296	Black-lored Tit	*Parus xanthogenys*
	SITTIDAE	
297	Velvet-fronted Nuthatch	*Sitta frontalis*
298	Chestnut-bellied Nuthatch	*Sitta castanea*
	STURNIDAE	
299	Hill Myna	*Gracula religiosa*
300	Asian Pied Starling	*Sturnus contra*
301	Common Myna	*Acridotheres tristis*
302	Jungle Myna	*Acridotheres fuscus*

Sl. No.	Common Name	Scientific Name
303	Bank Myna	*Acridotheres ginginianus*
304	Brahminy Starling	*Sturnus pagodarum*
305	Chestnut-tailed Starling	*Sturnus malabaricus*
306	Rosy Starling	*Sturnus roseus*
	PYCNONOTIDAE	
307	Black-crested Bulbul	*Pycnonotus melanicterus*
308	Red-whiskered Bulbul	*Pycnonotus jocosus*
309	Red-vented Bulbul	*Pycnonotus cafer*
310	White-browed Bulbul	*Pycnonotus luteolus*
	CISTICOLIDAE	
311	Zitting Cisticola	*Cisticola juncidis*
312	Grey-breasted Prinia	*Prinia hodgsonii*
313	Ashy Prinia	*Prinia socialis*
314	Plain Prinia	*Prinia inornata*
315	Jungle Prinia	*Prinia sylvatica*
	ZOSTEROPIDAE	
316	Oriental White-eye	*Zosterops palpebrosa*
	SYLVIDAE	
317	Pale-footed Bush Warbler	*Cettia pallidipes*
318	Spotted Bush Warbler	*Bradypterus thoracicus*
319	Clamorous Reed Warbler	*Acrocephalus stentoreus*
320	Thick-billed Warbler	*Acrocephalus aedon*
321	Blyth's Reed Warbler	*Acrocephalus dumetorum*
322	Paddy-field Warbler	*Acrocephalus Agricola*
323	Bristled Grassbird	*Chaetornis striatus*
324	Common Tailor Bird	*Orthotomus sutorius*
325	Orphean Warbler	*Sylvia hortensis*
326	Lesser Whitethroat	*Sylvia curruca*
327	Common Chiffchaff	*Phylloscopus collybita*
328	Greenish Leaf Warbler	*Phylloscopus trochiloides*
329	Hume's Warbler	*Phylloscopus humei*
330	Tickell's Leaf Warbler	*Phylloscopus affinis*
331	Eastern Crowned Warbler	*Phylloscopus coronatus*
332	Golden-spectacled Warbler	*Phylloscopus burkii*
333	Puff-throated Babbler	*Pellorneum ruficeps*
334	Brown-cheeked Fulvetta	*Alcippe poioicephala*
335	Jungle Babbler	*Turdoides striatus*
336	Rufous-capped Babbler	*Stachyris ruficeps*
337	Striped Tit Babbler	*Macronous gularis*
338	Common Babbler	*Turdoides caudatus*
339	Indian Scimitar Babbler	*Pomatorhinus horsfieldii*
340	Tawny-bellied Babbler	*Dumetia hyperythra*
341	Yellow-eyed Babbler	*Chrysomma sinense*

Sl. No.	Common Name	Scientific Name
	ALAUDIDAE	
342	Rufous-winged Bushlark	Mirafra assamica
343	Crested Lark	*Galerida cristata*
344	Oriental Skylark	*Alauda gulgula*
345	Ashy-crowned Sparrow Lark	*Eremopterix grisea*
346	Rufous-tailed Lark	*Ammomanes phoenicurus*
347	Indian Bushlark	*Mirafra erythroptera*
348	Jerdon's Bushlark	*Mirafra affinis*
	NECTARINIIDAE	
349	Purple-rumped Sunbird	*Nectarinia zeylonica*
350	Purple Sunbird	*Nectarinia asiatica*
351	Crimson Sunbird	*Aethopyga siparaja*
352	Little Spiderhunter	*Arachnothera longirostra*
	DICAEIDAE	
353	Thick-billed Flowerpecker	*Dicaeum agile*
354	Pale-billed Flowerpecker	*Dicaeum erythrorhynchos*
	MOTACILLIDAE	
355	White Wagtail	*Motacilla alba*
356	Yellow Wagtail	*Motacilla flava*
357	Citrine Wagtail	*Motacilla citreola*
358	Grey Wagtail	*Motacilla cinerea*
359	White-browed Wagtail	*Motacilla maderaspatensis*
360	Forest Wagtail	*Dendronanthus indicus*
361	Paddy-field Pipit	*Anthus rufulus*
362	Olive-backed Pipit	*Anthus hodgsonii*
363	Tree Pipit	*Anthus trivialis*
364	Blyth's Pipit	*Anthus godlewskii*
	PASSERIDAE	
365	House Sparrow	*Passer domesticus*
366	Chestnut-shouldered Petronia	*Petronia xanthocollis*
367	Baya Weaver	*Ploceus philippinus*
368	Black-breasted Weaver	*Ploceus benghalensis*
369	Streaked Weaver	*Ploceus manyar*
	HIRUNDINIDAE	
370	Dusky Crag Martin	*Hirundo concolor*
371	Northern House Martin	*Delichon urbica*
372	Barn Swallow	*Hirundo rustica*
373	Wire-tailed Swallow	*Hirundo smithii*
374	Red-rumped Swallow	*Hirundo daurica*
375	Streak-throated Swallow	*Hirundo fluvicola*
	ESTRILDIDAE	
376	Black-headed Munia	*Lonchura Malacca*
377	Scaly-breasted Munia	*Lonchura punctulata*

Sl. No.	Common Name	Scientific Name
378	White-rumped Munia	*Lonchura striata*
379	Black-throated Munia	*Lonchura kelartii*
380	Indian Silverbill	*Lonchura malabarica*
381	Red Avadavat	*Amandava amandava*
	FRINGILLIDAE	
382	Common Rosefinch	*Carpodacus erythrinus*
	EMBERIZIDAE	
383	Crested Bunting	*Melophus lathami*
384	Grey-necked Bunting	*Emberiza buchanani*
385	White-capped Bunting	*Emberiza stewarti*

 # No direct sighting in past three years (Nayak and Naik 2014)

** Unconfirmed record (Manoj V. Nair pers. com.)*

Appendix-VIII: List of Amphibians and Reptiles Recorded in Similipal Biosphere Reserve.

AMPHIBIANS

Order: Anura (Frogs and Toads)

Family: Bufonidae Gray, 1825

1. *Duttaphrynus scaber* (Schneider,1799) (Odia: Luni benga)
2. *Duttaphrynus stomaticus* (Lütken, 1864) (Odia: Katha benga)
3. *Duttaphrynus melanostictus* (Schneider,1799) (Odia: Sinduria benga)

Family: Dicroglossidae Anderson, 1871

4. *Euphlyctis cyanophlyctis* (Schneider,1799) (Odia: Pani benga)
5. *Fejervarya orissaensis* (Dutta,1997) (Odia: Pani benga)
6. *Minervarya syhadrensis* (Annandale, 1919) (Odia: Pani benga)
7. *Minervarya kalinga* (Raj, Dinesh, Das, Dutta, Kar, and Mohapatra, 2018) (Odia: Pani benga)
8. *Hoplobatrachus crassus* (Jerdon, 1853) (Odia: Cheli benga)
9. *Hoplobatrachus tigerinus* (Daudin, 1802) (Odia: Brahmani benga)
10. *Sphaerotheca breviceps* (Schneider, 1799) (Odia: Mati pota benga)
11. *Spaerotheca dobsonii* (Boulenger, 1882) (Odia: Mati pota matia benga)
12. *Sphaerotheca rolandae* (Dubois,1983) (Odia: Chhota mati pota benga)

Family: Microhylidae Gunther, 1858

13. *Microhyla mymensinghensis* Hasan, Islam, Kuramoto, Kurabayashi, and Sumida, 2014 (Odia: Chuin benga)
14. *Uperodon taprobanicus* (Parker, 1934) (Parker, 1934) (Odia: Sinduria phutka benga)
15. *Uperodon variegatus* (Stoliczka,1872) (Odia: Suneli benga)
16. *Uperodon globulosus* (Günther, 1864) (Odia: Kolathia benga)
17. *Uperodon systoma* (Schneider, 1799) (Odia: Chitra benga)

Family: Ranidae Rafinesque-Schmaltz, 1814

18. *Hydrophylax bahuvistara* Padhye, Jadhav, Modak, Nameer & Dahanukar, 2015 (Odia: Nali benga)
19. *Hylarana tytleri* Theobald, 1868 (Odia: Nali benga)

Family: Rhacophoridae Hoffman, 1932

20. *Chirixalus simus* Annandale, 1915 (Odia: Chuin buda benga)
21. *Raorchestes sanctisilvaticus* (Das and Chanda, 1997) (Odia: Similipal buda benga)
22. *Polypedates maculatus* Gray,1834 (Odia: Katha benga; Akhi dian benga)
23. *Polypedates teraiensis* Dubois,1987 (Odia: Akhi dian benga)

REPTILES

Order: Testudines (Turtles and Tortoises)

Family: Geoemydidae Theobald, 1868 (Fomerly Bataguridae Gray, 1869)

1. *Pangshura tecta* (Gray,1831) (Odia: Nali beka katha kainchha)
2. *Pangshura tentoria* (Gray, 1834) (Odia: Katha kainchha)
3. *Melanochelys tricarinata* (Blyth,1856) (Odia: Tini garia pahadi kainchha)
4. *Melanochelys trijuga indopeninsularis* (Annandale, 1913) (Odia: Kala/ Pahadi kainchha)

Family: Testudinidae Gray, 1825

5. *Indotestudo elongata* (Blyth,1853) (Odia: Haladia katha kainchha)

Family: Trionychidae Bell, 1828

6. *Lissemys punctata punctata* (Bonnaterre,1789) (Odia: Pani/Pankua kainchha)
7. *Lissemys punctata andersoni* (Webb, 1980) (Odia: Pani/Pankua kainchha)

Order: Crocodylia (Crocodiles)

Family: Crocodylidae Gray, 1825

8. *Crocodylus palustris* Lesson,1831 (Odia: Magara kumbhira)

Order: Sauria (Lizards)

Family: Agamidae Gray, 1825

9. *Calotes versicolor* (Daudin, 1802) (Odia: Endua)
10. *Psammophilus blanfordanus* (Stoliczka,1871) (Odia: Endua)
11. *Sitana spinaecephalus* Deepak, Vyas & Giri, 2016 (Odia: Mati endua)

Family: Chamaeleonidae Gray, 1825

12. *Chamaeleo zeylanicus* Laurenti,1768 (Odia: Bahurupi; Pohola endua; Maka sapa)

Family: Eublepharidae Boulenger, 1883

13. *Eublepharis hardwickii* Gray,1827 (Odia: Kalakuta; Kalakuta sapa)

Family: Gekkonidae Gray, 1825

14. *Cyrtodactylus nebulosus* (Beddome,1870) (Odia: Kalakuta)
15. *Hemidactylus brookii* Gray,1845 (Odia: Jhitipiti)
16. *Hemidactylus flaviviridis* Rüppell,1835 (Odia: Jhitipiti)
17. *Hemidactylus frenatus* Schlegel, 1836 (Odia: Jhitipiti)
18. *Hemidactylus leschenaultii* Dumeril & Bibron,1836 (Odia: Jhitipiti)
19. *Hemidactylus* sp. (Odia: No specific name)

Family: Lacertidae Cope, 1864

20. *Ophisops jerdonii* Blyth,1853 (Odia: Champei neula)

Family: Scincidae Gray, 1825

21. *Lygosoma albopunctata* (Gray,1846) (Odia: Champei sapa)
22. *Lygosoma punctata* (Gmelin,1799) (Odia: Nali lanzia champei sapa)
23. *Eutropis beddomei* (Jerdon,1870) (Odia: Champei neula)
24. *Eutropis carinata* (Schneider,1801) (Odia: Champei neula)
25. *Eutropis allapallensis* (Schmidt,1926) (Odia: Champei neula)
26. *Eutropis macularia* (Blyth,1853) (Odia: Champei neula)
27. *Sphenomorphus maculatus* (Blyth,1853) (Odia: Champei neula)
28. *Sepsophis punctatus* (Beddome,1846) (Odia: No specific name)

Family: Varanidae Gray, 1827

29. *Varanus bengalensis* (Daudin, 1802) (Odia: Matia godhi)
30. *Varanus flavescens* (Hardwicke & Gray, 1827) (Odia: Sorisia godhi)

Order: Squamata (Snakes)

Family: Boidae Gray, 1842

31. *Eryx johnii* (Russell, 1801) (Odia: Domundia sapa)
32. *Eryx conicus* (Schneider,1801) (Odia: Boda sapa)

Family: Pythonidae Fitzinger, 1826

33. *Python molurus molurus* (Linnaeus,1758) (Odia: Ajagara sapa)

Family: Colubridae Cope, 1893

Subfamily: Ahaetuliinae

34. *Ahaetulla nasuta* (Lacepede,1789) (Odia: Laudankia sapa)
35. *Ahaetulla laudankia* Deepak, Narayanan, Sarkar, Dutta & Mohapatra, 2019 (Odia: Laudankia sapa)
36. *Ahaetulla anomala* (Annandale, 1906) (Odia: Laudankia sapa)
37. *Chrysopelea ornata* (Shaw,1802) (Odia: Udanta sapa)
38. *Dendrelaphis tristis* (Daudin,1803) (Odia: Kanala; Kauchia)

Subfamily: Colubrinae

39. *Argyrogena fasciolata* (Shaw,1802) (Odia: No specific name)
40. *Boiga flaviviridis* Vogel & Ganesh, 2013 (Odia: No specific name)
41. *Boiga forsteni* (Dumeril, Bibron & Dumerill,1854) (Odia: Katakatia naga)
42. *Boiga trigonata* (Schneider,1802) (Odia: Pahadia chiti sapa)
43. *Coelognathus helena* (Daudin,1803) (Odia: Donger/Pahadia chiti)
44. *Coelognathus helena nigriangularis* Mohapatra, Schulz, Hofmann & Dutta, 2016 (Odia: Donger/ Pahadia chiti)
45. *Coelognathus radiatus* (Boie,1827) (Odia: Pahadia chiti)
46. *Lycodon aulicus* (Linnaeus,1758) (Odia: Kaudia chiti sapa)

47. *Lycodon jara* (Shaw,1802) (Odia: No specific name)
48. *Lycodon striatus* (Shaw,1802) (Odia: Kaudia chiti)
49. *Oligodon arnensis* (Shaw,1802) (Odia: Matia hara sapa)
50. *Oligodon taeniolatus* (Jerdon, 1853) (Odia: No specific name)
51. *Oligodon* cf *cyclurus* (Cantor, 1839) (Odia: No specific name)
52. *Ptyas mucosus* (Linnaeus,1758) (Odia: Dhamana sapa)

Subfamily: Natricinae

53. *Amphiesma stolatum* (Linnaeus,1758) (Odia: Matibiradi/Matiali sapa)
54. *Rhabdophis plumbicolor* (Cantor,1839) (Odia: Sabuja dhanda)
55. *Rhabdophis subminiatus* (Schlegel, 1837) (Odia: Nali beka dhanda)
56. *Fowlea piscator* (Schneider, 1799) (Odia: Pani dhanda; Dhanda sapa)

Family: Homalopsidae Gyiophis Murphy & Voris, 2014

Subfamily: Homalopsinae

57. *Enhydris enhydris* (Schneider,1799) (Odia: Jala ganthia)

Family: Lamprophiidae Fitzinger, 1843

58. *Psammodynastes pulverulentus* (Boie,1827) (Odia: No specific name)

Family: Sibynophiidae Dunn, 1928

59. *Sibynophis sagittarius* (Cantor,1839) (Odia: Dhulia naga)

Family: Elapidae Boie, 1827

60. *Bungarus caeruleus* (Schneider,1801) (Odia: Chiti sapa)
61. *Bungarus fasciatus* (Schneider,1801) (Odia: Rana sapa)
62. *Naja kaouthia* Lesson,1831 (Odia: Tampa sapa)
63. *Naja naja* (Linnaeus,1758) (Odia: Naga/Gokhara sapa)
64. *Ophiophagus hannah* (Cantor,1836) (Odia: Ahiraja)

Family: Typhlopidae Boulenger, 1891

65. *Grypotyphlops acutus* (Dumeril & Bibron,1844) (Odia: Teli sapa)
66. *Indotyphlops braminus* (Daudin,1803) (Odia: Domundia/Telia sapa)
67. *Argyrophis diardii* (Schlegel, 1839) (Odia: Telia sapa)

Family: Viperidae Bonaparte, 1840

68. *Daboia russelii* (Shaw & Nodder,1797) (Odia: Chandan boda)
69. *Echis carinatus* (Schneider,1801) (Odia: Dhulia boda)
70. *Trimeresurus gramineus* (Shaw, 1802) (Odia: Fatka sapa; Gendamundia boila)

Appendix-IX: List of Fishes Recorded in Similipal Biosphere Reserve. (Baliarsingh *et al.*, 2013)

CLASS: PISCES

Scientific Name	*Local Name*
Order: Osteoglossiformes	
Family: **Notopteridae**	
1. *Notopterusnotopterus* (Pallas)	Flai
2. *Chitalachitala* (Hamilton)	Chitala
Order: Cypriniformes	
Family: **Cyprinidae**	
3. *Amblypharyngodonmola* (Hamilton)	Pathari
4. *Bariliusbagra* (Hamilton)	Jhalli
5. *Danio rerio* (Hamilton)	Jaradi
6. *Catlacatla* (Hamilton)	Bhakur
7. *Cirrhinusfulungee* (Sykes)	Dumala
8. *Cirrhinusmrigala* (Hamilton)	Mira
9. *Crossocheiluslatius* (Hamilton)	Kalabatuli
10. *Cyprinuscarpio* (Linnaeus)	Mirkha
11. *Devarioaequipinnatus* (McCelland)	Kulia
12. *Esomusdanricus* (Hamilton)	Jhai
13. *Garragotyala* (Gray)	Patharachatta
14. *Garramullya* (Sykes)	Gentu
15. *Labeobata* (Hamilton)	Pohala
16. *Labeocalbasu* (Hamilton)	Kalabanisi
17. *Labeorohita* (Hamilton)	Rohi
18. *Laubuca sp.*	Jodda
19. *Osteobramavigorsii* (Sykes)	Chilanti
20. *Puntius amphibious* (Valenciennes)	Kerandi
21. *Puntius conchonius* (Hamilton)	Pitakerandi
22. *Pantiussarana* (Hamilton)	Sema
23. *Pantiussophore* (Hamilton)	PatiaKerandi
24. *Puntius ticto* (Hamilton)	Kujikerandi
25. *Puntius sp. 1*	Kerandi
26. *Puntius sp. 2*	Kerandi
27. *Rasboradaniconius* (Hamilton)	Jilo
28. *Rasbora sp.*	Mino
29. *Salmophasiabacaila* (Hamilton)	Baunsapatri
30. *Tor putitora* (Hamilton)*	Mahaseer
31. *Tor tor* (Ham)*	Mahaseer
Family: **Nemacheilidae**	
32. *Acanthocobitisbotia* (Hamilton)	Patharachatta
33. *Schistura sp. 1*	Patharachatta
34. *Schistura sp. 2*	Patharachatta

Scientific Name	*Local Name*
Family: **Cobitidae**	
35. *Lephidocephalichthysguntea* (Hamilton)	Jimani
36. *Lephidocephalichthysthermalis* (Val.)	Balijimani
37. *Lephidocephalichthys sp.*	Jimani
Order: Siluriformes	
Family: **Bagridae**	
38. *Sperataaor* (Hamilton)*	Singla
39. *Sperataseenghala* (Sykes)	Singhi
40. *Mystusbleekeri* (Day)	Kujikantia
41. *Mystuscavasius* (Hamilton)	Baikantia
42. *Mystusgulio* (Hamilton)	Kantia
43. *Mystusvittatus* (Bloch.)	Kantia
Family: **Siluridae**	
44. *Ompokbimaculatus* (Bloch)	Baisilia
45. *Wallago attu* (Schneider)	Balia
Family: **Schilbeidae**	
46. *Alia coila* (Hamilton)*	Putulli
47. *Eutropiichthysvacha* (Hamilton)*	Batbacha
Family: **Ambiycipitidae**	
48. *Amblycepsmangois* (Hamilton)*	Jhilikantia
Family: **Erethistidae**	
49. Hara sp.	-
Family: **Claridae**	
50. Clarias sp.	Magur
Family: **Heteropneustidae**	
51. *Heteropneustesfossilis* (Bloch)	Singi
Order: Beloniformes	
Family: **Belonidae**	
52. *Strongylurastrongylura* (Vanhasselt)	Gania
53. *Xenentodoncancila* (Hamilton)	Gangeitodi
Order: Synbranchiformes	
Family: **Synbranchidae**	
54. *Monopteruscuchia* (Hamilton)	Cuchia
Family: **Mastacembelidae**	
55. *Macrognathusaculeatus* (Bloch)	Todi
56. *Mastacembeluspancalus* (Hamilton)	Gomitodi
Order: Perciformes	
Family: **Badidae**	
57. *Badisbadis* (Hamilton)	Kalileep

Scientific Name	*Local Name*
Family: **Cichlidae**	
58. Oreochromismossambicus (Peters)	Tilapia
Family: **Anabantidae**	
59. Anabas cobojius (Hamilton)	Kau
60. Anabas testudineus (Bloch)	Raja kau
Family: **Belontidae**	
61. Trichogasterfasciata (Bloch and Schneider)	Raja kau
62. Trichogasterlatius (Hamilton)	Raja kau

Scientific Name	*Local Name*
Family: **Channidae**	
63. Channamarulius (Hamilton)	Sahala
64. Channapunctata (Bloch)	Gadisa
65. Channastriata (Bloch)	Seola
66. Channa sp.	-

Appendix-X: Invertebrate Fauna Recorded in Similipal Biosphere Reserve. (Ramakrishna *et al.*, 2006).

Phylum ARTHROPODA

Class CRUSTACEA

Decapoda (Crabs and Prawns)

Suborder BRACHYURA (Crabs)

Family POTAMONIDAE

1. **Paratelphusa jaquemonti* Ratham

Suborder MACRURA (Prawns)

Class ARACHNIDA

Scorpionida (Scorpions)

Family BUTHIDAE

2. *Lychas* (*Endotrichus*) *laevifrons* Pocock, 1897

Family ISCHNURIDAE

3. **Hormurus australasiae* (Fabr., 1775)

Family SCORPIONIDAE

4. **Heterometrus* (*Gigantometrus*) *swammerdami* Simon, 1872
5. *Heterometrus sp.*

Araneae (Spiders)

Family ARANEIDAE

6. *Nephila maculata* (Fabricius, 1793)
7. *Argiope catenulate* (Dolecchall, 1859)
8. *Argiope pulchella* Thorell, 1881
9. **Tinus pirauridae*
10. **Hersilia savigny* (Lucas)
11. **Nephila maculate* (Fabricius)
12. **Neoscona punnasis* Tikadar

Family SALTICIDAE

13. *Plexippus paykullii* (Audouin, 1825)
14. *Zygoballus narmadaensis* Tikader, 1975

Family CLUBIONIDAE

15. *Oedignatha microsculata* Reimoser

Family OXYOPORIDAE ·

16. **Oxyopes sp.*

Family THOMISIDAE

17. **Thomisus sp.*

Family LYCOSIDAE

18. **Hipassa sp.*

Family SALTICIDAE

19. *Plexippus paykullii* (Audouin, 1825)
20. *Zygoballus narmadaensis* Tikader, 1975

Prostigmata (Ticks and Mites)

Family IXODIDAE

21. **Dermncentor auratus* Supino, 1897

Family TERANYCHIDAE

22. *Schizotetranychus andropogoni* (Hirst, 1926)

Family PHYTOSEIIDAE

23. *Amblyseius coccineae* Gupta, 1975
24. *A. kalimpongensis* Gupta, 1970
25. *A. suknaensis* Gupta, 1970
26. *A. syzygii* Gupta, 1975
27. *Phytoseius macropilis* (Banks, 1909)
28. *P. Kapuri* Gupta, 1969
29. *Typhlodromus homalli* Gupta, 1970

Class DIPLOPODA (**Millipedes**)

30 & 31. 2 spp. of millipedes (Under identification)

Class CHILOPODA (**Centipedes**)

Family SCOLOPENDRIDAE

32. *Cormocephalus dentipes* Pocock
33. *Rhysia longipes longipes* Newport

Class INSECTA: DICTYOPTERA (**Blattids**)

Family DEROCALYMMIDAE

34. *Trichoblatta sericea* Saussure, 1863

Family EPILAMPRIDAE

35. *Stictolampra plicata* (Navas, 1904)
36. *Haanina auriculata* (Brunner, 1865)

Family PYCNOSCELIDIDAE

37. *Pycnoscelus surinamensis* (Linnaeus, 1758)

Class INSECTA: ORTHOPTERA

(Grasshoppers, Crickets, Kotydids)

Family TETRIGIDAE

38. *Coptotettix fossulatus* Bolivar, 1987
39. *Hedotettix gracilis* (De Haan, 1842)
40. *Ergatettix dorsifera* (Walker, 1871)

Suborder ENSIFERA

Superfamily GRYLLOIDEA

Family GRYLLIDAE

41. *Gryllus bimaculatus* De Geer, 1773
42. *Teleogryllus testaceus* (Walker, 1869)
43. *Teleogryllus longipennis* (Saussure, 1877)
44. *Modicogryllus confirmatus* (Walker, 1859)

Family TRIGONIDIIDAE

45. *Trigonidium humbertianum* (Saussure, 1878)

Family PYRGOMORIPHIDAE

46. *C. oxypterus* (Blanchard, 1836)
47. *Atractomorpha crenulata* Fabricius, 1739

Family ACRIDIDAE

48. *Acrida exaltata* (Walker, 1859)
49. *Phlaeoba infumata* Brunner, 1893
50. *Pterno bimaculata* (Thunberg, 1815)
51. *Clonacris sila Rehn*, 1944
52. *Eucoptacra praemorsa* (Stal, 1860)
53. *Chondracris rosea* (Degeer, 1773)
54. *C. robustus* (Serville, 1839)
55. *Tylotropidus varicernis* (Walker, 1870)
56. *Eyprepocnemis rosen* Uvrov, 1942
57. *Gerenia bengalensis* Bhowmik and Haldar, 1983
58. *Diabolocatantops sukhdae* Bhowmik, 1985

Class INSECTA: DERMAPTERA (Ear wigs)

Family PYGIDICRANIDAEDAE

Subfamily DIPLATYINAE

59. *Diplatys tikaderi* Srivastava
60. *Diplatys chowdhuryi*

Family LABIDURIDAE

Subfamily NALINAE

61. *Nala lividipes* (Dufour, 1820)

Class INSECTA: ISOPTERA (**Termites**)

Family TERMITIDAE

62. *Speculitermes cyclops* Wasmann, 1902
63. *Microcerotermes annandalei* Silvestri, 1923
64. *Speculitermes sinhalensis* Roonwal and Sen-Sarma, 1960
65. *Odontotermes assmuthi* Holmgren, 1913
66. **Danaus genutia* (Cramer, 1779)
67. *O. bellahunisensis* Holmgren and Holmgren, 1917
68. *O. microdentatus* Roonwal and Sen-Sarma, 1960

Class INSECTA: EPHEMEROPTERA (**May flies**)

Family EPHEMERIDAE

69. *Ephemera* sp.

Family HETAGENIDAEDAE

70. *Rhithrogena parva* (Ulmer, 1969)

Class INSECTA: ODONATA (**Dragon flies**)

Family LIBELLULIDAE

71. **Potamarcha obscura* (Rambur, 1842)
72. *Potamarcha congener* (Rambur)
73. *Pantala flavescens* (Fabricius, 1798)
74. **Trithemis festiva* (Schneider, 1936)

Family COENAGRIONIDAE

75. **Cercion malayanum* Selys
76. **Cercion calamorum dyeri* Fraser
77. **Onychargia atrocyana* Selys
78. **Aciagrion* sp.

Family LESTIDAE

79. **Lestes viridulus* Ramber
80. **Disparoneura* sp.

Family AESHNIDAE

81. **Anax imperator* Leach

Class INSECTA: HEMIPTERA (**Bugs**)

82. **Nepa cinerea*

Class INSECTA: LEPIDOPTERA (**Butterflies**)

Family PAPILIONIDAE (**Swallow tails**)

83. *Graphium doson eleius* (Frushstorfer, 1939)
84. **Graphium antiphates naira* (Moore, 1939)
85. *Graphium nomius nomiu* (Esper, 1939)
86. **Papilio clytia clytia* (Linnaeus, 1758)
87. **Papilio polytes* Linnaeus
88. **Princeps nephelus nephelus* (Boisduval, 1836)
89. *P. polymnestor polymnestor* (Cramer, 1775)
90. *P. polytes romulus* (Cramer, 1775)
91. *P. chaon chaon* Westwood
92. **P. demoleus demoleus* (Linnaeus, 1758)
93. **Pachlipata aristolochiae* Fabricius

Family PIERIDAE (White and Yellows)

94. **Delias eucharis* Drury
95. **Eurema hecabe* (Linnaeus, 1881)
96. **Eurema brigitta* (Wallace, 1867)
97. **Catopsilia pyranthe* (Linnaeus, 1758)

Family NYMPHALIDAE (Brush-footed butterflies)

98. **Neptis hylas* Moore, 1772
99. **Junonia lemonias* Linnaeus, 1912
100. **Junonia hierta* Fabricius, 1798
101. **Junonia orithya* Butler, 1885
102. **Junonia almana* Linnaeus, 1758
103. **Danaus genutia* (Cramer, 1779)
104. **Danaus chrysippus* (Linnaeus, 1758)
105. **Euploea core* (Cramer, 1780)
106. **Charaxes polyxena imna* (Bulter)
107. **Euthalia lepidea miyana* Frushtorfer, 1913
108. **Euthalia garuda anagama* Frushtorfer
109. *Cyrestis thyodamus* Boisduval, 1836
110. **Hypolimnas misippus* (Linnaeus, 1764)
111. **Hypolimnas bolina* (Linnaeus, 1764)
112. **Precis almana almana* (Linnaeus, 1758)
113. **Precis atlitis* (Linnaeus, 1763)
114. **Precis hierata hierata* (Fabricius, 1798)
115. **Phalanta Phalantha Phalantha* (Drury, 1773)
116. **Mycalesis perseus tubitha* (Fabricius, 1947)
117. *Ypthima ceylonica hubneri* Kirby, 1871
118. *Ergolis ariadnae* Linnaeus
119. *Melanitis leda ismena* Linnaeus
120. *Vagrans egista sinha* (Kollar)
121. *Tanaecia lepidea* Butler

Family LYCAENIDAE (**Blues**)

123. *Zizina otis* Murray, 1874

Family HESPRIIDAE (**Skippers**)

124. *Badamia exclamationis* Fabricius

Class INSECTA: COLEOPTERA (**Beetles)**

Family CHRYSOMELIDAE

125. *Monalepta signata* (Oliver, 1808)
126. *Aphthona nigrilabris* Duviver, 1892
127. **Phyllotretu chotanica* Duviver, 1892

Class INSECTA: HYMENOPTERA (**Bees)**

Family APIDAE

128. **Apis indica*

Class INSECTA: DIPTERA (**Mosquitoes, Flies**)

Family TABANIDAE

129. *Chrysops pellucidus* Fabricius, 1805

Family SYRPHIDAE

130. *Phytomia errans* Fabricius, 1787

Family SEPSIDAE

131. *Sepsis indica* Wiedemann, 1824

Family MUSCIDAE

132. *Orthelia coerulea* Wiedemann, 1819

Family CALLIPHORIDAE

133. *Hemipyrellia pulchra* Wiedemann, 1830

Note: While double ** preceding species name indicates first report of distributional record from Odisha/ Similipal. A single * preceding species name implies first record from Similipal.

Appendix-XI: List of Butterflies Recorded in Two Selected Study Sites in the Fringe Area of Similipal Biosphere Reserve with Protection Status under Indaian Wildlife (Protection) Act, 1972 (Payra *et al.*, 2016).

Family Papilionidae

1. *Graphium sarpedon* (Common Bluebottle)
2. *Graphium doson* (Common Jay)
3. *Graphium eurypylus* (Great Jay)
4. *Graphium agamemnon* (Tailed Jay)
5. *Graphium* nomius (Spot Swordtail)
6. *Graphium* antiphates (Fivebar Swordtail)
7. *Papilio polymnestor* (Blue Mormon)
8. *Papilio crino* (Common Banded Peacock)
9. *Papilio nephelus* (Yellow Helen)
10. *Papilio polytes* (Common Mormon)
11. *Papilio demoleus* (Lime Butterfly)
12. *Papilio clytia* (Common Mime)
13. *Atrophaneura hector* (Crimson Rose) **Sch I (PartIV)**
14. *Atrophaneura aristolochiae* (Common Rose)

Family Pieridae

15. *Cepora nerissa* (Common Gull)
16. *Belenois aurota* (Pioneer)
17. *Delias eucharis* (Common Jezebel)
18. *Appias libythea* (Striped Albatross) **Sch IV**
19. *Leptosia nina* (Psyche)
20. *Pareronia hippia* (Common Wanderer)
21. *Catopsilia pomona* (Common/Lemon Emigrant)
22. *Catopsilia pyranthe* (Mottled Emigrant)
23. *Gandaca harina* (Tree Yellow)
24. *Eurema brigitta* (Small Grass Yellow)
25. *Eurema laeta* (Spotless Grass Yellow)
26. *Eurema andersonii* (Onespot Grass Yellow)
27. *Eurema blanda* (Threespot Grass Yellow)
28. *Eurema hecabe* (Common Grass Yellow)

Family Lycaenidae

29. *Abisara echerius* (Plum Judy)
30. *Spalgis epeus* (Apefly)
31. *Poritia hewitsoni* (Common Gem) **Sch II (Part II)**
32. *Curetis thetis* (Indian Sunbeam)
33. *Curetis acuta* (Toothed Sunbeam)
34. *Petrelaea dana* (Dingy Lineblue)
35. *Prosotas nora* (Common Lineblue)
36. *Prosotas dubiosa* (Tailless Lineblue)
37. *Nacaduba berenice* (Rounded Sixlineblue)
38. *Prosotas noreia* (White-tipped Lineblue) **Sch I (Part IV)**
39. *Caleta decidia* (Angled Pierrot)
40. *Caleta elna* (Elbowed Pierrot)
41. *Jamides bochus* (Dark Cerulean)
42. *Jamides celeno* (Common Cerulean)
43. *Catochrysops strabo* (Forgetmenot)
44. *Lampides boeticus* (Pea Blue) **Sch II (Part II)**
45. *Leptotes plinius* (Zebra Blue)
46. *Castalius rosimon* (Common Pierrot)
47. *Zizeeria karsandra* (Dark Grass Blue)
48. *Zizeeria maha* (Pale Grass Blue)
49. *Zizina otis* (Lesser Grass Blue)
50. *Zizula hylax* (Tiny Grass Blue)
51. *Everes lacturnus* (Indian Cupid)
52. *Neopithecops zalmora* (Quaker)
53. *Megisba malaya* (Malayan)
54. *Acytolepis puspa* (Common Hedge Blue)
55. *Chilades lajus* (Lime Blue)
56. *Chilades pandava* (Plains Cupid)
57. *Freyeria trochylus* (Grass Jewel)
58. *Anthene emolus* (Ciliate Blue)
59. *Anthene lycaenina* (Pointed Ciliate Blue) **Sch II (Part II)**
60. *Spindasis vulcanus* (Common Silverline)
61. *Spindasis lohita* (Long-banded Silverline) **Sch II (Part II)**
62. *Arhopala atrax* (Indian Oakblue)
63. *Arhopala amantes* (Large Oakblue)
64. *Amblypodia anita* (Purple Leaf Blue)
65. *Loxura atymnus* (Yamfly)
66. *Rathinda amor* (Monkeypuzzle)
67. *Zeltus amasa* (Fluffy Tit)
68. *Rapala varuna* (Indigo Flash) **Sch II (Part II)**
69. *Rapala manea* (Slate Flash)
70. *Rapala iarbus* (Indian Red Flash)

Family Nymphalidae

71. *Tirumala limniace* (Blue Tiger)
72. *Danaus genutia* (Common Tiger)

73. *Danaus chrysippus* (Plain Tiger)
74. *Euploea core* (Common Indian Crow)
75. *Euploea klugii* (King Crow)
76. *Melanitis leda* (Common Evening Brown)
77. *Elymnias hypermnestra* (Common Palmfly)
78. *Mycalesis perseus* (Common Bushbrown)
79. *Orsotriaena medus* (Nigger)
80. *Ypthima huebneri* (Common Fourring)
81. *Ariadne ariadne* (Angled Castor)
82. *Ariadne merione* (Common Castor)
83. *Phalanta phalantha* (Common Leopard)
84. *Vagrans egista* (Vagrant)
85. *Junonia hierta* (Yellow Pansy)
86. *Junonia orithya* (Blue Pansy)
87. *Junonia lemonias* (Lemon Pansy)
88. *Junonia almana* (Peacock Pansy)
89. *Junonia atlites* (Grey Pansy)
90. *Junonia iphita*(Chocolate Soldier)
91. *Kallima inachus* (Orange Oakleaf)
92. *Hypolimnas bolina* (Great Eggfly)
93. *Hypolimnas misippus* (Danaid Eggfly) **Sch II (Part II)**
94. *Cyrestis cocles* (Marbled Map) **Sch II (Part II)**
95. *Cyrestis thyodamas* (Common Map)
96. *Neptis clinia* (Clear Sailer)
97. *Neptis hylas* (Common Sailer)
98. *Phaedyma columella* (Short-banded Sailer)
99. *Neptis jumbah* (Chestnut-streaked Sailer)
100. *Pantoporia hordonia* (Common Lascar)
101. *Lasippaviraja* (Yellowjack Sailer)
102. *Athyma perius* (Common Sergeant)
103. *Moduza procris* (Commander)
104. *Euthalia nais* (Baronet)
105. *Euripus consimilis* (Painted Courtesan) **Sch II (Part II)**
106. *Euthalia monina* (Powdered Baron)
107. *Euthalia aconthea* (Baron)
108. *Euthalia lubentina* (Gaudy Baron) **Sch IV**
109. *Polyura athamas* (Common Nawab)
110. *Polyura eudamippus* (Great Nawab)
111. *Charaxes psaphon* (Plain Tawny Rajah)
112. *Charaxes solon* (Black Rajah)
113. *Acraea violae* (Tawny Coster)
114. *Tanaecia lepidea* (Grey Count) **Sch II (Part II)**

Family Hesperiidae

115. *Hasora chromus* (Common Banded Awl)
116. *Badamia exclamationis* (Brown Awl)
117. *Bibasis sena* (Orangetail Awl) **Sch II (Part II)**
118. *Celaenorrhinus leucocera* (Common Spotted Flat)
119. *Sarangesa dasahara* (Common Small Flat)
120. *Coladenia indrani* (Tricolour Pied Flat)
121. *Tagiades japetus* (Common Snow Flat)
122. *Tagiades gana* (Suffused Snow Flat)
123. *Caprona ransonnetii* (Golden Angle)
124. *Tagiades litigiosa* (Water Snow Flat)
125. *Iambrix salsala* (Chestnut Bob)
126. *Notocrypta paralysos* (Common Banded Demon)
127. *Suastus gremius* (Indian Palm Bob)
128. *Matapa aria* (Common Redeye)
129. *Oriens goloides* (Ceylon Dartlet)
130. *Telicota bambusae* (Dark Palm Dart)
131. *Parnara guttata* (Straight Swift)
132. *Pelopidas* sp
133. *Pelopidas mathias* (Small Branded Swift)
134. *Baoris farri* (Paintbrush Swift) **Sch IV**
135. *Caltoris kumara* (Blank Swift)
136. *Cupitha purreea* (Wax Dart)

Appendix-XII: List of Odonates (Dragonflies and Damselflies) Recorded in Buffer Areas of Similipal Biosphere Reserve (Das *et al.*, 2012).

Family Coenagrionidae

1. *Agriocnemis pygmea* (Rambur, 1842)
2. *Agriocnemis lacteola* (Selys, 1877)
3. *Ceriagrion coromandelianum* (Fabr. 1798)
4. *Ceriagrion cerinorubellum* (Brauer, 1865)
5. *Ceriagrion olivaceum* Laidlaw, 1914)
6. *Ischnura aurora* (Brauer, 1865)
7. *Ischnura senegalensis* (Ramb, 1842)
8. *Pseudagrion microcephalum* (Rambur, 1842)
9. *Pseudagrion rubriceps* (Selys, 1876)
10. *Onychargia atrocyana* (Selys, 1865)
11. *Aciagrion pallidum* (Selys, 1891)

Family Protoneuridae

12. *Disparoneura quadrimaculata* (Rambur, 1842)
13. *Caconeura ramburi* (Fraser, 1922)

Family Lestidae

14. *Lestes umbrinus* (Selys, 1891
15. *Lestes viridulus* (Rambur, 1842)

Family Calopterygidae

16. *Neurobasis chinensis* (Linnaeus, 1758)
17. *Vestalis gracilis* (Rambur, 1842)
18. *Vestalis apicalis* (Selys, 1873)

Family Platycnemididae

19. *Copera ciliate* (Selys, 1863)
20. *Copera vittata* Selys, 1863
21. *Copera marginipes* (Rambur, 1842)

Family Chlorocyphidae

22. *Libellago Lineata* (Burmeister, 1839)
23. *Rhinocypha bisignata* Hagen (Selys, 1853)

Family Gomphidae

24. *Ictinogomphus rapax* (Rambur, 1842)
25. *Paragomphus lineatus* (Selys, 1850)

Family Aeshnidae

26. *Anax guttatus* (Burmeister, 1839)
27. *Anax immaculifrons* (Rambur, 1842)

Family Libellulidae

28. *Acisoma panorpoides* (Rambur, 1842)
29. *Aethriamanta brevipennis* (Rambur, 1842)
30. *Brachydiplax sobrina* (Rambur, 1842)
31. *Brachythemis contaminata* (Fabr.1793)
32. *Bradinopyga geminata* (Rambur, 1842)
33. *Cratilla lineata* (Foerster, 1903)
34. *Crocothemis servilia* (Drury, 1770)
35. *Diplacodes trivialis* (Rambur, 1842)
36. *Diplacodes nebulosa* Fabricius, 1793
37. *Neurothemis fulvia* (Drury, 1773)
38. *Neurothemis tullia* (Drury, 1773)
39. *Orthetrum pruinosum* (Rambur, 1842)
    Acad. J. Entomol., 5 (1): 54-61, 2012
40. *Orthetrum sabina* (Drury, 1770)
41. *Orthetrum triangulare* (Selys, 1878)
42. *Orthetrum glaucum* (Brauer, 1865)
43. *Orthetrum luzonicum* (Brauer, 1868)
44. *Orthetrum triangulare* (Selys, 1878)
45. *Orthetrum taeniolatum* (Schneider, 1845)
46. *Palpopleura sexmaculata* (Fabricius, 1787)
47. *Pantala flavescens* (Fabr. 1798)
48. *Potamarcha congener* (Rambur, 1842)
49. *Rhodothemis rufa* (Rambur, 1842)
50. *Rhyothemis variegate* (Linn. 1763)
51. *Tholymis tillarga* (Fabr. 1798)
52. *Tramea limbata* (Desjardins, 1832)
53. *Tramea basilaris* (Kirby, 1889)
54. *Trithemis aurora* (Burmeister, 1839)
55. *Trithemis festiva* (Rambur, 1842)
56. *Trithemis pallidinervis* (Selys, 1889)
57. *Neurothemis intermedia* (Rambur, 1842)
58. *Urothemis signata* (Rambur, 1842)

Index

R

S

T

U

V

W